VOLUME THREE HUNDRED AND THIRTY FIVE

INTERNATIONAL REVIEW OF CELL AND MOLECULAR BIOLOGY

Transcriptional Gene Regulation in Health and Disease

INTERNATIONAL REVIEW OF CELL AND MOLECULAR BIOLOGY

VOLUME THREE HUNDRED AND THIRTY FIVE

INTERNATIONAL REVIEW OF CELL AND MOLECULAR BIOLOGY

Transcriptional Gene Regulation in Health and Disease

Edited by

FRIEDEMANN LOOS

Centre de Recherche des Cordeliers, Paris, France

ELSEVIER

ACADEMIC PRESS

An imprint of Elsevier

Academic Press is an imprint of Elsevier
50 Hampshire Street, 5th Floor, Cambridge, MA 02139, United States
525 B Street, Suite 1800, San Diego, CA 92101-4495, United States
The Boulevard, Langford Lane, Kidlington, Oxford OX5 1GB, United Kingdom
125 London Wall, London, EC2Y 5AS, United Kingdom

First edition 2018

Notices
Knowledge and best practice in this field are constantly changing. As new research and experience broaden our understanding, changes in research methods, professional practices, or medical treatment may become necessary.

Practitioners and researchers must always rely on their own experience and knowledge in evaluating and using any information, methods, compounds, or experiments described herein. In using such information or methods they should be mindful of their own safety and the safety of others, including parties for whom they have a professional responsibility.

To the fullest extent of the law, neither the Publisher nor the authors, contributors, or editors, assume any liability for any injury and/or damage to persons or property as a matter of products liability, negligence or otherwise, or from any use or operation of any methods, products, instructions, or ideas contained in the material herein.

ISBN: 978-0-12-812339-3
ISSN: 1937-6448

For information on all Academic Press publications
visit our website at https://www.elsevier.com/books-and-journals

Publisher: Zoe Kruze
Acquisition Editor: Ashlie M. Jackman
Editorial Project Manager: Fenton Coulthurst
Production Project Manager: James Selvam
Cover Designer: Matthew Limbert

Typeset by SPi Global, India

CONTENTS

CONTRIBUTORS

Tobias M. Backes
Saarland University Medical Faculty, Homburg, Germany

Atze J. Bergsma
Molecular Stem Cell Biology, Department of Clinical Genetics; Department of Pediatrics; Center for Lysosomal and Metabolic Diseases, Erasmus Medical Center, Rotterdam, The Netherlands

Mike Broeders
Molecular Stem Cell Biology, Department of Clinical Genetics; Department of Pediatrics; Center for Lysosomal and Metabolic Diseases, Erasmus Medical Center, Rotterdam, The Netherlands

Ruaidhrí J. Carmody
Centre for Immunobiology, Institute of Infection, Immunity, and Inflammation, University of Glasgow, Glasgow, United Kingdom

Andrea Lesch
Saarland University Medical Faculty, Homburg, Germany

Tokameh Mahmoudi
Erasmus University Medical Center, Rotterdam, The Netherlands

Jennifer P. Mitchell
Rheumatoid Arthritis Pathogenesis Centre of Excellence, Centre for Immunobiology, Institute of Infection, Immunity, and Inflammation, University of Glasgow, Glasgow, United Kingdom

Enrico Ne
Erasmus University Medical Center, Rotterdam, The Netherlands

Robert-Jan Palstra
Erasmus University Medical Center, Rotterdam, The Netherlands

W.W.M. Pim Pijnappel
Molecular Stem Cell Biology, Department of Clinical Genetics; Department of Pediatrics; Center for Lysosomal and Metabolic Diseases, Erasmus Medical Center, Rotterdam, The Netherlands

Delphine Quénet
Robert Larner College of Medicine, University of Vermont, Burlington, VT, United States

Oliver G. Rössler
Saarland University Medical Faculty, Homburg, Germany

Sandra Rubil
Saarland University Medical Faculty, Homburg, Germany

Gerald Thiel
Saarland University Medical Faculty, Homburg, Germany

Ans T. van der Ploeg
Department of Pediatrics; Center for Lysosomal and Metabolic Diseases, Erasmus Medical Center, Rotterdam, The Netherlands

Erik van der Wal
Molecular Stem Cell Biology, Department of Clinical Genetics; Department of Pediatrics; Center for Lysosomal and Metabolic Diseases, Erasmus Medical Center, Rotterdam, The Netherlands

Kazuya Yamagata
Faculty of Life Sciences, Kumamoto University, Kumamoto, Japan

Tatsuya Yoshizawa
Faculty of Life Sciences, Kumamoto University, Kumamoto, Japan

CHAPTER ONE

Histone Variants and Disease

Delphine Quénet[1]
Robert Larner College of Medicine, University of Vermont, Burlington, VT, United States
[1]Corresponding author: e-mail address: delphine.quenet@uvm.edu

Contents

Abstract

In eukaryotes, the genome is organized into a complex nucleoprotein structure called chromatin. Despite the simplicity of its monomer, DNA and two copies of four histones, the existence of histone variants opens possibilities of multiple chromatin landscapes and fine-tune regulation of molecular mechanisms for the regulation of gene

ISSN 1937-6448
https://doi.org/10.1016/bs.ircmb.2017.07.006

expression and maintenance of genome stability. However, any defects in these combinations may contribute to disease development and/or progression. Here, I review human histone variants and their chaperones, and discuss how they contribute to pathological conditions.

1. INTRODUCTION

In humans, the genome (2 m of DNA) is organized into a complex nucleoprotein structure, called chromatin, and confined into a 2- to 10-μm nucleus, indicating the need for a high level of compaction. However, efficient DNA-dependent mechanisms, including transcription, replication, and DNA repair, require chromatin decondensation and free DNA, suggesting a precise regulation between chromatin packaging and accessibility. The organization of DNA into nucleosomes constitutes the first step of chromatin condensation. Further folding into 30 nm fiber originates higher order chromatin structures up to the optimum compact metaphase chromosome in mitosis (Woodcock and Dimitrov, 2001). In interphase cells, two major chromatin domains are defined based on their level of condensation and gene expression, namely, euchromatin and heterochromatin. The transcriptionally permissive euchromatin displays a relaxed organization, whereas the low level of expression at heterochromatin is maintained by a condensed structure (Li and Reinberg, 2011; Zhou et al., 2011). This condensation is mediated by various factors including histone variants, their posttranslational modifications (PTMs), and their chaperones, which are the focus of this review.

2. OVERVIEW OF NUCLEOSOME STRUCTURE

The basic unit of chromatin is the nucleosome that consists of ~147 bp of DNA wrapped 1.75 times in a left helix manner around an octamer of histone proteins. The histone octamer is composed of two heterodimers H3–H4 associated as a tetramer and surrounds on both sides with a H2A–H2B dimer (Luger et al., 1997). This structure is stabilized by a fifth histone, histone H1 or linker histone, that binds at the entry and exit of the nucleosome (Woodcock et al., 2006).

Histones H3, H4, H2A, and H2B are organized in cluster throughout the genome and are expressed in a replication-dependent manner. Ranging from 11 to 15 kDa, these basic proteins share similar structure consisting in a lysine-rich N-terminal tail, a globular histone fold domain, and a C-terminal tail. Their targeting, loading, and deposition onto chromatin occur during DNA replication and are mediated by chaperone proteins. These histone-binding proteins play an essential role in the turnover of histone and the maintenance of chromatin organization over the cell cycle. Despite the extensive investigation on this topic, not all chaperones have been identified. This gap of knowledge limits our understanding of histone recycling at the chromatin during DNA-dependent mechanisms, and consequences of defects in histone–chaperone interaction.

3. POSTTRANSLATIONAL MODIFICATIONS OF HISTONES

The unstructured N-terminal tail of histone comprises ~30 residues that extends outside of the nucleosome and is a site of extensive PTMs (Table 1; Bannister and Kouzarides, 2011). These PTMs participate to the regulation of chromatin condensation and DNA-dependent mechanisms (Zentner and Henikoff, 2013). Most extensive studied PTMs are methylation, acetylation, and phosphorylation (Table 1). Yet, 18 different PTMs on at least 400 residues have been described so far, indicating the large possibility of PTM combination to regulate molecular and cellular processes (Bannister and Kouzarides, 2011; Huang et al., 2014).

3.1 Acetylation

Acetylation is the predominant PTM catalyzed by histone acetyltransferase (HAT) enzymes that transfer the acetyl moiety from acetyl-CoA to the ε-amino group of lysine (K) residue. This PTM is reversed by histone deacetylase (HDAC) enzymes. Acetylation is often associated with transcription activation mediated by at least two nonexclusive mechanisms: (i) by neutralizing the basic charge of the ε-amino group, destabilizing histone–DNA interactions, and increasing nucleosome mobility and accessibility of the transcription machinery to promoters and (ii) by recruiting bromodomain-containing proteins that specifically recognize acetylated residue (Choi and Howe, 2009; Wang et al., 2008b; Yang, 2004). For instance, transcribed units are found acetylated on H3K14, H3K23, H4K12, and H4K16, and the double acetylation of histone H4 promotes the recruitment of the transcription factor IID (TFIID), leading to the assembly and stabilization of the transcription preinitiation complex (Naar et al., 2001; Wang et al., 2008a).

Table 1 Examples of Posttranslational Modifications of Histones

Histone Modification	Change	Modified Residue	Writer	Eraser	Reader Domain	References
Acetylation	Neutralization of positive charge	K	HAT	HDAC	Bromodomain	DesJarlais and Tummino (2016)
Methylation	No charge change, increase of volume, and hydrophobicity	K and R	KMT and RMT	KDM	Chromodomain, Tudor domain, PHD finger domain, MBT domain, WD40r domain, PWWP, BAH, Ankyrin, ADD, zinc-finger CW domain	DesJarlais and Tummino (2016)
Phosphorylation	Addition of negative charge	S, T, Y, and H	PK	PP	14-3-3, BRCT	Rossetto et al. (2012)
ADP-ribosylation	Addition of volume and negative charge	E, K, and S	PARP	PARG, MacroD1/D2, TARG1	Macro domain, WWE domain, PBZ	Barkauskaite et al. (2015)

Modified residue: *E*, glutamic acid; *H*, histidine; *K*, lysine; *R*, arginine; *S*, serine; *T*, threonine; *Y*, tyrosine. Writer and eraser: *HAT*, histone acetyltransferase; *HDAC*, histone deacetylase; *KDM*, lysine demethylase; *KMT*, lysine methyltransferase; *PARP*, poly(ADP-ribose) polymerase, *PARPG*, poly(ADP-ribose) glycohydrolase; *PK*, protein kinase; *PP*, protein phosphatase; *RMT*, arginine methyltransferase; *TARG1*, terminal ADP-ribose protein glycohydrolase. Reader domain: *ADD*, ATRX–DNMT3–DNMT3L, ADD_{ATRX}; *BAH*, bromo adjacent homology; *BRCT*, BRCA1 C terminus domain; *MBT*: malignant brain tumor; *PBZ*, PAR-binding zinc finger; *PHD*, plant homeodomain.

3.2 Methylation

Based on the transfer of methyl group from a molecule of *S*-adenosyl methionine to lysine or arginine (R) residues, methylation is catalyzed by lysine methyltransferase (KMT) or arginine methyltransferase (RMT) and histone demethylase (Shi and Whetstine, 2007; Trievel, 2004). This PTM displays several configurations: mono- (me), di- (me2, symmetric and asymmetric), and tri- (me3) methylation, which have different outcomes in the regulation of chromatin structure and function (Taverna et al., 2007). One example is the methylation of histone H3 on lysine 4 (H3K4). At promoters of housekeeping genes in embryonic stem cells, H3K4 is dimethylated (H3K4me2) and trimethylated (H3K4me3), providing a chromatin environment prone to transcription (Meissner et al., 2008). On the other hand, H3K4me3 in combination with H3K27me3 at developmental genes represses their expression. Methylation adds volume and hydrophobicity to the modified residue, regulating nucleosome stability and establishing a platform for proteins recognizing methylated histones, via domains such as chromo, Tudor, or PHD (plant homeodomain) finger (Adams-Cioaba and Min, 2009; Kim et al., 2006).

3.3 Phosphorylation

Catalyzed by protein kinase (PK) and protein phosphatase (PP), histone phosphorylation is the addition of phosphate group from an ATP molecule to a serine (S), threonine (T), tyrosine (Y), or histidine (H) residue. This PTM plays an important function in the DNA damage response, transcription regulation, mitosis/meiosis, and chromatin organization.

One of the best-defined role is in mitosis. In M phase, Aurora B phosphorylates serine 10 on histone H3 (H3S10), preventing the binding of the heterochromatin protein HP1α via its chromodomain to di- and trimethylated lysine 9 of histone H3 (H3K9me2/3) (Fischle et al., 2005; Hirota et al., 2005). This phosphorylation and the removal of HP1α initiate chromatin condensation during mitosis.

3.4 ADP-Ribosylation

ADP-ribosylation is a reaction catalyzed by most of members of the poly (ADP-ribose) polymerase (PARP) family, that is composed of 17 proteins in human (Barkauskaite et al., 2015). With the exception of PARP9 and PARP13, PARP proteins transfer a single mono-ADP-ribose group from NAD^+ to their target proteins on arginine, aspartate, glutamate, lysine, and serine residues. PARP1–5 can add additional ADP-ribose units to generate a long negatively charged polymer (PAR, up to 200 units) (D'Amours et al., 1999). The degradation of the polymer may involve several enzymatic activities. Poly(ADP-ribose) glycohydrolase (PARG) cleaves the *O*-glycosidic ribose bond between the first ADP-ribose and the PAR chain (Brochu et al., 1994; Slade et al., 2011). Then, the final mono-ADP-ribose is cleaved by MacroD1 and MacroD2 (Jankevicius et al., 2013; Rosenthal et al., 2013; Sharifi et al., 2013). TARG1 (terminal ADP-ribose glycohydrolase 1/C6orf130) can either release the PAR or the final mono-ADP-ribose by a nucleophile attack (Sharifi et al., 2013).

Three domains recognize and bind to PAR chain: the poly(ADP-ribose)-binding zinc finger (PBZ), WWE domain, and the macrodomain (Barkauskaite et al., 2015). The macrodomain is found in numerous proteins, including PARP9/14/15 and the histone variant macroH2A. Interestingly, PARG, MacroD1, MacroD2, and TARG1 also contain a macrodomain which evolves to become their catalytic site.

This PTM is well described for its role in promoting DNA repair (Martin-Hernandez et al., 2017). During this process, ADP-ribosylation of chromatin leads to chromatin reorganization, recruitment of the DNA repair machinery, and regulation of gene expression (Tallis et al., 2014). It is noteworthy that ADP-ribosylation also occurs in the absence of DNA damage, suggesting a broad function of this PTM in physiological and pathological conditions.

3.5 Writer, Eraser, and Reader

Three types of proteins or protein complexes are involved in the dynamic and function of PTMs: writers, erasers, and readers (Fig. 1; Table 1). Writers place the epigenetic modifications on histones, whereas erasers remove them. These two factors are divided into classes based on the PTM they regulate (e.g., HDAC vs HAT for acetylation, PARP vs PARG for ADP-ribosylation). Readers contain domain(s) able to recognize one or a combination of PTMs (e.g., chromodomain for methylation, bromodomain for acetylation; Table 1), and further relay their signal for an appropriate cellular response by recruiting additional chromatin-modifying complexes, affecting chromatin organization, compaction, and transcription. Like for histone variants and PTMs, there are extensive investigations on these three types of proteins (i.e., writers, erasers, and readers) to comprehend their molecular function.

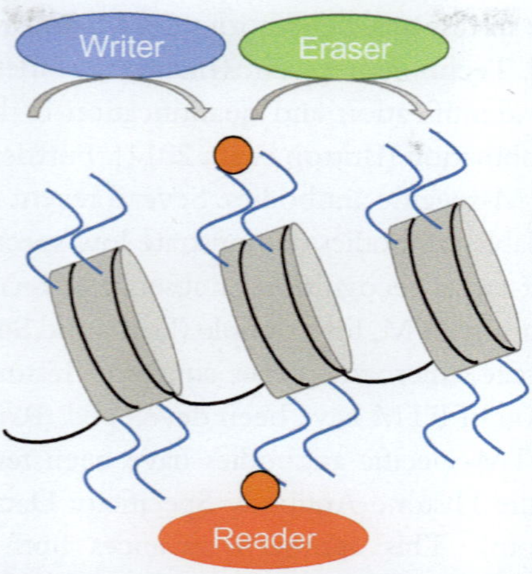

Fig. 1 Actors of the plasticity of the chromatin. Writer is a protein or protein complex that adds a posttranslational modification (*orange circle*) to a histone, whereas the eraser removes this modification. Due to the presence of a specific binding domain, the reader recognizes the posttranslational modification to relay this cellular signal.

To reach this goal, small molecules that either inhibit catalytic activity of writer and eraser, or interfere with reader binding have been generated, and therapeutic use has been proposed. Hence, clinical trials have been successful for some of them, leading to their approval by the Food and Drug Administration (FDA), like the HDAC inhibitor verinostat for the treatment of cutaneous T cell lymphoma, or the PARP inhibitor olaparib for the treatment of advanced ovarian cancer associated with defective BRCA genes (Konecny and Kristeleit, 2016; West and Johnstone, 2014). Thus, the study of PTMs and writers/erasers/readers will eventually provide knowledge on chromatin organization and function, as well as unveil new epigenetic therapeutic options.

3.6 Solely vs Combination of PTM

A specific PTM may not have meaning on its own or may display apparently opposite function. One extensively studied example is the phosphorylation H3S10 (Nowak and Corces, 2004). On one hand, this PTM marks condensed chromosomes during mitosis and meiosis (Gurley et al., 1978). On the other hand, H3S10 phosphorylation is associated with histone acetylation to promote chromatin decondensation and transcription activation (Chadee et al., 1999; Mahadevan et al., 1991). Thus, to comprehend PTM function in normal DNA-dependent mechanisms and the consequences of their presence or absence in diseases, neighboring PTMs and writer/eraser/reader have to be described. Technologies such as mass spectrometry have remarkably contributed to the identification and quantification of histone PTMs, and analysis of their combination (Britton et al., 2011). Further functional analysis usually relies on PTM-specific antibodies. Several recent studies have shown that numerous available antibodies demonstrate low specificity and/or sensitivity, leading to off-target recognition or absence of recognition due to the presence of neighboring PTM, for example (Fuchs and Strahl, 2011). To rule out this technical issue, microarrays that consist of histone peptide carrying one or a combination of PTM have been developed (Rothbart et al., 2015). So far, over 100 PTM-specific antibodies have been tested, and data have been compiled in the Histone Antibody Specificity Database (http://www.histoneantibodies.com). This approach advances both the validation of PTM-specific antibodies and the characterization of readers (Rothbart et al., 2012, 2015). The combination of these methods with biochemical, molecular, and cellular techniques contributes to understand how the PTMs and writer/eraser/reader regulate chromatin structure and cellular pathways, and to propose models that describe simple or complex functional interaction.

4. HISTONE VARIANTS

With the exception of histone H4, all histones comprise histone variants (Table 2) (Malik and Henikoff, 2003; Wiedemann et al., 2010). In contrast to previously described histones, histone variants are expressed over the cell cycle and are more likely incorporated into the genome in a replication-independent manner, such as during transcription (Marzluff and Duronio, 2002). Encoded by enclustered genes, these proteins may differ from the histones by only few residues (e.g., H3.3 differs from H3.1 in 5 amino acids), or large domains (e.g., compared to H2A, macroH2A presents an additional macrodomain of $\sim$210 amino acids). These differences define both their genomic distribution by selecting their chaperones, and their function (Fig. 2). As their counterparts, histone variants can be modified, recruiting readers and increased the repertoire of fine-tune regulations of DNA-dependent mechanisms. This section focuses on variant of the histone octamer, H3 and H2A.

4.1 The Human Histone H3 Family

The human histone H3 family comprises six histone variants: H3.3, the centromeric CENP-A, the primate-specific H3.X and H3.Y, and the two testis-specific histones H3.t and H3.5 (Filipescu et al., 2013).

Table 2 List of Known Human H3 and H2A Histone Variants and Their Function

Histone (Residues)	Major Cellular Location or Tissue Expression	Function	Chaperone
H3.3 (136)	Euchromatin and heterochromatin	Transcription	HIRA and DAXX/ ATRX
CENP-A (140)	Centromeres	Epigenetic mark of centromere	HJURP
H3.X (147) and H3.Y (136)	Brain	Transcription in response to stress stimuli	
H3t (136) and H3.5 (135)	Testis	Chromatin organization during spermatogenesis	H3t: NAP2?
H2A.X (143)		DNA damage repair/sex chromosome inactivation (meiosis)	
H2A.Z (128)	Euchromatin	Transcription/DNA damage repair	P400 and SRCAP
H2A.Bbd (115)	Testis and brain	Gene expression	FACT?
macroH2A (372)	Heterochromatin	Gene expression/X chromosome inactivation	

ATRX, α-thalassemia/mental retardation syndrome X-linked; *DAXX*, death domain-associated protein 6; *FACT*, facilitates chromatin transcription; *HIRA*, histone cell cycle regulator; *HJURP*, Holliday junction-recognizing protein; *NAP2*, nucleosome assembly protein; *SRCAP*, Snf2-related CREBBP activator protein.

4.1.1 The H3.3 Histone Variant

Encoded by two difference genes, namely, H3F3A and H3F3B, H3.3 differs by 4–5 amino acids (S31A, A87S, I89V, G90M, and S96C) from H3.1 (Albig et al., 1995; Szenker et al., 2011; Tagami et al., 2004). Its deposition at specific genomic domains is replication-dependent and -independent (Ahmad and Henikoff, 2002). This specific spatial and temporal distribution is determined by two histone chaperone complexes, HIRA (histone cell cycle regulator) and DAXX/ATRX (death domain-associated protein 6/α-thalassemia/mental retardation syndrome X-linked) complexes (Lewis et al., 2010; Tagami et al., 2004).

H3.3 roles comprise regulation of transcription, development, gametogenesis, and cell differentiation to cite few of them. H3.3 function in gene

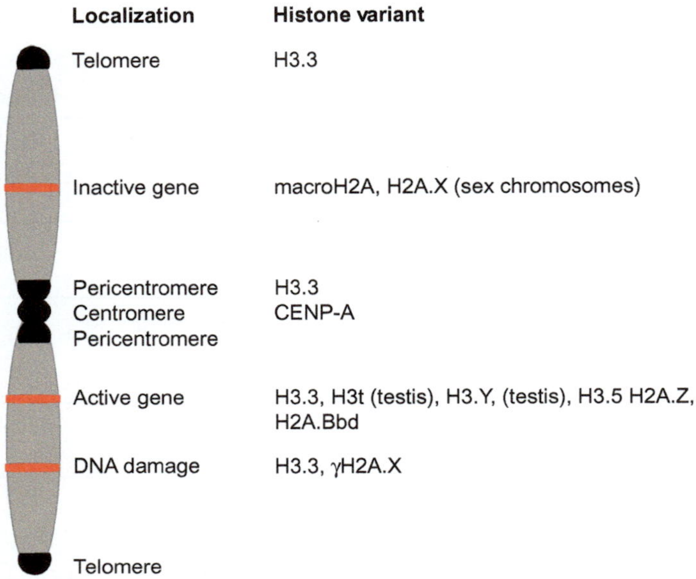

Fig. 2 Chromosome localization of histone variants.

expression is convoluted, as it may induce or repress transcription, and depends on the epigenetic context. At actively transcribed genes, H3.3 is associated with the presence of dimethylated H3K4 and RNA polymerase II, priming chromatin in an open state for an efficient transcription (Ahmad and Henikoff, 2002). Upon DNA damage, H3.3 is loaded at the break site, contributing to transcription reactivation after DNA damage repair (Adam et al., 2013). The deposition of H3.3 at heterochromatin domains (e.g., telomeres, pericentromeres, retroviral elements), as well as at imprinted genes, silences their expression, maintains repressive marks, and for the latter, supports proper allelic expression (Elsasser et al., 2015; Voon et al., 2015). Bivalent promoters are chromatin regions marked by both H3K27me3 and H3K4me3, such as developmentally regulated genes. In mouse embryonic stem cells, H3.3 represses transcription level of these genes by establishing and maintaining epigenetic mark, namely, H3K27me3 via the recruitment of PRC2 (polycomb repressive complex 2) (Banaszynski et al., 2013). Depletion of H3.3 is accompanied by the loss of PRC2 and H3K27me3, but not H3K4me3 at bivalent promoters, leading to abnormal differentiation. Altogether, H3.3 is not a solely marker of gene expression, but a histone variant whose function is defined by neighboring epigenetic marks and chromatin-binding proteins.

4.1.2 The Centromeric CENP-A

The centromere is a structurally and functionally specialized chromosome domain. In yeast, centromeres are genetically defined, whereas in most eukaryotes, they are epigenetically characterized by the presence of nucleosomes containing the H3 histone variant CENP-A (McKinley and Cheeseman, 2016). Centromeric chromatin is the docking site for a 40+ protein complex called kinetochore, which is the attachment site of the spindle microtubules during mitosis. Thus, CENP-A plays an essential role is the correct segregation of chromosomes.

Human CENP-A is encoded by a single gene that evolved rapidly with ~50% of amino acid identity with H3.1, H3.2, and H3.3, and whose disruption leads to lethality (Howman et al., 2000; Schueler et al., 2010). CENP-A deposition onto centromeres is mediated by a unique mechanism. During DNA replication, CENP-A molecules are equally distributed between the two DNA strands, leading to the dilution by half of CENP-A pool (Jansen et al., 2007). Synthesized in G2 phase, neo-CENP-A are loaded onto centromeres in late mitosis–early G1 by the chaperone HJURP (Holliday junction-recognizing protein) (Dunleavy et al., 2009; Foltz et al., 2009).

Defined as a constitutive heterochromatin domain (like pericentromeres and telomeres), centromeric chromatin consists of interspersed domains of CENP-A- and H3-containing nucleosomes (Blower et al., 2002; Sullivan and Karpen, 2004). Its heterochromatic feature is counter-balanced by the dimethylation of H3K4 and H3K36, two transcriptionally permissive marks (Bergmann et al., 2011; Blower et al., 2002; Sullivan and Karpen, 2004). Alterations of this organization lead to kinetochore defects and chromosome missegregation, and are observed in some diseases.

4.1.3 Primate-Specific Histones H3.X and H3.Y

H3.X and H3.Y are primate-specific histone H3 variants, preferentially expressed in human brain (Wiedemann et al., 2010). Evolutionary derived from H3.3, H3.Y seems to prefer forming an heterotypic nucleosome with H3.3 around transcription starting sites (TSS) in human cells (Kujirai et al., 2016). In vitro and in vivo, the H3.Y nucleosome forms a relaxed chromatin conformation with flexible DNA ends (Kujirai et al., 2017). H3.Y has been found to change expression of genes (upregulation and downregulation) involved in cell cycle control in response to stress stimuli, leading to the reduction of cell growth (Wiedemann et al., 2010). To date, no specific diseases have been associated with H3.X or H3.Y. Future research, including the identification of proteins interacting with these two histone variants by mass spectrometry and the analysis of their expression in normal and cancer tissues will unveil new function.

4.1.4 Testis-Specific Histones H3t and H3.5

Specifically expressed in testis, H3t plays a major role in the early stages of spermatogenesis by replacing the canonical histone H3 in spermatogonia. The absence of H3t alters the expression levels of meiosis-related genes, impairing entry into meiosis and leading to the absence of sperm, a medical condition called azoospermia (Ueda et al., 2017). The assumption is H3t loading occurs during DNA replication in the solely mitotic cells in testis, namely, spermatogonia. The instability of H3t nucleosome compared to H3.1 nucleosome may participate to chromatin reorganization processes occurring during spermatogenesis, such as the histone–protamine exchange (Tachiwana et al., 2010).

So far, two studies have described the role of human H3.5 histone variant (Schenk et al., 2011; Urahama et al., 2016). Resulting from the duplication of the H3.3 gene H3F3B, H3.5 is expressed in human seminiferous tubules and is accumulated around TSS. Like H3t, the instability of H3.5 nucleosomes and their detection during the first step of spermatogenesis suggest a potential role in the chromatin remodeling process required for sperm maturation.

4.2 The Human Histone H2A Family

The human histone H2A family is composed of histone variants H2A.X, H2A.Z, H2A.Bbd, and macroH2A (Bonisch and Hake, 2012).

4.2.1 The Histone Variant H2A.X

Defined by a SQ[E/D]P motif in the C-terminal domain, H2A.X is well characterized for its signaling role in the DNA damage response. Upon double-strand break occurrence, H2A.X is phosphorylated at serine 139 (labeled γH2A.X) by ATM (Ataxia Telangiectasia Mutated) kinase protein and forms foci spreading on each side of the damage (Kinner et al., 2008; Rogakou et al., 1998). This event promotes DNA repair by contributing to chromatin accessibility, and containing repair factors adjacent to the break.

Beyond this role in DNA damage response, H2A.X is involved in the maintenance of chromatin structure and development. γH2A.X, accumulated at mitotic kinetochores, contributes to the recruitment of MDC1, a mediator of DNA damage checkpoint (Eliezer et al., 2014). With ATM, H2A.X and MDC1 activate the formation of the mitotic spindle assembly checkpoint complex SAC, which regulates mitotic progression. The signal activating this ATM/γH2A.X/MDC1 pathway has not been identified, and the presence of endogenous DNA damages at centromeres remains plausible.

Another relevant role of H2A.X is in the establishment of an authentic epigenetic environment for sex chromosome. During meiosis in male mammals, H2A.X is accumulated with macroH2A and PARP2 in a subnuclear domain called sex body, whose function is to repress gene expression of sex chromosomes (Baarends et al., 1999; Dantzer et al., 2006; Turner et al., 2001). Loss of H2A.X leads to male infertility due to the abnormal formation of the sex body, alteration of transcriptional silencing of sex chromosomes (called MSCI—Meiotic Sex Chromosome Inactivation), and defects of meiotic recombination, indicating a major function of H2A.X in germ cells (Celeste et al., 2002; Fernandez-Capetillo et al., 2003). In somatic cells, H2A.X is transiently located within the territory of the inactivated chromosome X in late S phase (Chadwick and Lane, 2005). With the DNA damage factor BRCA1, H2A.X participates to the reestablishment of epigenetic markers characterizing inactivated X chromosome (e.g., histone hypoacetylation, enrichment of macroH2A) following DNA replication. Altogether, these two roles contribute to the maintenance of genome stability to prevent development defects and genomic instability.

4.2.2 The Histone Variant H2A.Z

H2A.Z is encoded by two different genes (i.e., H2AFZ and H2AFV) that generate two proteins, H2A.Z.1 and H2A.Z.2, differing by three residues and presenting 60% of similarity with canonical H2A (Eirin-Lopez et al., 2009). The two isoform transcripts are expressed in various human tissues, with some displaying either a higher level of H2A.Z.1 (e.g., brain) or H2A.Z.2 (e.g., liver and kidney) (Dryhurst et al., 2009). These differences associated with the lethality of H2A.Z.1 knockout mice strongly suggest that H2A.Z isoforms may have nonredundant function (Faast et al., 2001).

Ubiquitously expressed, H2A.Z nucleosomes are involved in all DNA-dependent mechanisms, from DNA repair to transcription by both stabilizing and destabilizing chromatin, and by regulating heterochromatin/euchromatin status and chromosome segregation (Billon and Cote, 2013; Zlatanova and Thakar, 2008). For instance, both H2A.Z histones are found associated with phosphorylated H3S10, supporting their role in chromatin condensation and during mitosis (Dryhurst et al., 2009). On the other hand, H2A.Z-mediated transcription is characterized by the presence of H2A.Z and trimethylated H3K4 nucleosomes near TSS (John et al., 2008). In vitro, the absence of H2A.Z-nucleosome at the +1 position of TSS results in transcription inhibition (Weber et al., 2014). In addition, H2A.Z is acetylated at euchromatin domains, but not heterochromatin (Venkatasubrahmanyam et al., 2007). As a binding site of the double-bromodomain-containing protein BRD2, acetylated H2A.Z regulates gene expression in hyperacetylated H4 environment (Draker et al., 2012). Heterotypic H2A/H2A.Z or homotypic H2A.Z/H2A.Z nucleosomes display specific genomic patterns all over the genome, supporting the major function of H2A.Z in gene expression (Weber et al., 2010).

4.2.3 The Histone Variant H2A.Bbd

Like CENP-A, H2A.Bbd (Barr Body Deficient) is a rapidly evolving histone variant that shares ~50% sequence identity with H2A (Chadwick and Willard, 2004; Gonzalez-Romero et al., 2008; Malik and Henikoff, 2003). Primarily expressed in testis and to a lower level in brain, H2A.Bbd is enriched in active regions of genome but excluded from the inactive X chromosome (Soboleva et al., 2011). The potential role of H2A.Bbd in sperm is to activate gene expression after fertilization. In contrast to canonical H2A, H2A.Bbd does not have the property to compact nucleosome, due to the absence of glutamate residues that usually form an acidic patch on the surface of H2A/H2B dimer and contribute to the higher order chromatin structure (Kalashnikova et al., 2013). Further biochemical analyzes have shown that H2A.Bbd nucleosomes are less stable than those containing H2A and are more rapidly exchanged, supporting H2A.Bbd role in active transcription (Bao et al., 2004; Gautier et al., 2004). The downregulation of H2A.Bbd in human cells induces both up- and downregulation in gene expression, as well as alteration of mRNA splicing with increased retention of intron and exon inclusion (Tolstorukov et al., 2012). This last function is supported by the binding of protein components of spliceosome to H2A.Bbd chromatin. Finally, the unstable H2A.Bbd nucleosome has also been described at sites of DNA synthesis and DNA damage, where the histone variant is transiently incorporated, suggesting H2A.Bbd may have a general function in maintaining an unstable chromatin to facilitate DNA-dependent mechanisms (Sansoni et al., 2014).

4.2.4 The Histone Variant macroH2A

macroH2A is encoded by two different genes H2AY and H2AFY2, generating three different histones: macroH2A1.1 and macroH2A1.2 from splicing events of H2AY, and macroH2A2 from H2AFY2 (Rasmussen et al., 1999). In contrast to other histones and histone variants, macroH2A contains a 25-kDa nonhistone domain at its C-terminal region called macrodomain that extends outside of the core nucleosome to form a docking site for poly(ADP-ribosyl)ated proteins (Han et al., 2011; Posavec et al., 2013). It is noteworthy that macrodomain of macroH2A1.1, but not macroH2A1.2 or macroH2A2, has been described as reader of NAD^+-derived metabolites PAR and *O*-acetyl catalyzed by PARP and sirtuin proteins, respectively (Kustatscher et al., 2005). This selectivity results from structural changes that may explain specific functions of these three histone variants.

macroH2A is preferentially accumulated at repressive chromatin domains, including inactive X chromosome and regulatory regions of pluripotency genes (Costanzi and Pehrson, 1998; Pasque et al., 2012). At inactive X chromosome, macroH2A colocalizes with Xist RNA, H3K27me3, and PARP1, to maintain the heterochromatic structure of this chromosome (Chadwick and Willard, 2004; Nusinow et al., 2007a,b; Ouararhni et al., 2006; Timinszky et al., 2009). Furthermore, macroH2A1.1 can form transcriptionally active domains (Chen et al., 2014). In the presence of PARP1, macroH2A1 regulates CBP (cAMP response element-binding protein (CREB)-binding protein)-dependent acetylation of H2BK12 and H2BK120 in primary human cells, leading to the organization of an acetylated and macroH2A-enriched chromatin. This regulation pathway is lost in cancer cells. Thus, macroH2A1 is a positive and negative transcription regulator, whose function may be modulated by PARP protein.

5. HISTONE CHAPERONES

Chaperones are histone-binding proteins that modulate the structure of chromatin by assisting the assembly and disassembly of the nucleosome (Elsasser and D'Arcy, 2013). Via an ATP-dependent mechanism, chaperones spatially and temporally regulate the localization of histones and indirectly PTMs associated to histones. Consequently, the absence or mutation or overexpression of chaperone can affect chromatin structure and DNA-dependent mechanisms, leading to genome instability. So far, the in vivo chaperone activity of three histone variants has been described and is summarized in this section.

5.1 The Two Chaperones of H3.3 Histones: HIRA and ATRX/DAXX Complexes

H3.3 can be either associated with euchromatin or heterochromatin. These specific localizations are dependent on HIRA and DAXX/ATRX chaperones, which indirectly control H3.3 function. DAXX/ATRX complex interacts with a unique H3.3 motif (residues 87–90) and deposits this histone variant at pericentromeres, telomeres, and repetitive elements (Liu et al., 2012). On the other hand, HIRA complex mediates H3.3 localization at bivalent promoters, active promoters, and transcribed gene bodies (Hamiche and Shuaib, 2013).

Until recently, the deposition of H3.3 mediated by the multiprotein HIRA complex was obscure. New biochemical structural studies have revealed that HIRA complex specifically recognizes and interacts with H3.3 via its UBN1 (ubinuclein-1) subunit (Ricketts and Marmorstein, 2016; Ricketts et al., 2015). In a similar mechanism than DAXX/ATRX, UBN1 through its Hpc2-related domain (HRD) binds preferentially H3.3 G90 residue. Mutation of this specific residue abolishes the interaction (Ricketts et al., 2015).

DAXX/ATRX and HIRA complexes function at loading the same histone variant in different chromatin context, highlighting the crosstalk between histone variant, chaperone, and PTMs for the correct targeting and functionality of chromatin regions. DAXX/ATRX-dependent deposition leads to the accumulation of both H3.3 and H3K9me3 at pericentromeres, whereas H3.3 loading is mediated by H3K27me3 or H3K36me3 enrichment at bivalent promoters and transcribed genes, respectively (Chen et al., 2013; Goldberg et al., 2010; Lewis et al., 2010; Voon et al., 2015). In addition, H3.3 chaperones can directly interact with chromatin remodelers, such as HIRA with PRC2 for the silencing of developmental genes (Banaszynski et al., 2013). Taken together, DAXX/ATRX and HIRA complexes are major contributors of H3.3-dependent gene regulation and genome stability.

5.2 The Centromeric Chaperone HJURP

HJURP has been initially described as DNA damage response factor (Kato et al., 2007). Upon DNA double-strand breaks, HJURP is recruited at the damage, where it interacts with two proteins of the MRN complex, hMSH5 and NBS1, for an effective homologous recombination pathway. HJURP is more likely involved in ATM pathway, a major phosphorylation cascade that regulates the choice between cell cycle arrest and DNA repair or apoptosis for the maintenance of chromosome stability (Kato et al., 2007). How HJURP exactly contributes to this molecular pathway is still elusive.

Its chaperone function has been more extensively explored. HJURP is the only described chaperone for CENP-A (Dunleavy et al., 2009; Foltz et al., 2009). It mediates the deposition of CENP-A at centromeres at late mitosis–early G1, by a unique mechanism requiring the dimerization of HJURP, activity of Mis18 complex, and noncoding RNAs (Barnhart et al., 2011; Foltz et al., 2009; McNulty et al., 2017; Quenet and Dalal, 2014; Zasadzinska et al., 2013). The integrity of this mechanism is modulated by the expression level of both CENP-A and HJURP. Misregulation (overexpression or downregulation) of CENP-A or HJURP altered the ratio between these two proteins, leading to mitotic defects and genomic instability (Lacoste et al., 2014; Mishra et al., 2011).

5.3 The Two Chaperones of H2A.Z: Anp32e and YL1

H2A.Z is deposited in a replication-independent manner by two chaperones, Anp32e and YL1 (Latrick et al., 2016; Mao et al., 2014; Obri et al., 2014). Anp32e is a component of the p400/Tip60 complex (Mao et al., 2014; Obri et al., 2014). This protein contributes to the remodeling of H2A.Z-containing chromatin, especially in response to DNA damage (Alatwi and Downs, 2015; Gursoy-Yuzugullu et al., 2015). The preferential binding of Anp32e to H2A.Z–H2B dimers compared to H2A–H2B dimers results in specific recognition of the extended H2A.Z α-helix at the C-terminal domain (Mao et al., 2014). This interaction leads to the removal of H2A.Z from chromatin. Deposition of H2A.Z–H2B dimer is mediated by YL1, a subunit of SRCAP (Snf2-related CREBBP activator protein) (Latrick et al., 2016; Liang et al., 2016). YL1 recognizes H2A.Z via its C-terminal α-helix (like Anp32e) and the acidic patch.

The two transcription elongation factors FACT (facilitates chromatin transcription) and Spt6 are histone chaperones that assist nucleosome assembly and disassembly during transcription (Avvakumov et al., 2011; Burgess and Zhang, 2013). A recent study using budding yeast as a model has identified both FACT and Spt6 as regulator of H2A.Z localization (Jeronimo et al., 2015). The absence of the two chaperones is associated with aberrant H2A.Z loading at gene bodies that leads to cryptic transcription due to ubiquitous recruitment of the ATP-dependent chromatin remodeling complex SWR-C. In contrast, in wild-type conditions, FACT and Spt6 counteract the deposition of H2A.Z excess. Interestingly, the ectopic localization of the centromeric variant CENP-A at euchromatin is also prevented by FACT in budding yeast, suggesting a potential role for this chaperone in the removal of histones at actively transcribed genes (Deyter and Biggins, 2014). The correct localization of histone variant is not only dependent on an accurate deposition but also on preventing pervasive recruitment that will alter chromatin structure and dysregulate gene expression, potentially contributing to diseases development.

5.4 Other Potential Histone Chaperones

Identifying testis-specific chaperones can be challenging, due to the low quantity of biological material. The in vivo chaperone catalyzing the replacement of H3 by H3t and H3.5 is unknown. However, in vitro studies have revealed that the testis-specific chaperone NAP2 (nucleosome assembly protein), but not the ubiquitous NAP1, promotes nucleosome assembly with H3t/H4 (Tachiwana et al., 2008). This in vitro specificity suggests the NAP2 may be the H3t chaperone. Further studies are in need to validate this hypothesis.

To date, no chaperone has been described with certitude for macroH2A, yet ATRX is a potential candidate. ATRX, but not DAXX, is interacting with macroH2A (Ratnakumar et al., 2012). In addition, in human erythroleukemia cells, the absence of ATRX is accompanied by the increase of macroH2A and the decrease of H3.3 at telomeres (Ratnakumar et al., 2012). Still, it is unclear whether the accumulation of macroH2A results from the loss of H3.3 or an active ATRX-dependent deposition. It is foreseen that the potential ATRX chaperone function for macroH2A will be further investigated.

FACT is an important factor during chromatin remodeling event that participates to the assembly/disassembly of several types of nucleosomes. As previously mentioned, ectopic localization of CENP-A and H2A.Z is prevented by the chaperone activity of FACT (Deyter and Biggins, 2014; Jeronimo et al., 2015). Upon DNA damages, FACT facilitates H2A.X nucleosome exchange (Heo et al., 2008). This example reveals that FACT is not only a proofreading chaperone and can be required for the correct process of DNA-dependent mechanism, such as DNA repair.

To analyze nucleosome structure, a common technique is to purify recombinant histones and then to mix them with DNA in the presence or not of a chaperone. Most of the time, this in vitro assay is performed in the presence of the chaperone NAP1. This approach provides structural characteristics of the nucleosome, as well as a postulate regarding the identity of the chaperone. Thus, H3.Y- and H2A.Bbd nucleosomes have been successfully reconstituted in vitro in the presence of NAP1 (Kujirai et al., 2016; Okuwaki et al., 2005). Without doubt, future research will characterize still unknown in vivo chaperones of histone variants, especially for H3.5 and H3.X. A successful strategy is the immunoprecipitation of the histone variant followed by mass spectrometry to identify associated proteins. Potential obstacles include the low abundance of histone variant, the specificity of the antibody used for the purification step, the high percentage of similarity between histones, and the choice of tested cellular conditions (e.g., after DNA damage, at a specific step of the cell cycle, during development). Like H3.3, a specific histone variant may be loaded by several chaperones for regulation purpose, and reciprocally, a chaperone may load or remove different histone variants, as described for FACT in budding yeast. Altogether, identifying chaperone identity is in need to better comprehend the function of the histone variants and consequences of altered functions.

6. HISTONE VARIANTS, CHAPERONES, AND DISEASES

6.1 Somatic Mutations of H3.3 Histone Variants and Their Chaperones

The identification of high-frequency mutations of H3.3 proteins in pediatric glioblastoma has led to a broad effort to inventory all H3.3 mutations, as well as other histones in various cancer types (Table 3; Maze et al., 2014). Currently, somatic H3.3 mutations have been found associated with pediatric brain tumors, and five classes of bone and cartilage neoplasms (Behjati et al., 2013; Gessi et al., 2013; Gielen et al., 2013; Khuong-Quang et al., 2012; Schwartzentruber et al., 2012; Wu et al., 2012a).

In the majority of chondroblastoma (95% of tested samples), lysine 36 of H3.3 is substituted with methionine in either H3F3A or H3F3B genes (Behjati et al., 2013). On the other hand, only mutations in H3F3A gene leading to Glycine 34 substitution with Tryptophane or Leucine have been described in giant cell tumor of bone. Interestingly, for both tumors, the presence of the mutation has been reported predominantly in cells from the stromal compartment. While H3F3A and H3F3B genes display different exonic and intronic DNA sequences, the amino acid sequence of encoded H3.3 is identical. The observation of a preferred mutation for chondroblastoma (H3F3B) vs giant cell tumor of bone (H3F3A) indicates an absence of redundancy of the two H3.3 proteins and/or a specific tissue expression of H3F3

Table 3 List of H3.3 Mutations and Their Associated Diseases

Gene	Residue	Disease	References
H3F3A	K27M	Pediatric diffuse intrinsic pontine glioma	Khuong-Quang et al. (2012), Wu et al., (2012a), and Sturm et al. (2012)
H3F3A	K27M or G34R/V	Pediatric glioblastoma	Gessi et al. (2013)
H3F3A	G34W ≫ G34L	Giant cell tumor of bone	Behjati et al. (2013)
H3F3B ≫ H3F3A	K36M	Chondroblastoma	Behjati et al. (2013)
H3F3B	K36M	Clear cell chondrosarcoma	Behjati et al. (2013)
H3F3B	K36M	Conventional chondrosarcoma	Behjati et al. (2013)
H3F3A and H3F3B	G34R	Osteosarcoma	Behjati et al. (2013)

Residue: G, glycine; K, lysine; L, leucine; M, methionine; R, arginine; V, valine; W, tryptophan.
Adapted from Maze, I., Noh, K.M., Soshnev, A.A., Allis, C.D., 2014. Every amino acid matters: essential contributions of histone variants to mammalian development and disease. Nat. Rev. Genet. 15, 259–271.

genes. To date, these differential expression profiles have been described in normal human tissues, but not in tumors (Frank et al., 2003).

Direct consequences of H3.3 mutations are the loss of a specific PTM (i.e., acetylation or methylation) and/or of reader interaction, leading to the alteration of chromatin structure and function, especially gene expression in physiological and pathological conditions (Maze et al., 2014). Hence, in pediatric brain tumors, H3K27M mutation is associated with a decrease level of di- and trimethylation (Chan et al., 2013; Lewis et al., 2013). This drop does not result from the absence of recruitment of EZH2, the catalytic subunit of PRC2, but the inhibition of its methyltransferase activity.

Recognition of histone variant by their chaperone can also be affected by mutation in the chaperone. So far, no mutation of HIRA complex has been reported in tumors, suggesting the H3.3 localization at actively transcribed genes is not altered. In addition, H3.3 mutations described (K27, G34, and K36) do not play a role in the expression of euchromatic genes. These data indicate that the oncogenic property of H3.3 is more likely associated with its localization at heterochromatin and its role in the maintenance of genome integrity.

Mutations in DAXX have been described in pediatric glioblastoma multiforme (Schwartzentruber et al., 2012). DAXX mutation (observed in 14/42 samples) leads to an increase of alternative lengthening of telomeres and genomic instability, potentially due to the loss of H3.3 at telomeres (Schwartzentruber et al., 2012). Because ATRX forms a heterodimer with DAXX, it is not surprising that mutations of the histone chaperone have also been found (observed in 2/42 samples), but to a lower extend.

Somatic mutations in both ATRX and DAXX have also been recounted in pancreatic neuroendocrine tumors (Marinoni et al., 2014). Detailled analysis revealed the nature of these mutations (i.e., insertion, deletion, and nonsense/missense mutation), and their preferential localization. In ATRX, most of missense mutations are within the amino-terminus ADD (ATRX–DNMTT3–DNMT3L) domain and the C-terminal helicase domain. The former is a recognition and binding site of the heterochromatic mark H3K9me3, clarifying how DAXX/ATRX/H3.3 complex is recruited at heterochromatin (Eustermann et al., 2011; Iwase et al., 2011).

Similarly, ADD domain of ATRX has been found mutated in ~50% of patients diagnosed with α-thalassemia mental retardation syndrome (Gibbons et al., 1995a,b; Picketts et al., 1996). The observation that mutations in ATRX are associated with diseases as different as cancer and mental retardation syndrome indicates the central role of chaperone in genome integrity. Investigating whether H3.3 histone loss contributes to the development of these diseases is in high interest to propose potential epigenetic treatment.

6.2 CENP-A and HJURP in Cancers

The overexpression of CENP-A and its chaperone HJURP has been described in numerous cancers, including colorectal, breast, and lung cancers (Hu et al., 2010; McGovern et al., 2012; Qiu et al., 2013; Wu et al., 2012b; Zhang et al., 2016). The overexpression of HJURP in colon, gastric, lung, and brain cancers is an indicator of poor prognostic (de Tayrac et al., 2013; Hu et al., 2010; Kato et al., 2007; Valente et al., 2013; Zhang et al., 2016). For some tumors, such as lung and breast cancers, a co-upregulation is observed, without comprehending the function of this expression level in tumorigenesis.

When only CENP-A is overexpressed, this histone variant is deposited all over the genome by two molecular mechanisms dependent on HJURP and DAXX/ATRX chaperones (Athwal et al., 2015; Lacoste et al., 2014).

In these conditions in HeLa cells, centromeric localization is still dependent on HJURP, whereas the excess of CENP-A binds to DAXX/ATRX and is incorporated at ectopic loci, such as chromosome arms (Lacoste et al., 2014). To a lesser extent, this interaction is also observed when CENP-A is not overexpressed in HeLa cells, but the histone variant is only observed at the centromere, suggesting a proofreading mechanism for ectopic CENP-A removal (Lacoste et al., 2014). Such molecular mechanism has not yet been shown in human. Nonetheless, in yeast, the chromatin remodeling factor FACT and the E3 ubiquitin ligase Psh1 remove CENP-A which is subsequently degraded by the proteasome (Collins et al., 2004; Deyter and Biggins, 2014; Kitagawa et al., 2014).

It is unknown why CENP-A prefers DAXX/ATRX as back-up chaperone among the other H3 histone chaperones. CENP-A is the most distant H3 variant, and the peptide motif required for DAXX/ATRX binding to H3.3 (residues 87–90: AAIG) is not conserved in CENP-A (residues: QALL). However, most of these residues are nonpolar amino acids (A, I, G, and L), suggesting they may share characteristics. Further investigations are in need to describe the recognition and binding mechanisms of DAXX/ATRX with CENP-A.

Ectopic deposition of CENP-A may generate dicentric chromosomes and neocentromeres (i.e., ectopic centromeres that may originate from non-centromeric regions of chromosomes), which are associated with increased frequency of chromosome missegregation, DNA damages, and aneuploidy (Crasta et al., 2012; Janssen et al., 2011; Santaguida and Amon, 2015; Wu et al., 2012b). To date, more than 100 neocentromeres isolated from patients mostly affected by developmental disabilities have been described (Alonso et al., 2010; Klein et al., 2012; Marshall et al., 2008). With the exception of a higher frequency at 3q, 13q, and 15q regions, no common characteristics among neocentromeres have been identified (Burrack and Berman, 2012; Liehr et al., 2010; Marshall et al., 2008). The formation of neocentromere is still a rare event that may occur on already existing surnumerary chromosomes. One hypothesis is neocentromere formation is a cellular mechanism to stabilize the genome in tumors (Burrack and Berman, 2012). To date, it is unclear whether CENP-A or HJURP overexpression is a cause of neocentromere formation and genome instability in cancer.

Recently, a "centromeric and kinetochore gene expression score" (CES) has been proposed to evaluate chromosomal instability due to aberrant centromere and kinetochore function (Zhang et al., 2016). This CES comprises the expression level of CENP-A and numerous factors involved in the targeting, loading, and deposition of CENP-A, including HJURP, Mis18B, and CENP-K/L/M/N. Interestingly, a high CES correlates with genomic instability and poor survival prognostic in tested tumors (e.g., breast and lung cancers). In addition, a high CES can predict a good response to chemotherapy and radiotherapy. This observation is not surprising regarding growing evidence that increasing genome instability can be a therapeutic approach (Bakhoum et al., 2015).

6.3 γH2A.X in Aging

Defect in DNA damage repair is a common feature of aged cells that lead to accumulation of unrepaired DNA breaks and increased frequency of mutations and genomic instability (Burgess et al., 2012). Unsurprisingly, several human premature aging diseases are associated with defects in the DNA damage repair mechanism (e.g., *Xeroderma pigmentosum*, Werner syndrome, Cockayne syndrome).

The accumulation of γH2A.X at telomeres triggers persistent DNA damage response and senescence. Senescence will indirectly cause aging by stopping irreversible cell division and DNA replication, increasing activity of senescence-associated β-galactosidase (SA-β-gal), altering chromatin organization and gene expression, and inducing the senescence-associated secretory phenotype (SASP) (Campisi and d'Adda di Fagagna, 2007). SASP proteins comprise cytokines, growth factors, and proteases that can have opposite effects: beneficial by contributing to tissue repair or deleterious by promoting inflammation and tumorigenesis (Freund et al., 2010). Thus, it is tempting to propose that γH2A.X contributes to any senescent-associated diseases by its continual phosphorylation that is activating DNA damage response.

H2A.X is also associated with cancer. H2AFX gene is localized on chromosome 11 (11q23) in a region frequently mutated or deleted in hematopoietic malignancies (Bassing et al., 2003). Moreover, accumulation of γH2A.X has been described in numerous human cancer cell lines, such as cervical and ovarian cancers (Banath et al., 2004; Yu et al., 2006). Due to its role in DNA damage response, it has been proposed that the quantification of γH2A.X signal may predict tumor sensitivity to chemo- or radiotherapy (Sedelnikova and Bonner, 2006). By targeting γH2A.X-expressing tumors, DNA damaging treatment will overwhelm DNA damage repair pathways and activate cell death. This strategy may be a simple approach to select patients who will more likely respond to therapy.

6.4 H2A.Z and YL1 in Cancers and Floating-Harbor Syndrome

H2A.Z overexpression has been described in several types of cancers including metastatic melanoma, bladder, prostate, and breast cancers (Baptista et al., 2013; Hua et al., 2008; Kim et al., 2013; Svotelis et al., 2010; Vardabasso et al., 2015). The limit of most of these studies is the absence of discrimination between the two isoforms H2A.Z.1 and H2A.Z.2. Two recent works have characterized specific functions for H2A.Z.1 and H2A.Z.2 in tumorigenesis.

In melanoma cells, H2A.Z.2 promotes proliferation by upregulating expression level of E2F target genes (Vardabasso et al., 2015). These transcription changes result from the H2A.Z.2 accumulation at promoters that stabilizes histone acetylation and the subsequent recruitment of the bromodomain and extraterminal motif (BET) protein BRD2 and the transcription factor E2F1. Interestingly, H2A.Z.2 downregulation in melanoma cells prevents chromatin acetylation at E2F target genes and blocks cell cycle progression at G1/S. H2A.Z.2-depleted cells are sensitive to the BET inhibitor JQ1, leading to cell death. Unfortunately, the short lifetime of JQ1 restricts its use to laboratory research; therefore, alternative drugs displaying similar activity are in need for clinical purpose. Altogether, H2A.Z.2 is a driver of cell proliferation and drug sensitivity in malignant melanoma.

The overexpression of H2A.Z.1 in patients diagnosed with liver cancer (HCC, hepatocellular carcinoma) has been proposed as a marker of poor prognostic (Yang et al., 2016). Like H2A.Z.2, H2A.Z.1 promotes cell proliferation by regulating transcription of cell cycle proteins. This H2A.Z isoform also modulates the epithelial–mesenchymal transition (EMT), a cellular mechanism that initiates metastasis. Downregulation of H2A.Z.1 induces G1/S arrest and apoptotic cell death, and inhibits EMT activation by altering the expression level of E-cadherin and fibronectin in a TGFβ-dependent manner. H2A.Z.1 role as mediator of cell proliferation and transformation makes this histone variant a promising target for liver cancer therapy.

Studying H2A.Z chaperones may also have a potential interest, like Anp32e which is overexpressed in gastric cancers and associated with poor prognostic in myeloma (Tsukamoto et al., 2008; Walker et al., 2010). The other H2A.Z chaperone, SRCAP is a diagnostic marker for the Floating-Harbor syndrome (also known as Pelletier–Leisti syndrome) (Hood et al., 2012). This autosomal dominant disease is mainly characterized by delayed bone mineralization and growth deficiency (Leisti et al., 1975). In this syndrome, SRCAP, the ATPase subunit of SRCAP, is truncated due to genetic mutation, potentially affecting the integrity of the complex and perturbing H2A.Z deposition. This hypothesis is supported by the decreased loading of H2A.Z and acetylation level in lung adenocarcinoma cell lines depleted of SRCAP (Wong et al., 2007). Thus, it is tempting to speculate that Floating-Harbor syndrome derives to a certain degree from an inactive H2A.Z chaperone activity.

Floating-Harbor syndrome and Rubinstein–Taybi syndrome share phenotypic features, such as short stature and anomalous thumbs. The major cause of Rubinstein–Taybi syndrome is mutations in CBP and p300 genes. CBP interacts with the chaperone SRCAP to induce steroid receptor-mediated transcription (Monroy et al., 2003). The functional interaction between these two proteins and their genetic mutations may explain the clinical overlap between the two syndromes.

6.5 H2A.Bbd Overexpression in Hodgkin's Lymphoma

H2A.Bbd and macroH2A often display opposed characteristics, such as their localization, and their potential involvement in cancer (Chadwick and Willard, 2004; Tolstorukov et al., 2012). H2A.Bbd RNA has been found overexpressed in Hodgkin's lymphoma cells, whereas macroH2A suppresses melanoma progression (Kapoor et al., 2010; Winkler et al., 2012). How H2A.Bbd contributes to tumorigenesis is not well understood. Overexpression of H2A.Bbd in mouse embryonic fibroblasts is associated with shortened S phase and sensitivity to DNA damages (Sansoni et al., 2014). These phenotypes are observed in Hodgkin's lymphoma cells overexpressing H2A.Bbd, but not in cancer cells with low H2A.Bbd level. Consequently, this histone variant has been proposed as marker to predict success of radio- and chemotherapy for patient diagnosed with Hodgkin's lymphoma (Sansoni et al., 2014; Yeoh and Mikhaeel, 2011).

6.6 macroH2A in Cancer and Neurodegenerative Diseases

A general acceptance describes macroH2A isoforms as tumor suppressor; so, it is unsurprising that these histone variants are found downregulated in cancers (e.g., lung, colon; Sporn and Jung, 2012; Sporn et al., 2009). However, the situation is more complex for macroH2A1. Indeed, the ratio between macroH2A1.1 and macroH2A1.2, resulting from the preferential splicing for one isoform, determines the patient survival (Novikov et al., 2011). In lung cancer, a high ratio of macroH2A1.1 will have a better prognosis (Sporn et al., 2009). This observation validates the assumption that macroH2A1.1 acts as a tumor suppressor, whereas the role of macroH2A1.2 is context dependent (Cantarino et al., 2013). In HCC, both macroH2A1.1 and macroH2A1.2 are overexpressed (Borghesan et al., 2016). Treatment of cells with the DNA-demethylating agent, 5-aza-dC, induces senescence, whereas the same treatment in the absence of macroH2A1 inhibits cell proliferation, without inducing senescence. These data suggest that macroH2A1 is a modulator of senescence and a potential mark to predict prognosis and progression of HCC.

macroH2A has also been found associated with neurodegenerative diseases, especially Alzheimer's and Huntington diseases (Hu et al., 2011; Mueller et al., 2005). macroH2A is described as a biomarker of Huntington disease (Hu et al., 2011). In this disorder, macroH2A is overexpressed and contributes to the development of phenotypes associated with the loss of htt (huntingtin) protein. Upon DNA damage, macroH2A has a dual function in chromatin remodeling at sites of DNA damages and recruitment of DNA repair proteins in a PARP-dependent manner. In Huntington disease, mutant htt induces DNA double-strand breaks. By trapping the DNA repair protein Ku70, mutant htt prevents the activation of the DNA-dependent protein kinase (DNA-PK) and the repair by nonhomologous end joining (NHEJ), leading to the accumulation of damages (Enokido et al., 2010). A positive feedback loop will be activated, with the recruitment of PARP1, macroH2A, and DNA repair proteins involved in NHEJ, leading to Huntington diseases pathology.

Not all studies distinguish the two variants macroH2A1.1 and macroH2A1.2, challenging the understanding of their respective function in cancer and neurodegenerative diseases.

7. CONCLUDING REMARKS

Identification and characterization of histone variants are still ongoing with three new histone H3 variant recently characterized (Taguchi et al., 2017). It is fantastic to observe this large panel of features, such as localization, PTM, and function, for proteins with high sequence similarity. Thus, the variety of diseases associated with the upregulation or downregulation of histone variants is not surprising (Tables 3 and 4). It also demonstrates the importance of comprehending these epigenetics components to propose personalized therapeutic option. Future studies will unveil the intricate functional interaction between histone variant, chaperone, and PTM to improve diagnostic and treatment of diseases.

Table 4 Diseases Associated With Change of Expression or Mutation of Histone Variants. Only Examples Presented in the Review Are Listed

Histone Variant	Type of Alteration	Disease	References
CENP-A	Overexpression	Colorectal cancer, breast cancer, lung cancer	Hu et al. (2010), McGovern et al. (2012), Qiu et al. (2013), Wu et al. (2012b), and Zhang et al. (2016)
H3t	Downregulation	Azoospermia	Ueda et al. (2017)
H2A.X	Mutation or deletion	Hematopoietic malignancies	Bassing et al. (2003)
	Accumulation (γH2A.X)	Cervical cancer, ovarian cancer	Banath et al. (2004) and Yu et al. (2006)
H2A.Z	Overexpression	Metastatic melanoma, bladder cancer, prostate cancer, breast cancer, liver cancer	Baptista et al. (2013), Hua et al. (2008), Kim et al. (2013), Svotelis et al. (2010), Yang et al. (2016), and Vardabasso et al. (2015)
H2A.Bbd	Overexpression	Hodgkin's lymphoma	Chadwick and Willard (2004) and Tolstorukov et al. (2012)
macroH2A	Downregulation	Melanoma, lung cancer, colon cancer, Alzheimer's disease, Huntington disease	Sporn and Jung (2012), Sporn et al. (2009), Hu et al. (2011), and Mueller et al. (2005)
macroH2A.1	Overexpression	Hepatocellular carcinoma	Borghesan et al. (2016)

ACKNOWLEDGMENTS

I apologize to my colleagues for work that could not be cited owing to space limitations. Work in my laboratory is supported by start-up funding from the Robert Larner College of Medicine, University of Vermont and from the American Cancer Society—Institutional Research Grant (126773-IRG-14-196-01-IRG).

REFERENCES

Adam, S., Polo, S.E., Almouzni, G., 2013. Transcription recovery after DNA damage requires chromatin priming by the H3.3 histone chaperone HIRA. Cell 155, 94–106.

Adams-Cioaba, M.A., Min, J., 2009. Structure and function of histone methylation binding proteins. Biochem. Cell Biol. 87, 93–105.

Ahmad, K., Henikoff, S., 2002. The histone variant H3.3 marks active chromatin by replication-independent nucleosome assembly. Mol. Cell 9, 1191–1200.

Alatwi, H.E., Downs, J.A., 2015. Removal of H2A.Z by INO80 promotes homologous recombination. EMBO Rep. 16, 986–994.

Albig, W., Bramlage, B., Gruber, K., Klobeck, H.G., Kunz, J., Doenecke, D., 1995. The human replacement histone H3.3B gene (H3F3B). Genomics 30, 264–272.

Alonso, A., Hasson, D., Cheung, F., Warburton, P.E., 2010. A paucity of heterochromatin at functional human neocentromeres. Epigenetics Chromatin 3, 6.

Athwal, R.K., Walkiewicz, M.P., Baek, S., Fu, S., Bui, M., Camps, J., Ried, T., Sung, M.H., Dalal, Y., 2015. CENP-A nucleosomes localize to transcription factor hotspots and subtelomeric sites in human cancer cells. Epigenetics Chromatin 8, 2.

Avvakumov, N., Nourani, A., Cote, J., 2011. Histone chaperones: modulators of chromatin marks. Mol. Cell 41, 502–514.

Baarends, W.M., Hoogerbrugge, J.W., Roest, H.P., Ooms, M., Vreeburg, J., Hoeijmakers, J.H., Grootegoed, J.A., 1999. Histone ubiquitination and chromatin remodeling in mouse spermatogenesis. Dev. Biol. 207, 322–333.

Bakhoum, S.F., Kabeche, L., Wood, M.D., Laucius, C.D., Qu, D., Laughney, A.M., Reynolds, G.E., Louie, R.J., Phillips, J., Chan, D.A., et al., 2015. Numerical chromosomal instability mediates susceptibility to radiation treatment. Nat. Commun. 6, 5990.

Banaszynski, L.A., Wen, D., Dewell, S., Whitcomb, S.J., Lin, M., Diaz, N., Elsasser, S.J., Chapgier, A., Goldberg, A.D., Canaani, E., et al., 2013. Hira-dependent histone H3.3 deposition facilitates PRC2 recruitment at developmental loci in ES cells. Cell 155, 107–120.

Banath, J.P., Macphail, S.H., Olive, P.L., 2004. Radiation sensitivity, H2AX phosphorylation, and kinetics of repair of DNA strand breaks in irradiated cervical cancer cell lines. Cancer Res. 64, 7144–7149.

Bannister, A.J., Kouzarides, T., 2011. Regulation of chromatin by histone modifications. Cell Res. 21, 381–395.

Bao, Y., Konesky, K., Park, Y.J., Rosu, S., Dyer, P.N., Rangasamy, D., Tremethick, D.J., Laybourn, P.J., Luger, K., 2004. Nucleosomes containing the histone variant H2A.Bbd organize only 118 base pairs of DNA. EMBO J. 23, 3314–3324.

Baptista, T., Graca, I., Sousa, E.J., Oliveira, A.I., Costa, N.R., Costa-Pinheiro, P., Amado, F., Henrique, R., Jeronimo, C., 2013. Regulation of histone H2A.Z expression is mediated by sirtuin 1 in prostate cancer. Oncotarget 4, 1673–1685.

Barkauskaite, E., Jankevicius, G., Ahel, I., 2015. Structures and mechanisms of enzymes employed in the synthesis and degradation of PARP-dependent protein ADP-ribosylation. Mol. Cell 58, 935–946.

Barnhart, M.C., Kuich, P.H., Stellfox, M.E., Ward, J.A., Bassett, E.A., Black, B.E., Foltz, D.R., 2011. HJURP is a CENP-A chromatin assembly factor sufficient to form a functional de novo kinetochore. J. Cell Biol. 194, 229–243.

Bassing, C.H., Suh, H., Ferguson, D.O., Chua, K.F., Manis, J., Eckersdorff, M., Gleason, M., Bronson, R., Lee, C., Alt, F.W., 2003. Histone H2AX: a dosage-dependent suppressor of oncogenic translocations and tumors. Cell 114, 359–370.

Behjati, S., Tarpey, P.S., Presneau, N., Scheipl, S., Pillay, N., Van Loo, P., Wedge, D.C., Cooke, S.L., Gundem, G., Davies, H., et al., 2013. Distinct H3F3A and H3F3B driver mutations define chondroblastoma and giant cell tumor of bone. Nat. Genet. 45, 1479–1482.

Bergmann, J.H., Rodriguez, M.G., Martins, N.M., Kimura, H., Kelly, D.A., Masumoto, H., Larionov, V., Jansen, L.E., Earnshaw, W.C., 2011. Epigenetic engineering shows H3K4me2 is required for HJURP targeting and CENP-A assembly on a synthetic human kinetochore. EMBO J. 30, 328–340.

Billon, P., Cote, J., 2013. Precise deposition of histone H2A.Z in chromatin for genome expression and maintenance. Biochim. Biophys. Acta 1819, 290–302.

Blower, M.D., Sullivan, B.A., Karpen, G.H., 2002. Conserved organization of centromeric chromatin in flies and humans. Dev. Cell 2, 319–330.

Bonisch, C., Hake, S.B., 2012. Histone H2A variants in nucleosomes and chromatin: more or less stable? Nucleic Acids Res. 40, 10719–10741.

Borghesan, M., Fusilli, C., Rappa, F., Panebianco, C., Rizzo, G., Oben, J.A., Mazzoccoli, G., Faulkes, C., Pata, I., Agodi, A., et al., 2016. DNA hypomethylation and histone variant macroH2A1 synergistically attenuate chemotherapy-induced senescence to promote hepatocellular carcinoma progression. Cancer Res. 76, 594–606.

Britton, L.M., Gonzales-Cope, M., Zee, B.M., Garcia, B.A., 2011. Breaking the histone code with quantitative mass spectrometry. Expert Rev. Proteomics 8, 631–643.

Brochu, G., Duchaine, C., Thibeault, L., Lagueux, J., Shah, G.M., Poirier, G.G., 1994. Mode of action of poly(ADP-ribose) glycohydrolase. Biochim. Biophys. Acta 1219, 342–350.

Burgess, R.J., Zhang, Z., 2013. Histone chaperones in nucleosome assembly and human disease. Nat. Struct. Mol. Biol. 20, 14–22.

Burgess, R.C., Misteli, T., Oberdoerffer, P., 2012. DNA damage, chromatin, and transcription: the trinity of aging. Curr. Opin. Cell Biol. 24, 724–730.

Burrack, L.S., Berman, J., 2012. Neocentromeres and epigenetically inherited features of centromeres. Chromosome Res. 20, 607–619.

Campisi, J., d'Adda di Fagagna, F., 2007. Cellular senescence: when bad things happen to good cells. Nat. Rev. Mol. Cell Biol. 8, 729–740.

Cantarino, N., Douet, J., Buschbeck, M., 2013. MacroH2A—an epigenetic regulator of cancer. Cancer Lett. 336, 247–252.

Celeste, A., Petersen, S., Romanienko, P.J., Fernandez-Capetillo, O., Chen, H.T., Sedelnikova, O.A., Reina-San-Martin, B., Coppola, V., Meffre, E., Difilippantonio, M.J., et al., 2002. Genomic instability in mice lacking histone H2AX. Science 296, 922–927.

Chadee, D.N., Hendzel, M.J., Tylipski, C.P., Allis, C.D., Bazett-Jones, D.P., Wright, J.A., Davie, J.R., 1999. Increased Ser-10 phosphorylation of histone H3 in mitogen-stimulated and oncogene-transformed mouse fibroblasts. J. Biol. Chem. 274, 24914–24920.

Chadwick, B.P., Lane, T.F., 2005. BRCA1 associates with the inactive X chromosome in late S-phase, coupled with transient H2AX phosphorylation. Chromosoma 114, 432–439.

Chadwick, B.P., Willard, H.F., 2004. Multiple spatially distinct types of facultative heterochromatin on the human inactive X chromosome. Proc. Natl. Acad. Sci. U. S. A. 101, 17450–17455.

Chan, K.M., Fang, D., Gan, H., Hashizume, R., Yu, C., Schroeder, M., Gupta, N., Mueller, S., James, C.D., Jenkins, R., et al., 2013. The histone H3.3K27M mutation in pediatric glioma reprograms H3K27 methylation and gene expression. Genes Dev. 27, 985–990.

Chen, P., Zhao, J., Wang, Y., Wang, M., Long, H., Liang, D., Huang, L., Wen, Z., Li, W., Li, X., et al., 2013. H3.3 actively marks enhancers and primes gene transcription via opening higher-ordered chromatin. Genes Dev. 27, 2109–2124.

Chen, H., Ruiz, P.D., Novikov, L., Casill, A.D., Park, J.W., Gamble, M.J., 2014. MacroH2A1.1 and PARP-1 cooperate to regulate transcription by promoting CBP-mediated H2B acetylation. Nat. Struct. Mol. Biol. 21, 981–989.

Choi, J.K., Howe, L.J., 2009. Histone acetylation: truth of consequences? Biochem. Cell Biol. 87, 139–150.

Collins, K.A., Furuyama, S., Biggins, S., 2004. Proteolysis contributes to the exclusive centromere localization of the yeast Cse4/CENP-A histone H3 variant. Curr. Biol. 14, 1968–1972.

Costanzi, C., Pehrson, J.R., 1998. Histone macroH2A1 is concentrated in the inactive X chromosome of female mammals. Nature 393, 599–601.

Crasta, K., Ganem, N.J., Dagher, R., Lantermann, A.B., Ivanova, E.V., Pan, Y., Nezi, L., Protopopov, A., Chowdhury, D., Pellman, D., 2012. DNA breaks and chromosome pulverization from errors in mitosis. Nature 482, 53–58.

D'Amours, D., Desnoyers, S., D'Silva, I., Poirier, G.G., 1999. Poly(ADP-ribosyl)ation reactions in the regulation of nuclear functions. Biochem. J. 342 (Pt. 2), 249–268.

Dantzer, F., Mark, M., Quenet, D., Scherthan, H., Huber, A., Liebe, B., Monaco, L., Chicheportiche, A., Sassone-Corsi, P., de Murcia, G., et al., 2006. Poly(ADP-ribose) polymerase-2 contributes to the fidelity of male meiosis I and spermiogenesis. Proc. Natl. Acad. Sci. U. S. A. 103, 14854–14859.

de Tayrac, M., Saikali, S., Aubry, M., Bellaud, P., Boniface, R., Quillien, V., Mosser, J., 2013. Prognostic significance of EDN/RB, HJURP, p60/CAF-1 and PDLI4, four new markers in high-grade gliomas. PLoS One 8, e73332,

DesJarlais, R., Tummino, P.J., 2016. Role of histone-modifying enzymes and their complexes in regulation of chromatin biology. Biochemistry 55, 1584–1599.

Deyter, G.M., Biggins, S., 2014. The FACT complex interacts with the E3 ubiquitin ligase Psh1 to prevent ectopic localization of CENP-A. Genes Dev. 28, 1815–1826.

Draker, R., Ng, M.K., Sarcinella, E., Ignatchenko, V., Kislinger, T., Cheung, P., 2012. A combination of H2A.Z and H4 acetylation recruits Brd2 to chromatin during transcriptional activation. PLoS Genet. 8, e1003047,

Dryhurst, D., Ishibashi, T., Rose, K.L., Eirin-Lopez, J.M., McDonald, D., Silva-Moreno, B., Veldhoen, N., Helbing, C.C., Hendzel, M.J., Shabanowitz, J., et al., 2009. Characterization of the histone H2A.Z-1 and H2A.Z-2 isoforms in vertebrates. BMC Biol. 7, 86.

Dunleavy, E.M., Roche, D., Tagami, H., Lacoste, N., Ray-Gallet, D., Nakamura, Y., Daigo, Y., Nakatani, Y., Almouzni-Pettinotti, G., 2009. HJURP is a cell-cycle-dependent maintenance and deposition factor of CENP-A at centromeres. Cell 137, 485–497.

Eirin-Lopez, J.M., Gonzalez-Romero, R., Dryhurst, D., Ishibashi, T., Ausio, J., 2009. The evolutionary differentiation of two histone H2A.Z variants in chordates (H2A.Z-1 and H2A.Z-2) is mediated by a stepwise mutation process that affects three amino acid residues. BMC Evol. Biol. 9, 31.

Eliezer, Y., Argaman, L., Kornowski, M., Roniger, M., Goldberg, M., 2014. Interplay between the DNA damage proteins MDC1 and ATM in the regulation of the spindle assembly checkpoint. J. Biol. Chem. 289, 8182–8193.

Elsasser, S.J., D'Arcy, S., 2013. Towards a mechanism for histone chaperones. Biochim. Biophys. Acta 1819, 211–221.

Elsasser, S.J., Noh, K.M., Diaz, N., Allis, C.D., Banaszynski, L.A., 2015. Histone H3.3 is required for endogenous retroviral element silencing in embryonic stem cells. Nature 522, 240–244.

Enokido, Y., Tamura, T., Ito, H., Arumughan, A., Komuro, A., Shiwaku, H., Sone, M., Foulle, R., Sawada, H., Ishiguro, H., et al., 2010. Mutant huntingtin impairs Ku70-mediated DNA repair. J. Cell Biol. 189, 425–443.

Eustermann, S., Yang, J.C., Law, M.J., Amos, R., Chapman, L.M., Jelinska, C., Garrick, D., Clynes, D., Gibbons, R.J., Rhodes, D., et al., 2011. Combinatorial readout of histone H3 modifications specifies localization of ATRX to heterochromatin. Nat. Struct. Mol. Biol. 18, 777–782.

Faast, R., Thonglairoam, V., Schulz, T.C., Beall, J., Wells, J.R., Taylor, H., Matthaei, K., Rathjen, P.D., Tremethick, D.J., Lyons, I., 2001. Histone variant H2A.Z is required for early mammalian development. Curr. Biol. 11, 1183–1187.

Fernandez-Capetillo, O., Mahadevaiah, S.K., Celeste, A., Romanienko, P.J., Camerini-Otero, R.D., Bonner, W.M., Manova, K., Burgoyne, P., Nussenzweig, A., 2003. H2AX is required for chromatin remodeling and inactivation of sex chromosomes in male mouse meiosis. Dev. Cell 4, 497–508.

Filipescu, D., Szenker, E., Almouzni, G., 2013. Developmental roles of histone H3 variants and their chaperones. Trends Genet. 29, 630–640.

Fischle, W., Tseng, B.S., Dormann, H.L., Ueberheide, B.M., Garcia, B.A., Shabanowitz, J., Hunt, D.F., Funabiki, H., Allis, C.D., 2005. Regulation of HP1-chromatin binding by histone H3 methylation and phosphorylation. Nature 438, 1116–1122.

Foltz, D.R., Jansen, L.E., Bailey, A.O., Yates 3rd, J.R., Bassett, E.A., Wood, S., Black, B.E., Cleveland, D.W., 2009. Centromere-specific assembly of CENP-A nucleosomes is mediated by HJURP. Cell 137, 472–484.

Frank, D., Doenecke, D., Albig, W., 2003. Differential expression of human replacement and cell cycle dependent H3 histone genes. Gene 312, 135–143.

Freund, A., Orjalo, A.V., Desprez, P.Y., Campisi, J., 2010. Inflammatory networks during cellular senescence: causes and consequences. Trends Mol. Med. 16, 238–246.

Fuchs, S.M., Strahl, B.D., 2011. Antibody recognition of histone post-translational modifications: emerging issues and future prospects. Epigenomics 3, 247–249.

Gautier, T., Abbott, D.W., Molla, A., Verdel, A., Ausio, J., Dimitrov, S., 2004. Histone variant H2ABbd confers lower stability to the nucleosome. EMBO Rep. 5, 715–720.

Gessi, M., Gielen, G.H., Hammes, J., Dorner, E., Muhlen, A.Z., Waha, A., Pietsch, T., 2013. H3.3 G34R mutations in pediatric primitive neuroectodermal tumors of central nervous system (CNS-PNET) and pediatric glioblastomas: possible diagnostic and therapeutic implications? J. Neurooncol 112, 67–72.

Gibbons, R.J., Picketts, D.J., and Higgs, D.R. 1995a. Syndromal mental retardation due to mutations in a regulator of gene expression. Hum. Mol. Genet. 4, 1705–1709.

Gibbons, R.J., Picketts, D.J., Villard, L., Higgs, D.R., 1995b. Mutations in a putative global transcriptional regulator cause X-linked mental retardation with alpha-thalassemia (ATR-X syndrome). Cell 80, 837–845.

Gielen, G.H., Gessi, M., Hammes, J., Kramm, C.M., Waha, A., Pietsch, T., 2013. H3F3A K27M mutation in pediatric CNS tumors: a marker for diffuse high-grade astrocytomas. Am. J. Clin. Pathol. 139, 345–349.

Goldberg, A.D., Banaszynski, L.A., Noh, K.M., Lewis, P.W., Elsaesser, S.J., Stadler, S., Dewell, S., Law, M., Guo, X., Li, X., et al., 2010. Distinct factors control histone variant H3.3 localization at specific genomic regions. Cell 140, 678–691.

Gonzalez-Romero, R., Mendez, J., Ausio, J., Eirin-Lopez, J.M., 2008. Quickly evolving histones, nucleosome stability and chromatin folding: all about histone H2A.Bbd. Gene 413, 1–7.

Gurley, L.R., D'Anna, J.A., Barham, S.S., Deaven, L.L., Tobey, R.A., 1978. Histone phosphorylation and chromatin structure during mitosis in Chinese hamster cells. Eur. J. Biochem. 84, 1–15.

Gursoy-Yuzugullu, O., Ayrapetov, M.K., Price, B.D., 2015. Histone chaperone Anp32e removes H2A.Z from DNA double-strand breaks and promotes nucleosome reorganization and DNA repair. Proc. Natl. Acad. Sci. U. S. A. 112, 7507–7512.

Hamiche, A., Shuaib, M., 2013. Chaperoning the histone H3 family. Biochim. Biophys. Acta 1819, 230–237.

Han, W., Li, X., Fu, X., 2011. The macro domain protein family: structure, functions, and their potential therapeutic implications. Mutat. Res. 727, 86–103.

Heo, K., Kim, H., Choi, S.H., Choi, J., Kim, K., Gu, J., Lieber, M.R., Yang, A.S., An, W., 2008. FACT-mediated exchange of histone variant H2AX regulated by phosphorylation of H2AX and ADP-ribosylation of Spt16. Mol. Cell 30, 86–97.

Hirota, T., Lipp, J.J., Toh, B.H., Peters, J.M., 2005. Histone H3 serine 10 phosphorylation by Aurora B causes HP1 dissociation from heterochromatin. Nature 438, 1176–1180.

Hood, R.L., Lines, M.A., Nikkel, S.M., Schwartzentruber, J., Beaulieu, C., Nowaczyk, M.J., Allanson, J., Kim, C.A., Wieczorek, D., Moilanen, J.S., et al., 2012. Mutations in SRCAP, encoding SNF2-related CREBBP activator protein, cause Floating-Harbor syndrome. Am. J. Hum. Genet. 90, 308–313.

Howman, E.V., Fowler, K.J., Newson, A.J., Redward, S., MacDonald, A.C., Kalitsis, P., Choo, K.H., 2000. Early disruption of centromeric chromatin organization in centromere protein A (Cenpa) null mice. Proc. Natl. Acad. Sci. U. S. A. 97, 1148–1153.
Hu, Z., Huang, G., Sadanandam, A., Gu, S., Lenburg, M.E., Pai, M., Bayani, N., Blakely, E.A., Gray, J.W., Mao, J.H., 2010. The expression level of HJURP has an independent prognostic impact and predicts the sensitivity to radiotherapy in breast cancer. Breast Cancer Res. 12, R18.
Hu, Y., Chopra, V., Chopra, R., Locascio, J.J., Liao, Z., Ding, H., Zheng, B., Matson, W.R., Ferrante, R.J., Rosas, H.D., et al., 2011. Transcriptional modulator H2A histone family, member Y (H2AFY) marks Huntington disease activity in man and mouse. Proc. Natl. Acad. Sci. U. S. A. 108, 17141–17146.
Hua, S., Kallen, C.B., Dhar, R., Baquero, M.T., Mason, C.E., Russell, B.A., Shah, P.K., Liu, J., Khramtsov, A., Tretiakova, M.S., et al., 2008. Genomic analysis of estrogen cascade reveals histone variant H2A.Z associated with breast cancer progression. Mol. Syst. Biol. 4, 188.
Huang, H., Sabari, B.R., Garcia, B.A., Allis, C.D., Zhao, Y., 2014. SnapShot: histone modifications. Cell 159 (458), e451.
Iwase, S., Xiang, B., Ghosh, S., Ren, T., Lewis, P.W., Cochrane, J.C., Allis, C.D., Picketts, D.J., Patel, D.J., Li, H., et al., 2011. ATRX ADD domain links an atypical histone methylation recognition mechanism to human mental-retardation syndrome. Nat. Struct. Mol. Biol. 18, 769–776.
Jankevicius, G., Hassler, M., Golia, B., Rybin, V., Zacharias, M., Timinszky, G., Ladurner, A.G., 2013. A family of macrodomain proteins reverses cellular mono-ADP-ribosylation. Nat. Struct. Mol. Biol. 20, 508–514.
Jansen, L.E., Black, B.E., Foltz, D.R., Cleveland, D.W., 2007. Propagation of centromeric chromatin requires exit from mitosis. J. Cell Biol. 176, 795–805.
Janssen, A., van der Burg, M., Szuhai, K., Kops, G.J., Medema, R.H., 2011. Chromosome segregation errors as a cause of DNA damage and structural chromosome aberrations. Science 333, 1895–1898.
Jeronimo, C., Watanabe, S., Kaplan, C.D., Peterson, C.L., Robert, F., 2015. The histone chaperones FACT and Spt6 restrict H2A.Z from intragenic locations. Mol. Cell 58, 1113–1123.
John, S., Sabo, P.J., Johnson, T.A., Sung, M.H., Biddie, S.C., Lightman, S.L., Voss, T.C., Davis, S.R., Meltzer, P.S., Stamatoyannopoulos, J.A., et al., 2008. Interaction of the glucocorticoid receptor with the chromatin landscape. Mol. Cell 29, 611–624.
Kalashnikova, A.A., Porter-Goff, M.E., Muthurajan, U.M., Luger, K., Hansen, J.C., 2013. The role of the nucleosome acidic patch in modulating higher order chromatin structure. J. R. Soc. Interface 10, 20121022.
Kapoor, A., Goldberg, M.S., Cumberland, L.K., Ratnakumar, K., Segura, M.F., Emanuel, P.O., Menendez, S., Vardabasso, C., Leroy, G., Vidal, C.I., et al., 2010. The histone variant macroH2A suppresses melanoma progression through regulation of CDK8. Nature 468, 1105–1109.
Kato, T., Sato, N., Hayama, S., Yamabuki, T., Ito, T., Miyamoto, M., Kondo, S., Nakamura, Y., Daigo, Y., 2007. Activation of Holliday junction recognizing protein involved in the chromosomal stability and immortality of cancer cells. Cancer Res. 67, 8544–8553.
Khuong-Quang, D.A., Buczkowicz, P., Rakopoulos, P., Liu, X.Y., Fontebasso, A.M., Bouffet, E., Bartels, U., Albrecht, S., Schwartzentruber, J., Letourneau, L., et al., 2012. K27M mutation in histone H3.3 defines clinically and biologically distinct subgroups of pediatric diffuse intrinsic pontine gliomas. Acta Neuropathol. 124, 439–447.
Kim, J., Daniel, J., Espejo, A., Lake, A., Krishna, M., Xia, L., Zhang, Y., Bedford, M.T., 2006. Tudor, MBT and chromo domains gauge the degree of lysine methylation. EMBO Rep. 7, 397–403.

Kim, K., Punj, V., Choi, J., Heo, K., Kim, J.M., Laird, P.W., An, W., 2013. Gene dysregulation by histone variant H2A.Z in bladder cancer. Epigenetics Chromatin 6, 34.
Kinner, A., Wu, W., Staudt, C., Iliakis, G., 2008. Gamma-H2AX in recognition and signaling of DNA double-strand breaks in the context of chromatin. Nucleic Acids Res. 36, 5678–5694.
Kitagawa, T., Ishii, K., Takeda, K., Matsumoto, T., 2014. The 19S proteasome subunit Rpt3 regulates distribution of CENP-A by associating with centromeric chromatin. Nat. Commun. 5, 3597.
Klein, E., Rocchi, M., Ovens-Raeder, A., Kosyakova, N., Weise, A., Ziegler, M., Meins, M., Morlot, S., Fischer, W., Volleth, M., et al., 2012. Five novel locations of Neocentromeres in human: 18q22.1, Xq27.1 approximately 27.2, Acro p13, Acro p12, and heterochromatin of unknown origin. Cytogenet. Genome Res. 136, 163–166.
Konecny, G.E., Kristeleit, R.S., 2016. PARP inhibitors for BRCA1/2-mutated and sporadic ovarian cancer: current practice and future directions. Br. J. Cancer 115, 1157–1173.
Kujirai, T., Horikoshi, N., Sato, K., Maehara, K., Machida, S., Osakabe, A., Kimura, H., Ohkawa, Y., Kurumizaka, H., 2016. Structure and function of human histone H3.Y nucleosome. Nucleic Acids Res. 44, 6127–6141.
Kujirai, T., Horikoshi, N., Xie, Y., Taguchi, H., Kurumizaka, H., 2017. Identification of the amino acid residues responsible for stable nucleosome formation by histone H3.Y. Nucleus 8, 239–248.
Kustatscher, G., Hothorn, M., Pugieux, C., Scheffzek, K., Ladurner, A.G., 2005. Splicing regulates NAD metabolite binding to histone macroH2A. Nat. Struct. Mol. Biol. 12, 624–625.
Lacoste, N., Woolfe, A., Tachiwana, H., Garea, A.V., Barth, T., Cantaloube, S., Kurumizaka, H., Imhof, A., Almouzni, G., 2014. Mislocalization of the centromeric histone variant CenH3/CENP-A in human cells depends on the chaperone DAXX. Mol. Cell 53, 631–644.
Latrick, C.M., Marek, M., Ouararhni, K., Papin, C., Stoll, I., Ignatyeva, M., Obri, A., Ennifar, E., Dimitrov, S., Romier, C., et al., 2016. Molecular basis and specificity of H2A.Z-H2B recognition and deposition by the histone chaperone YL1. Nat. Struct. Mol. Biol. 23, 309–316.
Leisti, J., Hollister, D.W., Rimoin, D.L., 1975. The Floating-Harbor syndrome. Birth Defects Orig. Artic. Ser. 11, 305.
Lewis, P.W., Elsaesser, S.J., Noh, K.M., Stadler, S.C., Allis, C.D., 2010. Daxx is an H3.3-specific histone chaperone and cooperates with ATRX in replication-independent chromatin assembly at telomeres. Proc. Natl. Acad. Sci. U. S. A. 107, 14075–14080.
Lewis, P.W., Muller, M.M., Koletsky, M.S., Cordero, F., Lin, S., Banaszynski, L.A., Garcia, B.A., Muir, T.W., Becher, O.J., Allis, C.D., 2013. Inhibition of PRC2 activity by a gain-of-function H3 mutation found in pediatric glioblastoma. Science 340, 857–861.
Li, G., Reinberg, D., 2011. Chromatin higher-order structures and gene regulation. Curr. Opin. Genet. Dev. 21, 175–186.
Liang, X., Shan, S., Pan, L., Zhao, J., Ranjan, A., Wang, F., Zhang, Z., Huang, Y., Feng, H., Wei, D., et al., 2016. Structural basis of H2A.Z recognition by SRCAP chromatin-remodeling subunit YL1. Nat. Struct. Mol. Biol. 23, 317–323.
Liehr, T., Kosyakova, N., Weise, A., Ziegler, M., Raabe-Meyer, G., 2010. First case of a neocentromere formation in an otherwise normal chromosome 7. Cytogenet. Genome Res. 128, 189–191.
Liu, C.P., Xiong, C., Wang, M., Yu, Z., Yang, N., Chen, P., Zhang, Z., Li, G., Xu, R.M., 2012. Structure of the variant histone H3.3-H4 heterodimer in complex with its chaperone DAXX. Nat. Struct. Mol. Biol. 19, 1287–1292.

Luger, K., Mader, A.W., Richmond, R.K., Sargent, D.F., Richmond, T.J., 1997. Crystal structure of the nucleosome core particle at 2.8 Å resolution. Nature 389, 251–260.
Mahadevan, L.C., Willis, A.C., Barratt, M.J., 1991. Rapid histone H3 phosphorylation in response to growth factors, phorbol esters, okadaic acid, and protein synthesis inhibitors. Cell 65, 775–783.
Malik, H.S., Henikoff, S., 2003. Phylogenomics of the nucleosome. Nat. Struct. Biol. 10, 882–891.
Mao, Z., Pan, L., Wang, W., Sun, J., Shan, S., Dong, Q., Liang, X., Dai, L., Ding, X., Chen, S., et al., 2014. Anp32e, a higher eukaryotic histone chaperone directs preferential recognition for H2A.Z. Cell Res. 24, 389–399.
Marinoni, I., Kurrer, A.S., Vassella, E., Dettmer, M., Rudolph, T., Banz, V., Hunger, F., Pasquinelli, S., Speel, E.J., Perren, A., 2014. Loss of DAXX and ATRX are associated with chromosome instability and reduced survival of patients with pancreatic neuroendocrine tumors. Gastroenterology 146 (453–460), e455,
Marshall, O.J., Chueh, A.C., Wong, L.H., Choo, K.H., 2008. Neocentromeres: new insights into centromere structure, disease development, and karyotype evolution. Am. J. Hum. Genet. 82, 261–282.
Martin-Hernandez, K., Rodriguez-Vargas, J.M., Schreiber, V., Dantzer, F., 2017. Expanding functions of ADP-ribosylation in the maintenance of genome integrity. Semin. Cell Dev. Biol. 63, 92–101.
Marzluff, W.F., Duronio, R.J., 2002. Histone mRNA expression: multiple levels of cell cycle regulation and important developmental consequences. Curr. Opin. Cell Biol. 14, 692–699.
Maze, I., Noh, K.M., Soshnev, A.A., Allis, C.D., 2014. Every amino acid matters: essential contributions of histone variants to mammalian development and disease. Nat. Rev. Genet. 15, 259–271.
McGovern, S.L., Qi, Y., Pusztai, L., Symmans, W.F., Buchholz, T.A., 2012. Centromere protein-A, an essential centromere protein, is a prognostic marker for relapse in estrogen receptor-positive breast cancer. Breast Cancer Res. 14, R72.
McKinley, K.L., Cheeseman, I.M., 2016. The molecular basis for centromere identity and function. Nat. Rev. Mol. Cell Biol. 17, 16–29.
McNulty, S.M., Sullivan, L.L., Sullivan, B.A., 2017. Human centromeres produce chromosome-specific and array-specific alpha satellite transcripts that are complexed with CENP-A and CENP-C. Dev. Cell 42, 226–240.
Meissner, A., Mikkelsen, T.S., Gu, H., Wernig, M., Hanna, J., Sivachenko, A., Zhang, X., Bernstein, B.E., Nusbaum, C., Jaffe, D.B., et al., 2008. Genome-scale DNA methylation maps of pluripotent and differentiated cells. Nature 454, 766–770.
Mishra, P.K., Au, W.C., Choy, J.S., Kuich, P.H., Baker, R.E., Foltz, D.R., Basrai, M.A., 2011. Misregulation of Scm3p/HJURP causes chromosome instability in Saccharomyces cerevisiae and human cells. PLoS Genet. 7, e1002303,
Monroy, M.A., Schott, N.M., Cox, L., Chen, J.D., Ruh, M., Chrivia, J.C., 2003. SNF2-related CBP activator protein (SRCAP) functions as a coactivator of steroid receptor-mediated transcription through synergistic interactions with CARM-1 and GRIP-1. Mol. Endocrinol. 17, 2519–2528.
Mueller, S.G., Weiner, M.W., Thal, L.J., Petersen, R.C., Jack, C., Jagust, W., Trojanowski, J.Q., Toga, A.W., Beckett, L., 2005. The Alzheimer's disease neuroimaging initiative. Neuroimaging Clin. N. Am. 15 (869–877), xi–xii.
Naar, A.M., Lemon, B.D., Tjian, R., 2001. Transcriptional coactivator complexes. Annu. Rev. Biochem. 70, 475–501.
Novikov, L., Park, J.W., Chen, H., Klerman, H., Jalloh, A.S., Gamble, M.J., 2011. QKI-mediated alternative splicing of the histone variant MacroH2A1 regulates cancer cell proliferation. Mol. Cell. Biol. 31, 4244–4255.

Nowak, S.J., Corces, V.G., 2004. Phosphorylation of histone H3: a balancing act between chromosome condensation and transcriptional activation. Trends Genet. 20, 214–220.

Nusinow, D.A., Hernandez-Munoz, I., Fazzio, T.G., Shah, G.M., Kraus, W.L., Panning, B., 2007a. Poly(ADP-ribose) polymerase 1 is inhibited by a histone H2A variant, MacroH2A, and contributes to silencing of the inactive X chromosome. J. Biol. Chem. 282, 12851–12859.

Nusinow, D.A., Sharp, J.A., Morris, A., Salas, S., Plath, K., Panning, B., 2007b. The histone domain of macroH2A1 contains several dispersed elements that are each sufficient to direct enrichment on the inactive X chromosome. J. Mol. Biol. 371, 11–18.

Obri, A., Ouararhni, K., Papin, C., Diebold, M.L., Padmanabhan, K., Marek, M., Stoll, I., Roy, L., Reilly, P.T., Mak, T.W., et al., 2014. ANP32E is a histone chaperone that removes H2A.Z from chromatin. Nature 505, 648–653.

Okuwaki, M., Kato, K., Shimahara, H., Tate, S., Nagata, K., 2005. Assembly and disassembly of nucleosome core particles containing histone variants by human nucleosome assembly protein I. Mol. Cell. Biol. 25, 10639–10651.

Ouararhni, K., Hadj-Slimane, R., Ait-Si-Ali, S., Robin, P., Mietton, F., Harel-Bellan, A., Dimitrov, S., Hamiche, A., 2006. The histone variant mH2A1.1 interferes with transcription by down-regulating PARP-1 enzymatic activity. Genes Dev. 20, 3324–3336.

Pasque, V., Radzisheuskaya, A., Gillich, A., Halley-Stott, R.P., Panamarova, M., Zernicka-Goetz, M., Surani, M.A., Silva, J.C., 2012. Histone variant macroH2A marks embryonic differentiation in vivo and acts as an epigenetic barrier to induced pluripotency. J. Cell Sci. 125, 6094–6104.

Picketts, D.J., Higgs, D.R., Bachoo, S., Blake, D.J., Quarrell, O.W., Gibbons, R.J., 1996. ATRX encodes a novel member of the SNF2 family of proteins: mutations point to a common mechanism underlying the ATR-X syndrome. Hum. Mol. Genet. 5, 1899–1907.

Posavec, M., Timinszky, G., Buschbeck, M., 2013. Macro domains as metabolite sensors on chromatin. Cell. Mol. Life Sci. 70, 1509–1524.

Qiu, J.J., Guo, J.J., Lv, T.J., Jin, H.Y., Ding, J.X., Feng, W.W., Zhang, Y., Hua, K.Q., 2013. Prognostic value of centromere protein-A expression in patients with epithelial ovarian cancer. Tumour Biol. 34, 2971–2975.

Quenet, D., Dalal, Y., 2014. A long non-coding RNA is required for targeting centromeric protein A to the human centromere. Elife 3, e03254,

Rasmussen, T.P., Huang, T., Mastrangelo, M.A., Loring, J., Panning, B., Jaenisch, R., 1999. Messenger RNAs encoding mouse histone macroH2A1 isoforms are expressed at similar levels in male and female cells and result from alternative splicing. Nucleic Acids Res. 27, 3685–3689.

Ratnakumar, K., Duarte, L.F., LeRoy, G., Hasson, D., Smeets, D., Vardabasso, C., Bonisch, C., Zeng, T., Xiang, B., Zhang, D.Y., et al., 2012. ATRX-mediated chromatin association of histone variant macroH2A1 regulates alpha-globin expression. Genes Dev. 26, 433–438.

Ricketts, M.D., Marmorstein, R., 2017. A molecular prospective for HIRA complex assembly and H3.3-specific histone chaperone function. J. Mol. Biol. 429, 1924–1933.

Ricketts, M.D., Frederick, B., Hoff, H., Tang, Y., Schultz, D.C., Singh Rai, T., Grazia Vizioli, M., Adams, P.D., Marmorstein, R., 2015. Ubinuclein-1 confers histone H3.3-specific-binding by the HIRA histone chaperone complex. Nat. Commun. 6, 7711.

Rogakou, E.P., Pilch, D.R., Orr, A.H., Ivanova, V.S., Bonner, W.M., 1998. DNA double-stranded breaks induce histone H2AX phosphorylation on serine 139. J. Biol. Chem. 273, 5858–5868.

Rosenthal, F., Feijs, K.L., Frugier, E., Bonalli, M., Forst, A.H., Imhof, R., Winkler, H.C., Fischer, D., Caflisch, A., Hassa, P.O., et al., 2013. Macrodomain-containing proteins are new mono-ADP-ribosylhydrolases. Nat. Struct. Mol. Biol. 20, 502–507.

Rossetto, D., Avvakumov, N., Cote, J., 2012. Histone phosphorylation: a chromatin modification involved in diverse nuclear events. Epigenetics 7, 1098–1108.

Rothbart, S.B., Lin, S., Britton, L.M., Krajewski, K., Keogh, M.C., Garcia, B.A., Strahl, B.D., 2012. Poly-acetylated chromatin signatures are preferred epitopes for site-specific histone H4 acetyl antibodies. Sci. Rep. 2, 489.

Rothbart, S.B., Dickson, B.M., Raab, J.R., Grzybowski, A.T., Krajewski, K., Guo, A.H., Shanle, E.K., Josefowicz, S.Z., Fuchs, S.M., Allis, C.D., et al., 2015. An interactive database for the assessment of histone antibody specificity. Mol. Cell 59, 502–511.

Sansoni, V., Casas-Delucchi, C.S., Rajan, M., Schmidt, A., Bonisch, C., Thomae, A.W., Staege, M.S., Hake, S.B., Cardoso, M.C., Imhof, A., 2014. The histone variant H2A.Bbd is enriched at sites of DNA synthesis. Nucleic Acids Res. 42, 6405–6420.

Santaguida, S., Amon, A., 2015. Short- and long-term effects of chromosome mis-segregation and aneuploidy. Nat. Rev. Mol. Cell Biol. 16, 473–485.

Schenk, R., Jenke, A., Zilbauer, M., Wirth, S., Postberg, J., 2011. H3.5 is a novel hominid-specific histone H3 variant that is specifically expressed in the seminiferous tubules of human testes. Chromosoma 120, 275–285.

Schueler, M.G., Swanson, W., Thomas, P.J., Program, N.C.S., Green, E.D., 2010. Adaptive evolution of foundation kinetochore proteins in primates. Mol. Biol. Evol. 27, 1585–1597.

Schwartzentruber, J., Korshunov, A., Liu, X.Y., Jones, D.T., Pfaff, E., Jacob, K., Sturm, D., Fontebasso, A.M., Quang, D.A., Tonjes, M., et al., 2012. Driver mutations in histone H3.3 and chromatin remodelling genes in paediatric glioblastoma. Nature 482, 226–231.

Sedelnikova, O.A., Bonner, W.M., 2006. GammaH2AX in cancer cells: a potential biomarker for cancer diagnostics, prediction and recurrence. Cell Cycle 5, 2909–2913.

Sharifi, R., Morra, R., Appel, C.D., Tallis, M., Chioza, B., Jankevicius, G., Simpson, M.A., Matic, I., Ozkan, E., Golia, B., et al., 2013. Deficiency of terminal ADP-ribose protein glycohydrolase TARG1/C6orf130 in neurodegenerative disease. EMBO J. 32, 1225–1237.

Shi, Y., Whetstine, J.R., 2007. Dynamic regulation of histone lysine methylation by demethylases. Mol. Cell 25, 1–14.

Slade, D., Dunstan, M.S., Barkauskaite, E., Weston, R., Lafite, P., Dixon, N., Ahel, M., Leys, D., Ahel, I., 2011. The structure and catalytic mechanism of a poly(ADP-ribose) glycohydrolase. Nature 477, 616–620.

Soboleva, T.A., Nekrasov, M., Pahwa, A., Williams, R., Huttley, G.A., Tremethick, D.J., 2011. A unique H2A histone variant occupies the transcriptional start site of active genes. Nat. Struct. Mol. Biol. 19, 25–30.

Sporn, J.C., Jung, B., 2012. Differential regulation and predictive potential of MacroH2A1 isoforms in colon cancer. Am. J. Pathol. 180, 2516–2526.

Sporn, J.C., Kustatscher, G., Hothorn, T., Collado, M., Serrano, M., Muley, T., Schnabel, P., Ladurner, A.G., 2009. Histone macroH2A isoforms predict the risk of lung cancer recurrence. Oncogene 28, 3423–3428.

Sturm, D., Witt, H., Hovestadt, V., Khuong-Quang, D.A., Jones, D.T., Konermann, C., Pfaff, E., Tönjes, M., Sill, M., Bender, S., et al., 2012. Hotspot mutations in H3F3A and IDH1 define distinct epigenetic and biological subgroups of glioblastoma. Cancer Cell 22, 425–437.

Sullivan, B.A., Karpen, G.H., 2004. Centromeric chromatin exhibits a histone modification pattern that is distinct from both euchromatin and heterochromatin. Nat. Struct. Mol. Biol. 11, 1076–1083.

Svotelis, A., Gevry, N., Grondin, G., Gaudreau, L., 2010. H2A.Z overexpression promotes cellular proliferation of breast cancer cells. Cell Cycle 9, 364–370.

Szenker, E., Ray-Gallet, D., Almouzni, G., 2011. The double face of the histone variant H3.3. Cell Res. 21, 421–434.

Tachiwana, H., Osakabe, A., Kimura, H., Kurumizaka, H., 2008. Nucleosome formation with the testis-specific histone H3 variant, H3t, by human nucleosome assembly proteins in vitro. Nucleic Acids Res. 36, 2208–2218.

Tachiwana, H., Kagawa, W., Osakabe, A., Kawaguchi, K., Shiga, T., Hayashi-Takanaka, Y., Kimura, H., Kurumizaka, H., 2010. Structural basis of instability of the nucleosome containing a testis-specific histone variant, human H3T. Proc. Natl. Acad. Sci. U. S. A. 107, 10454–10459.

Tagami, H., Ray-Gallet, D., Almouzni, G., Nakatani, Y., 2004. Histone H3.1 and H3.3 complexes mediate nucleosome assembly pathways dependent or independent of DNA synthesis. Cell 116, 51–61.

Taguchi, H., Xie, Y., Horikoshi, N., Maehara, K., Harada, A., Nogami, J., Sato, K., Arimura, Y., Osakabe, A., Kujirai, T., et al., 2017. Crystal structure and characterization of novel human histone H3 variants, H3.6, H3.7, and H3.8. Biochemistry 56, 2184–2196.

Tallis, M., Morra, R., Barkauskaite, E., Ahel, I., 2014. Poly(ADP-ribosyl)ation in regulation of chromatin structure and the DNA damage response. Chromosoma 123, 79–90.

Taverna, S.D., Ueberheide, B.M., Liu, Y., Tackett, A.J., Diaz, R.L., Shabanowitz, J., Chait, B.T., Hunt, D.F., Allis, C.D., 2007. Long-distance combinatorial linkage between methylation and acetylation on histone H3 N termini. Proc. Natl. Acad. Sci. U. S. A. 104, 2086–2091.

Timinszky, G., Till, S., Hassa, P.O., Hothorn, M., Kustatscher, G., Nijmeijer, B., Colombelli, J., Altmeyer, M., Stelzer, E.H., Scheffzek, K., et al., 2009. A macrodomain-containing histone rearranges chromatin upon sensing PARP1 activation. Nat. Struct. Mol. Biol. 16, 923–929.

Tolstorukov, M.Y., Goldman, J.A., Gilbert, C., Ogryzko, V., Kingston, R.E., Park, P.J., 2012. Histone variant H2A.Bbd is associated with active transcription and mRNA processing in human cells. Mol. Cell 47, 596–607.

Trievel, R.C., 2004. Structure and function of histone methyltransferases. Crit. Rev. Eukaryot. Gene Expr. 14, 147–169.

Tsukamoto, Y., Uchida, T., Karnan, S., Noguchi, T., Nguyen, L.T., Tanigawa, M., Takeuchi, I., Matsuura, K., Hijiya, N., Nakada, C., et al., 2008. Genome-wide analysis of DNA copy number alterations and gene expression in gastric cancer. J. Pathol. 216, 471–482.

Turner, J.M., Burgoyne, P.S., Singh, P.B., 2001. M31 and macroH2A1.2 colocalise at the pseudoautosomal region during mouse meiosis. J. Cell Sci. 114, 3367–3375.

Ueda, J., Harada, A., Urahama, T., Machida, S., Maehara, K., Hada, M., Makino, Y., Nogami, J., Horikoshi, N., Osakabe, A., et al., 2017. Testis-specific histone variant H3t gene is essential for entry into spermatogenesis. Cell Rep. 18, 593–600.

Urahama, T., Harada, A., Maehara, K., Horikoshi, N., Sato, K., Sato, Y., Shiraishi, K., Sugino, N., Osakabe, A., Tachiwana, H., et al., 2016. Histone H3.5 forms an unstable nucleosome and accumulates around transcription start sites in human testis. Epigenetics Chromatin 9, 2.

Valente, V., Serafim, R.B., de Oliveira, L.C., Adorni, F.S., Torrieri, R., Tirapelli, D.P., Espreafico, E.M., Oba-Shinjo, S.M., Marie, S.K., Paco-Larson, M.L., et al., 2013. Modulation of HJURP (Holliday junction-recognizing protein) levels is correlated with glioblastoma cells survival. PLoS One 8, e62200.

Vardabasso, C., Gaspar-Maia, A., Hasson, D., Punzeler, S., Valle-Garcia, D., Straub, T., Keilhauer, E.C., Strub, T., Dong, J., Panda, T., et al., 2015. Histone variant H2A.Z.2 mediates proliferation and drug sensitivity of malignant melanoma. Mol. Cell 59, 75–88.

Venkatasubrahmanyam, S., Hwang, W.W., Meneghini, M.D., Tong, A.H., Madhani, H.D., 2007. Genome-wide, as opposed to local, antisilencing is mediated redundantly by the euchromatic factors Set1 and H2A.Z. Proc. Natl. Acad. Sci. U. S. A. 104, 16609–16614.

Voon, H.P., Hughes, J.R., Rode, C., De La Rosa-Velazquez, I.A., Jenuwein, T., Feil, R., Higgs, D.R., Gibbons, R.J., 2015. ATRX plays a key role in maintaining silencing at interstitial heterochromatic loci and imprinted genes. Cell Rep. 11, 405–418.
Walker, B.A., Leone, P.E., Chiecchio, L., Dickens, N.J., Jenner, M.W., Boyd, K.D., Johnson, D.C., Gonzalez, D., Dagrada, G.P., Protheroe, R.K., et al., 2010. A compendium of myeloma-associated chromosomal copy number abnormalities and their prognostic value. Blood 116, e56–65.
Wang, Y.L., Faiola, F., Xu, M., Pan, S., Martinez, E., 2008a. Human ATAC is a GCN5/PCAF-containing acetylase complex with a novel NC2-like histone fold module that interacts with the TATA-binding protein. J. Biol. Chem. 283, 33808–33815.
Wang, Z., Zang, C., Rosenfeld, J.A., Schones, D.E., Barski, A., Cuddapah, S., Cui, K., Roh, T.Y., Peng, W., Zhang, M.Q., et al., 2008b. Combinatorial patterns of histone acetylations and methylations in the human genome. Nat. Genet. 40, 897–903.
Weber, C.M., Henikoff, J.G., Henikoff, S., 2010. H2A.Z nucleosomes enriched over active genes are homotypic. Nat. Struct. Mol. Biol. 17, 1500–1507.
Weber, C.M., Ramachandran, S., Henikoff, S., 2014. Nucleosomes are context-specific, H2A.Z-modulated barriers to RNA polymerase. Mol. Cell 53, 819–830.
West, A.C., Johnstone, R.W., 2014. New and emerging HDAC inhibitors for cancer treatment. J. Clin. Invest. 124, 30–39.
Wiedemann, S.M., Mildner, S.N., Bonisch, C., Israel, L., Maiser, A., Matheisl, S., Straub, T., Merkl, R., Leonhardt, H., Kremmer, E., et al., 2010. Identification and characterization of two novel primate-specific histone H3 variants, H3.X and H3.Y. J. Cell Biol. 190, 777–791.
Winkler, C., Steingrube, D.S., Altermann, W., Schlaf, G., Max, D., Kewitz, S., Emmer, A., Kornhuber, M., Banning-Eichenseer, U., Staege, M.S., 2012. Hodgkin's lymphoma RNA-transfected dendritic cells induce cancer/testis antigen-specific immune responses. Cancer Immunol. Immunother. 61, 1769–1779.
Wong, M.M., Cox, L.K., Chrivia, J.C., 2007. The chromatin remodeling protein, SRCAP, is critical for deposition of the histone variant H2A.Z at promoters. J. Biol. Chem. 282, 26132–26139.
Woodcock, C.L., Dimitrov, S., 2001. Higher-order structure of chromatin and chromosomes. Curr. Opin. Genet. Dev. 11, 130–135.
Woodcock, C.L., Skoultchi, A.I., Fan, Y., 2006. Role of linker histone in chromatin structure and function: H1 stoichiometry and nucleosome repeat length. Chromosome Res. 14, 17–25.
Wu, G., Broniscer, A., McEachron, T.A., Lu, C., Paugh, B.S., Becksfort, J., Qu, C., Ding, L., Huether, R., Parker, M., et al., 2012a. Somatic histone H3 alterations in pediatric diffuse intrinsic pontine gliomas and non-brainstem glioblastomas. Nat. Genet. 44, 251–253.
Wu, Q., Qian, Y.M., Zhao, X.L., Wang, S.M., Feng, X.J., Chen, X.F., Zhang, S.H., 2012b. Expression and prognostic significance of centromere protein A in human lung adenocarcinoma. Lung Cancer 77, 407–414.
Yang, X.J., 2004. Lysine acetylation and the bromodomain: a new partnership for signaling. Bioessays 26, 1076–1087.
Yang, H.D., Kim, P.J., Eun, J.W., Shen, Q., Kim, H.S., Shin, W.C., Ahn, Y.M., Park, W.S., Lee, J.Y., Nam, S.W., 2016. Oncogenic potential of histone-variant H2A.Z.1 and its regulatory role in cell cycle and epithelial-mesenchymal transition in liver cancer. Oncotarget 7, 11412–11423.
Yeoh, K.W., Mikhaeel, N.G., 2011. Role of radiotherapy in modern treatment of Hodgkin's lymphoma. Adv. Hematol. 2011, 258797.
Yu, T., MacPhail, S.H., Banath, J.P., Klokov, D., Olive, P.L., 2006. Endogenous expression of phosphorylated histone H2AX in tumors in relation to DNA double-strand breaks and genomic instability. DNA Repair (Amst.) 5, 935–946.

Zasadzinska, E., Barnhart-Dailey, M.C., Kuich, P.H., Foltz, D.R., 2013. Dimerization of the CENP-A assembly factor HJURP is required for centromeric nucleosome deposition. EMBO J. 32, 2113–2124.
Zentner, G.E., Henikoff, S., 2013. Regulation of nucleosome dynamics by histone modifications. Nat. Struct. Mol. Biol. 20, 259–266.
Zhang, W., Mao, J.H., Zhu, W., Jain, A.K., Liu, K., Brown, J.B., Karpen, G.H., 2016. Centromere and kinetochore gene misexpression predicts cancer patient survival and response to radiotherapy and chemotherapy. Nat. Commun. 7, 12619.
Zhou, V.W., Goren, A., Bernstein, B.E., 2011. Charting histone modifications and the functional organization of mammalian genomes. Nat. Rev. Genet. 12, 7–18.
Zlatanova, J., Thakar, A., 2008. H2A.Z: view from the top. Structure 16, 166–179.

CHAPTER TWO

NF-κB and the Transcriptional Control of Inflammation

Jennifer P. Mitchell*, Ruaidhrí J. Carmody[†,1]
*Rheumatoid Arthritis Pathogenesis Centre of Excellence, Centre for Immunobiology, Institute of Infection, Immunity, and Inflammation, University of Glasgow, Glasgow, United Kingdom
[†]Centre for Immunobiology, Institute of Infection, Immunity, and Inflammation, University of Glasgow, Glasgow, United Kingdom
[1]Corresponding author: e-mail address: ruaidhri.carmody@glasgow.ac.uk

Contents

Abstract

The NF-κB transcription factor was discovered 30 years ago and has since emerged as the master regulator of inflammation and immune homeostasis. It achieves this status by means of the large number of important pro- and antiinflammatory factors under its transcriptional control. NF-κB has a central role in inflammatory diseases such as rheumatoid arthritis, inflammatory bowel disease, and autoimmunity, as well as diseases comprising a significant inflammatory component such as cancer and atherosclerosis. Here, we provide an overview of the studies that form the basis of our understanding of the role of NF-κB subunits and their regulators in controlling inflammation. We also describe the emerging importance of posttranslational modifications of NF-κB in the regulation of inflammation, and highlight the future challenges faced by researchers who aim to target NF-κB transcriptional activity for therapeutic benefit in treating chronic inflammatory diseases.

International Review of Cell and Molecular Biology, Volume 335
ISSN 1937-6448
https://doi.org/10.1016/bs.ircmb.2017.07.007

1. INTRODUCTION

The nature of the cellular microenvironment is dynamic and rapidly changes with injury, infection, and other stresses that compromise the physiological integrity of the host. Immune responses to these threats must be readily inducible, tailored to specific stresses, and be tightly controlled in order to restore the homeostatic state. Nuclear factor-κB (NF-κB) is an evolutionarily conserved transcription factor that provides a means to achieve inducible, specific, and regulated immune responses. Its role as a master regulator of the inflammatory response stems from its critical role in regulating the expression of hundreds of immune relevant genes, in particular those encoding proinflammatory cytokines, chemokines, as well as other genes important for the development of the immune system. NF-κB is activated by a vast range of stimuli, including microbial components known as pathogen-associated molecular patterns and endogenous damage-associated molecular patterns which are recognized by pattern recognition receptors such as Toll-like receptors (TLRs) (Carmody and Chen, 2007; Feldman et al., 2015). Additionally, NF-κB lies downstream of antigen receptors, growth factor receptors, and cytokine receptors, including the tumor necrosis factor (TNF) receptor superfamily. NF-κB may also be activated by environmental stresses including ultra violet light, irradiation, and reactive oxygen species. Given its extensive influence in a diverse array ofbiological systems, failure to regulate NF-κB activation can have devastating consequences such as chronic inflammatory diseases, autoimmunity, cancer, neurodegenerative disorders, and cardiovascular disease. To reflect the potential for unwanted host damage, NF-κB is regulated by a multitude of elements at various stages throughout the activating pathway. The activation and signaling pathways have been reviewed in great depth before and we are indebted to these authors for their extensive descriptions (Hayden and Ghosh, 2008; Karin and Lin, 2002; Zhang et al., 2017). This review will focus its attention first on what role NF-κB plays in the development of inflammation by looking at genetic studies in both mice and humans where individual components of the signaling pathways are mutated. We then examine how NF-κB regulates inflammation in light of the diverse array of genes it controls and the seemingly small armory of regulatory proteins that control its activity. Finally, we look to the future and what challenges and opportunities there are within the field of NF-κB research and how the potential for manipulating the activity of this transcription factor can lead to novel therapeutic targets in the treatment of chronic inflammatory disease, cancer, and other pathologies.

1.1 Overview of NF-κB

Discovered 30 years ago by David Baltimore and Ranjan Sen, the transcription factor NF-κB has since been established as the key regulator of inducible gene expression in the immune system (Sen and Baltimore, 1986). Initially, NF-κB was reported as a specific DNA-binding activity in B cells that recognized an enhancer element in the gene encoding the immunoglobulin-κ light chain and was named to reflect these attributes. Subsequently, NF-κB was found to be expressed in almost all cell types, although the name NF-κB persisted (Baeuerle and Baltimore, 1988). It was quickly appreciated that NF-κB was an inducible transcription factor with key roles in regulating the development and homeostasis of the immune system, and the coordination of the inflammatory response. Despite being essential for inflammation, NF-κB's role in the transcription of the κ light chain remains elusive.

Although often referred to as a single transcription factor, NF-κB in mammals is in fact a family of five related proteins that combine to form dimers, which can both positively and negatively regulate gene transcription (see Fig. 1). The five subunits are RelA, also known as p65 (encoded by *Rela*), RelB (*Relb*), c-Rel (*c-Rel*), p105 (*Nfkb1*) and p100 (*Nfkb2*). p105 and p100 are larger precursor proteins which are further processed into the subunits, p50 and p52, respectively (Hayden and Ghosh, 2014). All of

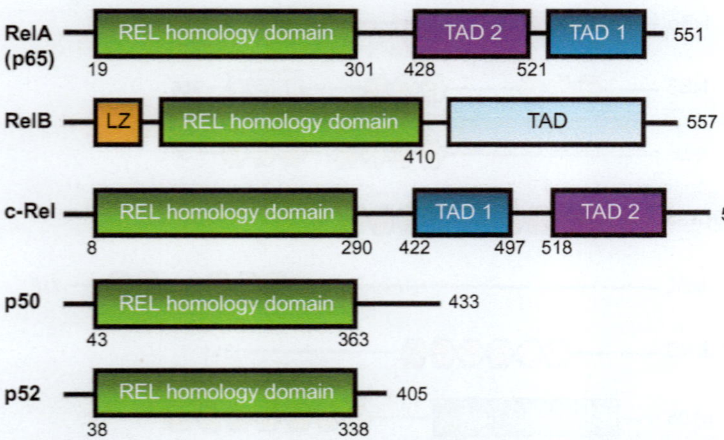

Fig. 1 Members of the NF-κB Rel protein family. The mammalian Rel protein family consists of five members: RelA (p65), RelB, c-Rel, p50, and p52. The p50 and p52 subunits are generated by the limited proteasomal processing of precursors p105 and p100, respectively. Numbers refer to amino acid sequence. *LZ*, leucine zipper; *TAD*, transactivating domain.

the subunits share a common, evolutionarily conserved 300 amino acid N-terminal Rel-homology domain (RHD) which contains DNA binding and dimerization activities (Fig. 1). The five NF-κB subunits can homo- and heterodimerize to form 15 possible subunit combinations with varying transcriptional and DNA binding properties. Additionally, p65, RelB, and c-Rel have a C-terminal transactivation domain (TAD) that enables these subunits to directly promote gene transcription. p50 and p52 on the other hand lack a TAD and so as homodimers lack any intrinsic transactivation properties. Thus, in the homodimeric form, p50 and p52 are thought to act predominantly as repressors of transcription through competition with transcriptionally active dimers for binding to specific sites in DNA (Hayden and Ghosh, 2008). However, when bound to a TAD-containing subunit as part of a heterodimer, the p50 and p52 subunits may positively regulate transcription.

The primary mechanism controlling NF-κB activation is through interaction with the inhibitor of NF-κB (IκB) family of proteins. In the resting state, NF-κB dimers are sequestered in the cytoplasm in an inactive state by members of the IκB family of proteins which include IκBα (*Nfkbia*), IκBβ (*Nfkbib*), IκBε (*Nfkbie*), and the NF-κB precursors p105 and p100 (Hayden and Ghosh, 2014). All IκB proteins possess a central ankyrin repeat domain (ANK) which mediates interaction with NF-κB dimers (Fig. 2). The most

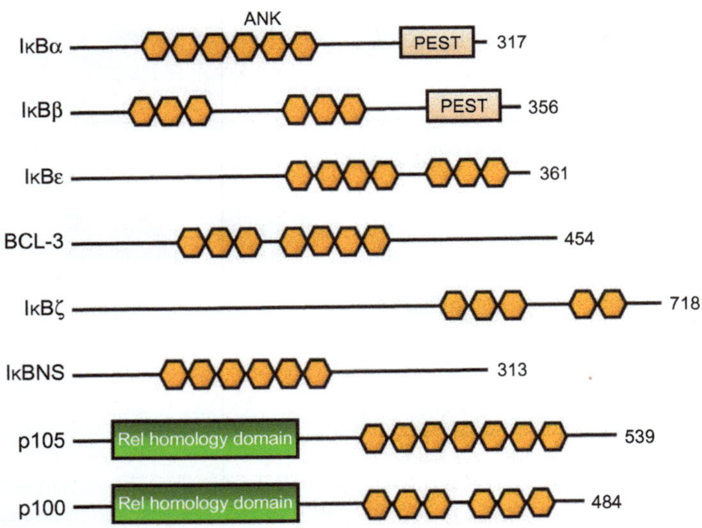

Fig. 2 The IκB protein family includes the prototypical members IκBα, IκBβ, IκBε; the atypical members Bcl-3, IκBζ, IκBNS, and IκBη; and the NF-κB precursors p105 and p100. *ANK*, ankryin repeat domains; *PEST*, PEST domain. Numbers refer to amino acid sequence.

extensively studied of these members is IκBα, deletion of which results in a severe inflammatory phenotype in mice leading to death by postnatal days 8–10. In the absence of a stimulus, IκBα binds NF-κB dimers in the cytoplasm, masking the nuclear localization signals of NF-κB. Of note, although IκBα:p65:p50 complexes may shuttle dynamically between the nucleus and cytoplasm, the equilibrium at the steady state is in the cytoplasm (Tam et al., 2000). Upon receiving an activating signal, the IκB proteins are phosphorylated by the IκB kinase (IKK) complex which triggers lysine48 (K48)-linked polyubiquitination of IκBα leading to proteasomal degradation of the IκB, thus freeing the NF-κB dimers which can now move into the nucleus where they can bind to specific sites in DNA to regulate gene transcription (Baker and Ghosh, 2010; Hayden and Ghosh, 2014; Scheidereit, 2006). The NF-κB-induced expression and resynthesis of IκBα is the primary means for terminating NF-κB transcriptional activity and constitutes a critical negative feedback loop (Hayden and Ghosh, 2008).

1.1.1 Activation of NF-κB

NF-κB activation occurs through either of two pathways termed the classical or canonical pathway, and the alternative or noncanonical pathway (Bonizzi, 2004) (Fig. 3). A key step in both of these pathways is the posttranslational modification of inhibitory IκB proteins. In both pathways this is controlled by the IKK complex. The IKK complex is composed of the kinases IKKα or IKKβ, as well as the regulatory subunit NEMO (NF-κB essential modifier), also known as IKKγ. Activation of the IKK complex is an essential step in the activation of NF-κB in both the classical and noncanonical pathways.

1.1.2 Classical Pathway

The classical pathway is the general model for NF-κB activation and occurs downstream of the majority of immunoreceptors including TLR, antigen receptors, cytokine, and growth factor receptors, including the TNFR superfamily. The classical pathway is dependent on IKKβ, which phosphorylates IκBα at positions S32 and S36. This targets IκBα for polyubiquitination by the $SCF^{\beta TrCP}$ ubiquitin ligase (at positions K21 and K22 on IκBα) and leads to the subsequently degradation of IκBα by the 26S proteasome (Wertz and Dixit, 2010). In this pathway, the liberated NF-κB dimers are largely composed of the p65, p50, and c-Rel subunits.

1.1.3 Noncanonical Pathway

The noncanonical pathway is triggered by a limited number of stimuli including B cell-activating factor (BAFF) and lymphotoxin β (LTβ), and

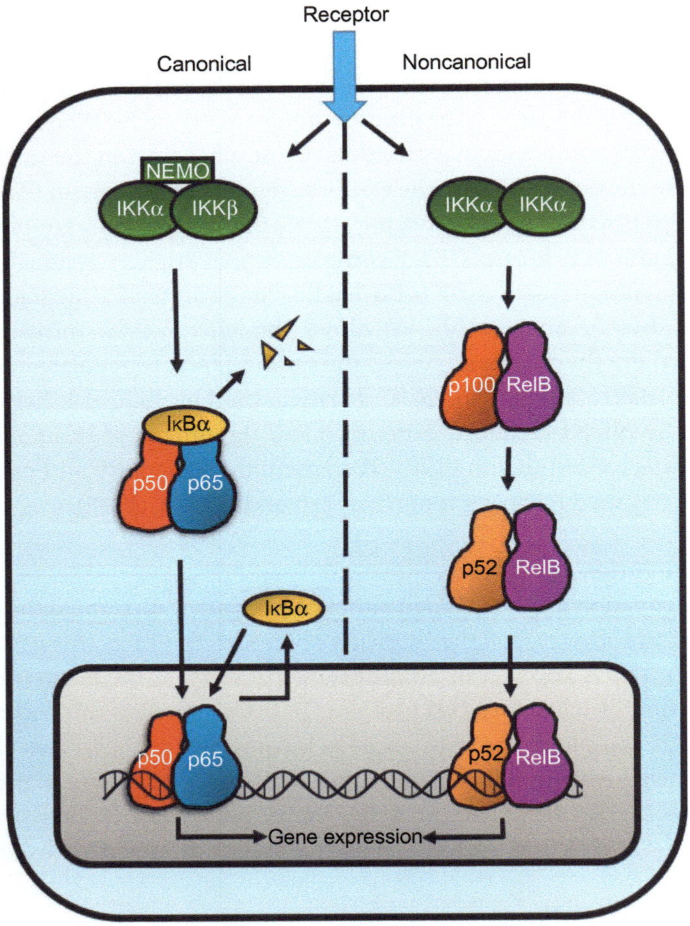

Fig. 3 Canonical and noncanonical pathways for activation of NF-κB. Upon receiving an activating signal, IκB proteins which sequester NF-κB dimers in the cytoplasm are phosphorylated by the IKK complex marking them for ubiquitination and proteasomal degradation. This frees the NF-κB dimers which can now move into the nucleus and activate gene expression. In the canonical pathway, the triggering stimuli include the binding of a TLR ligand which results in the phosphorylation and degradation of IκBα following which p65:p50 dimers move into the nucleus and bind to their DNA targets. The noncanonical pathway is induced by certain TNF family cytokines such as CD40L, BAFF, and LTβ and leads to the IKKα-mediated phosphorylation of p100 into p52 which is bound to RelB and is transcriptionally active.

has a major role during lymphoid organogenesis (Senftleben et al., 2001a). This pathway leads to activation of NF-κB dimers composed of the RelB and p52 subunits and is dependent on IKKα but independent of IKKβ and NEMO. NF-κB p100 is the key negative regulator of the noncanonical pathway and prevents the nuclear translocation of RelB in resting cells. Activation of the alternative pathway leads to the stabilization of NF-κB-inducing kinase (NIK) which phosphorylates and activates IKKα. Activated IKKα then phosphorylates p100 at two serine residues in the C-terminal (S176 and S180) triggering its polyubiquitination by the $SCF^{\beta TrCP}$ ubiquitin ligase. In this case, the ubiquitination of p100 triggers only partial proteasomal degradation to generate the p52 NF-κB subunit and the formation of free RelB:p52 dimers (Beinke and Ley, 2004). Whereas the classical pathway mediates a rapid response to activating stimuli, the alternative pathway is slower and its kinetics are dependent on the synthesis of new proteins which is in line with a role in organogenesis (Baltimore, 2011).

1.1.4 Oscillations of NF-κB Signaling

It has been observed that NF-κB subcellular localization can occur in oscillations between the cytoplasm and nucleus following activation. As detailed earlier, IκB proteins sequester NF-κB dimers in the cytoplasm of resting cells and upon receiving an activating signal, NF-κB is released and translocates to the nucleus. However, NF-κB also controls the transcription of IκBα (Sun et al., 1993) resulting in a negative feedback loop that causes oscillations in NF-κB activity leading to dynamic changes in gene expression that may depend on the number, duration, and intensity of such oscillations (Nelson et al., 2004). For example, treatment of cells with TNFα at time intervals that represented pulsatile inflammatory responses resulted in synchronized oscillations of translocation of NF-κB from the cytoplasm into the nucleus, with different stimulation intervals resulting in different patterns of gene expression (Ashall et al., 2009). Nuclear-cytoplasmic oscillations with a period of 100 min are observed for a range of TNFα concentrations, although as TNFα concentration decreases, a smaller fraction of the cells respond suggesting a threshold for activation (Turner et al., 2010).

2. THE ROLE OF NF-κB IN INFLAMMATION

NF-κB controls the expression of hundreds of biologically important genes that regulate cell function, cell death and survival, and proliferation (an ever-increasing list can be found at http://www.bu.edu/nf-kb/gene-resources/target-genes/). However, it is in the regulation of genes that control the immune response that NF-κB is of most physiological importance given that the phenotypes of mice models are predominantly immune in nature. NF-κB regulates gene transcription by binding to κB sites in the enhancer or promoter regions of target genes. κB sites are DNA sequences with a consensus of 5′-GGGRNWYCC-3′ (where N is any base, R is a purine, W is adenine or thymine, and Y is a pyrimidine) and are recognized and bound by the RHD of NF-κB dimers (Smale, 2012). NF-κB dimers can recognize variations of the κB sequence with different affinities which increases regulatory networks that emerge from just five protein subunits (Zhang et al., 2017). To discern in greater detail how such a modest transcription factor family can respond to a such a huge number of activators and then in turn activate the required patterns of gene expression in a number of different tissues, we can look to knockout studies carried out in mice and, more recently, to human genetic deficiencies that highlight the role of NF-κB subunits in the control of inflammation. NF-κB research has continuously been at the forefront of technological advances in biochemistry, molecular biology, and genetics and has benefited enormously from the early generation of transgenic mouse models of NF-κB subunits and regulatory components of the NF-κB activation pathway. These transgenic mice have provided a critical insight into the roles of the NF-κB subunits and their regulators in the inflammatory response. Moreover, they have highlighted the complexity of NF-κB and its regulation resulting from the apparent redundancies and interactions that exist between the various factors (Pasparakis et al., 2006). Thorough reviews of mouse knockout studies have been described previously (Gerondakis et al., 2006; Pasparakis et al., 2006) and so here we will focus on those aspects most pertinent to the regulation of inflammation (see Table 1 for an overview). It is also worth noting that continual progress in the study of human genetic diseases that affect the components of the NF-κB signaling pathway will significantly contribute to our future understanding of the regulation of NF-κB in human health and disease.

2.1 NF-κB Subunits

2.1.1 p105/p50

The deletion of *Nfkb1* in mice, which encodes for the p50 and p105 NF-κB proteins, is not fatal and results in viable adult development but with impaired innate and adaptive immune function (Artis et al., 2005; Cariappa et al., 2000; Sha et al., 1995). *Nfkb1*$^{-/-}$ mice are unable to clear *Listeria monocytogenes* (Sha et al., 1995) or *Citrobacter rodentium* infection (Dennis et al., 2008) and are more susceptible to infection with *Streptococcus pneumonia*, but show increased resistance to infection with murine encephalomyocarditis virus (Sha et al., 1995). *Nfkb1*$^{-/-}$ macrophages express increased levels of TNFα in response to TLR activation and are defective

Table 1 Mouse Model Phenotypes

Genotype	Fatal	Phenotype	References
NF-κB subunits			
Nfkb1$^{-/-}$	No	Impaired innate and adaptive immune function; Macrophages express increased levels of TNFα in response to TLR activation and are deficient in TLR-tolerance; NK cells exhibit enhanced proliferation and expression of IFNγ in vitro; Marginal zone and $CD5^+$ B cell numbers reduced; Differentiation to Th2 cells impaired; MAPK activation defective	Bohuslav et al. (1998), Tato et al. (2006), Grumont et al. (1998), Pohl et al. (2002), Artis et al. (2005) and Yang et al. (2012)
Nfkb1$^{\Delta CT/\Delta CT}$	No	Enlarged lymph nodes and spleen; Increased numbers of B cells; T cell proliferation and cytokine production impaired	Ishikawa et al. (1998)
Nfkb2$^{-/-}$	No	Impaired secondary lymphoid organ development; Impaired B cell maturation	Beinke and Ley (2004) and Caamano et al. (1998)
Nfkb2$^{\Delta CT/\Delta CT}$	No	Hyperproliferative defects	Ishikawa et al. (1997)
c-Rel$^{-/-}$	No	B cell proliferative and activation defects; Defective in Th1 responses and cytokine production; Reduced number of pDCs	Cheng et al. (2003), Grumont et al. (1998), Pohl et al. (2002), and Strasser et al. (1999)
Rela$^{-/-}$	Yes	Die on embryonic day 15 due to TNFα-induced fetal hepatocyte apoptosis; Rescued by deleting TNFα or TNFR1	Beg et al. (1995), Doi et al. (1999), and Alcamo et al. (2001)

Continued

Table 1 Mouse Model Phenotypes—cont'd

Genotype	Fatal	Phenotype	References
Rela$^{-/-}$ *Tnfr*$^{-/-}$	No	Die shortly after birth possibly due to increased susceptibility to infection; Impaired secondary lymphoid organ development	Alcamo et al. (2001) and Horwitz et al. (1997)
Relb$^{-/-}$	No	Severe inflammatory phenotype exhibited; Disrupted development of secondary lymphoid organs, germinal centers, and fDCs; Defects in humoral immunity;	Burkly et al. (1995), Yilmaz et al. (2003), Bonizzi et al. (2004), and Weih et al. (1997)
Nfkb1$^{-/-}$ *Nfkb2*$^{-/-}$	No	Impaired secondary lymphoid organ development	Lo et al. (2006)
Nfkb1$^{-/-}$ *Relb*$^{-/-}$	No	More severe organ inflammation than that seen in *relb*$^{-/-}$ mice	Weih et al. (1997)
Nfkb1$^{-/-}$ *c-Rel*$^{-/-}$	No	Increased severity in defects seen in single knockout mice	Pohl et al. (2002)
Nfkb1$^{-/-}$ *Rela*$^{-/-}$	Yes	Die on embryonic day 13	Horwitz et al. (1997)
IKK complex			
Ikka$^{-/-}$	Die at birth	Skin and skeletal defects	Cao et al. (2001)
Ikka$^{AA/AA}$ *knock-in*	No	Impaired secondary lymphoid organ development; Do not produce fDCs; Enhanced inflammatory response to infection	Bonizzi et al. (2004), Gerondakis et al. (2006), and Lawrence et al. (2005)
Ikkb$^{-/-}$	Yes	Die around embryonic day 13 due to TNFα-induced fetal hepatocyte apoptosis	Li et al. (1999)

Table 1 Mouse Model Phenotypes—cont'd

Genotype	Fatal	Phenotype	References
Nemo$^{-/-}$	Yes	Die around embryonic day 10 due to TNFα-induced fetal hepatocyte apoptosis	Kim et al. (2003)
Nemo$^{+/-}$	Females viable	Female mice exhibit an inflammatory skin disease after birth similar to the human condition incontinentia pigmenti	Makris et al. (2000)
Nemo$^{\Delta/EC}$	No	Spontaneous colitis develops following conditional deletion in intestinal epithelial cells; Chronic hepatitis leading to hepatocellular carcinoma when restricted to hepatocytes	Nenci et al. (2007) and Luedde et al. (2007)
IκB proteins			
Nfkbia$^{-/-}$	No	Die shortly after birth with severe granulopoeisis and inflammatory dermatitis; Impaired secondary lymphoid organ development	Beg et al. (1995)
Nfkbib$^{-/-}$	No	Minor immune defects; Increased expression of some cytokines; Resistant to LPS-induced septic shock and collagen-induced arthritis	Memet et al. (1999) and Rao et al. (2010)
Nfkbie$^{-/-}$	No	Minor immune defects; Increased expression of some cytokines	Memet et al. (1999)

Continued

Table 1 Mouse Model Phenotypes—cont'd

Genotype	Fatal	Phenotype	References
Bcl-3$^{-/-}$	No	Impaired secondary lymphoid organ development; Reduced numbers of follicular B cells but increased numbers of marginal zone B cells; Sensitive to some bacterial infections; Hyperresponsiveness to TLR stimulation and defective in TLR tolerance; Th2 cells secrete reduced levels of IL-4, IL-5, and IL-13; Activated T cells more susceptible to apoptosis; Increased sensitivity to autoimmune diabetes; Resistance to DSS-induced colitis	Franzoso et al. (1998), Gerondakis et al. (2006), Schwarz et al. (1997), Hovelmeyer et al. (2014), Zhang et al. (2013), Pene et al. (2011), Carmody et al. (2007b), Corn et al. (2005), Bauer et al. (2006), Ruan et al. (2010), and O'Carroll et al. (2013)
Bcl-3$^{-/-}$ *Nfkb2*$^{-/-}$	No	Die within a month of birth from multiorgan failure but no overt signs of autoimmunity	Zhang et al. (2007)
Nfkbiz$^{-/-}$	Yes	Rare survivors develop atopic dermatitis and conjunctivitis; Defective Th17 development and resistant to EAE	Yamamoto et al. (2004) and Okamoto et al. (2010)
Nfkbid$^{-/-}$	No	Intestinal inflammation	Kuwata et al. (2006)

in TLR tolerance, an important antiinflammatory mechanism of macrophages (Bohuslav et al., 1998) (see Section 3.3). Natural killer (NK) cells that are *Nfkb1* deficient, however, exhibit enhanced proliferation and increased expression of IFNγ in vitro when stimulated with *Toxoplasma gondii* which suggests that *Nfkb1* is a negative regulator of NK function (Tato et al., 2006). These phenotypes reflect the repressor role of p50 homodimers in limiting NF-κB-mediated gene transcription. In *Nfkb1*$^{-/-}$ mice marginal zone and

$CD5^+$ peritoneal B cell numbers are reduced, however, follicular B cell numbers are normal but turn over more quickly (Grumont et al., 1998; Pohl et al., 2002). These effects may be linked to BAFF-mediated B cell survival which although is a strong inducer of the noncanonical pathway of NF-κB activation may also activate p50 subunits (Claudio et al., 2002; Hatada et al., 2003). *Nfkb1*$^{-/-}$ B cells exhibit a poor proliferative response to lipopolysaccharide (LPS) stimulation which is a ligand for TLR4, although B cell receptor-induced proliferation is normal (Sha et al., 1995). *Nfkb1* is not essential for T cell maturation but it is indispensable for the development for Th2 responses involving the effector cytokines IL-4, IL-5, and IL-13 (Artis et al., 2005; Das et al., 2001). Deletion of the C-terminal ANK-containing region of p105 in mice (*Nfkb1*$^{\Delta CT/\Delta CT}$) results in mice that express p50 but not p105. This leads to increased levels of p50 homodimers in the nucleus of T cells, B cells, and macrophages, highlighting the role of p105 as an IκB protein (Gerondakis et al., 2006). *Nfkb1*$^{\Delta CT/\Delta CT}$ mice exhibit lymphoid infiltration into various organs and enlarged lymph nodes and spleen. Additionally, *Nfkb1*$^{\Delta CT/\Delta CT}$ B cells are hyperresponsive to activation and are present in increased numbers, however, *Nfkb1*$^{\Delta CT/\Delta CT}$ T cells are less capable of proliferation and their ability to produce cytokines diminished (Ishikawa et al., 1998).

It should also be noted that in addition to acting as a precursor to p50, and functioning as an IκB protein, p105 also serves an important NF-κB-independent role as a regulator of the mitogen-activated protein kinase (MAPK) pathway (Beinke and Ley, 2004). p105 forms a stoichiometric complex with Tpl-2 (MAP3K8), the apical kinase of the MAPK cascade activated by TLRs, TNFR, as well as cytokine receptors such as IL-17R (Gantke et al., 2012; Xiao et al., 2014). In resting cells, all cellular Tpl-2 is bound by p105, while Tpl-2 bound p105 represents only a small pool of the total cellular p105 protein. The interaction of p105 with Tpl-2 is required to keep Tpl-2 in an inactive but stable state. Activation of the MAPK pathway requires the IKKβ-induced phosphorylation of p105 and Tpl-2, which triggers the proteasomal degradation of p105 and the activation of Tpl-2 (Roget et al., 2012). In the absence of p105 there is little to no Tpl-2 detectable in cells as it is degraded rapidly and so *Nfkb1*$^{-/-}$ cells are defective in MAPK activation (Yang et al., 2012). Therefore, it is important to consider both the NF-κB and MAPK activation pathways when assessing the role of p105 in inflammation (Banerjee et al., 2006).

In humans, heterozygous mutations in *NFKB1* cause a reduction in p105 and p50 levels in resting cells and greatly diminish p105 phosphorylation following stimulation leading to common variable immunodeficiency (CVID). CVID presents as poor antibody responses and ineffective isotype class switching in patients (Boztug et al., 2016; Fliegauf et al., 2015).

2.1.2 p100/p52

As with *Nfkb1*$^{-/-}$ mice, *Nfkb2*$^{-/-}$ mice are viable but exhibit a number of immune deficiencies including abnormal secondary lymphoid organ development, impaired in B cell maturation, and abnormal T cell functionality (Beinke and Ley, 2004). *Nfkb2*$^{-/-}$ mice lack the perifollicular marginal zone in the spleen and B cell follicular areas are greatly diminished in both the spleen and lymph nodes (Caamano et al., 1998). This phenotype of *Nfkb2*$^{-/-}$ mice is shared with mice lacking other components of the non-canonical pathway of NF-κB activation including IKKα, NIK, and RelB as well as LTβ and LTβ-receptor (LTβR) knockout mice which activate the noncanonical pathway to induce the expression of chemokines crucial for lymphoid organogenesis (Beinke and Ley, 2004; Bonizzi, 2004). *Nfkb2*$^{-/-}$ mice also display defects in other aspects of noncanonical-dependent pathways including BAFF-mediated B cell survival, and *Nfkb2*$^{-/-}$ mice have poor B cell survival rates (Gerondakis and Strasser, 2003). In vitro activation of *Nfkb2*$^{-/-}$ T cells does not reveal any defect in cell proliferation or cytokine production (Franzoso et al., 1998). However, the Th1 responses of *Nfkb2*$^{-/-}$ mice appear to vary depending on the challenging pathogen. These mice are particularly susceptible to challenge by *Leishmania major* due to their inability to mount a normal IFNγ response but this is attributed to diminished CD40-dependent IL-12 production by *Nfkb2*$^{-/-}$ macrophages as opposed to an inability of *Nfkb2*$^{-/-}$ T cells to mount a Th1 response (Speirs et al., 2002). Challenge of *Nfkb2*$^{-/-}$ mice by *T. gondii* also results in increased susceptibility due to an inefficient IFNγ response; however, in this case it appears to be due to a Fas-dependent reduction in the number of T cells (Caamano et al., 2000). *Nfkb2*$^{\Delta CT/\Delta CT}$ mice, which express p52 but not p100 due to deletion of the C-terminal ANK-containing domainof p100 demonstrates the importance of p100 as an IκB protein. *Nfkb2*$^{\Delta CT/\Delta CT}$ mice exhibit many hyperproliferative-associated abnormalities such as lymphocytic infiltrates into organs, enlarged lymph nodes, and agranulocytosis (Ishikawa et al., 1997). These defects are consistent with the induction of p52-containing dimers by TNFR superfamily ligands (Beinke and Ley, 2004). The hyperproliferative defects seen in mice which lack p100 support a role for p52 in controlling cell proliferation (Gerondakis et al., 2006). This is further supported by the observations that some human lymphomas have similar alterations to the *Nfkb2*$^{\Delta CT/\Delta CT}$ mice (Courtois and Gilmore, 2006). In human genetic diseases, heterozygous mutations in *NFKB2* are present in an autosomal-dominant CVID where patients present with chronic respiratory infections and autoimmune infiltration of the scalp causing baldness, among other symptoms (Chen et al., 2013).

2.1.3 c-Rel

Studies of *c-Rel*$^{-/-}$ mice have identified an important role for this subunit in B cells, T cells, macrophages, and dendritic cells (DCs). c-Rel is required for B cell proliferation induced by LPS and B cell receptor activation, where it coordinates the cell cycle transition from G1- to S-phase by promoting the expression of E2F3a which is critical for cycle progression (Cheng et al., 2003; Grumont et al., 1998). c-Rel also promotes B cell survival by upregulating the prosurvival Bcl-2 family members *A1* and *Bcl-xl*. *c-Rel*$^{-/-}$ B cells are also defective in isotype switching (Pohl et al., 2002).

In T cells, c-Rel is not required for positive or negative selection but it does play an essential role in the control of CD4$^+$ and CD8$^+$ immunity (Strasser et al., 1999). In a model of autoimmune encephalomyelitis, *c-Rel*$^{-/-}$ mice are defective in Th1 but not Th2 responses. c-Rel regulates the development of Th1 cells through the c-Rel-dependent production of IL-12 by professional antigen-presenting cells (APCs) (Hilliard et al., 2002; Mason et al., 2004). Of note is the finding that c-Rel is not essential for the generation of Th1 cells and IL-12 production by APCs in mice challenged by *T. gondii* (Mason et al., 2004). In this case, *c-Rel*$^{-/-}$ mice had increased susceptibility to infection associated with decreased T cell activation, proliferation, and production of interferon gamma (IFNγ). In this model of infection the reduced T cell response did not appear to result from a defect in Th1 differentiation but rather from defective clonal expansion of Th1 effector cells *c-Rel*$^{-/-}$ mice. Furthermore, c-Rel has a role in the production of cytokines by T cells. In CD4$^+$ T cells, c-Rel is essential for proinflammatory cytokine priming of immune responses and promotes the expression of IFNγ and IL-2 by naïve T cells (Banerjee et al., 2005). When CD4$^+$ T cells are activated, c-Rel promotes the production of GM-CSF and IL-2, and directly regulates IL-2 transcription via chromatin remodeling on the *Il2* promoter (Chen et al., 2005a; Gerondakis et al., 1996). *c-Rel*$^{-/-}$ mice exhibit reduced numbers of type I IFN producing plasmacytoid DCs (pDCs) but show no abnormalities in the development of conventional DCs (cDCs) nor the expression of costimulatory factors (O'Keeffe et al., 2005). However, c-Rel is essential for the APC priming of antigen-specific cytotoxic T cell responses (Mintern et al., 2002) and for the expression of the p35 subunit of IL-12 and the p19 subunit of IL-23 in DCs (Carmody et al., 2007a; Grumont et al., 2001; Mintern et al., 2002). In macrophages, c-Rel is required for the LPS-induced expression of the IL-12 p40 subunit (Sanjabi et al., 2000). The reasons for the different requirements for regulation of IL-12 subunit production by c-Rel across different APCs is not fully understood, however, it may relate to the role of cDCs and macrophages as essential producers of IL-12 in acute and sustained immune responses (Gerondakis et al., 2006).

2.1.4 RelA (p65)

The p65 (*Rela*) subunit is the only NF-κB subunit that results in embryonic lethality upon deletion. *Rela*$^{-/-}$ mice die on embryonic day 15 due to TNFα-induced fetal hepatocyte apoptosis (Beg and Baltimore, 1996). *Rela*$^{-/-}$ lethality may be rescued by also deleting TNFα (Doi et al., 1999) or TNF-receptor1 (Alcamo et al., 2001). *Rela*$^{-/-}$ *Tnfr1*$^{-/-}$ mice are born without major overt abnormalities, however, shortly after birth die at increased rates possibly due to greater susceptibility to infection (Alcamo et al., 2001). The protective effect of p65 against TNFα-induced cell death is also observed in other cell types including macrophages (Beg and Baltimore, 1996), B cells (Prendes et al., 2003), and T cells (Senftleben et al., 2001b). In addition, p65 also prevents apoptosis induced by double-stranded RNA which is a TLR9 ligand (Li et al., 2001). The antiapoptotic role of RelA results from its positive regulation of the expression of a number of antiapoptotic and prosurvival genes including c-FLIP, A20, and GADD45β (Dutta et al., 2006). Analysis of *Rela*$^{-/-}$ *Tnfr1*$^{-/-}$ mice revealed that p65 has an important role in secondary lymphoid organ development. In this case, it is the expression of p65 in radio-resistant stromal cells rather than hemopoietic cells that appear to be responsible. In most cells, p65 comprises the predominant NF-κB transcriptional activity and consequently it is responsible for the expression of a large number of proinflammatory mediators that promote leukocyte recruitment by innate immune cells, T cell responses, and B cell isotype switching (Alcamo et al., 2002; Horwitz et al., 1997).

2.1.5 RelB

The predominant form of RelB appears to be as heterodimers with p50 and p52 (Hayden and Ghosh, 2014). *Relb*$^{-/-}$ mice exhibit a severe inflammatory phenotype comprising T cell infiltrates in multiple organs, inflammatory dermatitis, T cell-dependent myeloid hyperplasia, and splenomegaly (Burkly et al., 1995). *Relb*$^{-/-}$ mice also exhibit disrupted development of secondary lymphoid organs and lack Peyer's patches, splenic germinal centers, and follicular DCs (fDCs) (Yilmaz et al., 2003). In addition, RelB is indispensable for the proper development of the marginal zone and its infiltration by macrophages and B cells (Weih and Caamano, 2003). The defects in secondary lymphoid organ development observed in *Relb*$^{-/-}$ mice are due to the importance of p52/RelB-dependent chemokine expression induced by LTβ activation of the noncanonical pathway (Bonizzi et al., 2004). This is further supported by the pattern of RelB expression which is highest in the thymus, lymph nodes, and Peyer's patches (Gerondakis et al., 2006).

Relb$^{-/-}$ mice also revealed the selective regulation of myeloid DC lineages by RelB, which is thought to result from both defective stromal cells and impaired hemopoietic stem cells (Wu et al., 1998). The function of *Relb*$^{-/-}$ B cells in vitro appears to be normal, however, *Relb*$^{-/-}$ mice exhibit defects in humoral immunity, which may result from impaired $CD4^+$ T cell responses due to defective APC or T cell function (Weih et al., 1997). T cell expression of RelB is essential for the proliferation of single positive T cells in the latter stages of development and is required in mature $CD4^+$ T cells for normal Th1 function and IFNγ production (Guerin et al., 2002). *Relb*$^{-/-}$ mice also exhibit impaired Th1 cell differentiation, possibly due to decreased expression of Stat4 (Corn et al., 2005). In human studies, an autosomal recessive RelB deficiency has been observed in three patients from the same family, who all presented with combined immunodeficiency (CID) from a young age, experiencing frequent infections and severe autoimmune skin diseases (Sharfe et al., 2015).

2.1.6 Double NF-κB Subunit Knockout Mice

The studies of mice with individual knockout mutations for each of the genes that encode the five subunits of NF-κB shed much light on their unique roles in the control of inflammation. However, given the heterodimerization and homodimerization properties of each NF-κB subunit, the knockout of individual subunits may not fully capture the complexity of NF-κB dimer function. This is substantiated by studies of mice deficient in more than one NF-κB subunit.

The generation of *Nfkb1*$^{-/-}$*Nfkb2*$^{-/-}$ mice demonstrated a joint role for these subunits in coordinating the LTβ-dependent formation of secondary lymphoid organs (Lo et al., 2006). The phenotype seen in

Nfkb1$^{-/-}$ *Nfkb2*$^{-/-}$ mice is similar to that of mice lacking LTβR and is more severe than that exhibited by mice with single gene knockout. This is further supported by the demonstration that LTβ activates both p50- and p52-containing dimers (Gerondakis et al., 2006). Similarly, *Nfkb1* and *Relb* double knockout mice exhibit more severe organ inflammation than that seen in single *Relb*$^{-/-}$ mice (Weih et al., 1997). These double knockout mice experience inflammatory infiltrates that are entirely devoid of B cells, indicating a critical impairment in B cell development. Mice with both *Nfbk1* and *c-Rel* knocked out display similar overall defects as the single knockout mice but with increased severity, for example, the reduction in the $CD5^+$ peritoneal B cell population in *Nfkb1*$^{-/-}$ mice is even further diminished in *Nfkb1*$^{-/-}$ *c-Rel*$^{-/-}$ mice (Pohl et al., 2002). The embryonic death observed in *Rela*$^{-/-}$ mice occurs at an even earlier stage of development in *Nfkb1*$^{-/-}$ *Rela*$^{-/-}$ mutants (Horwitz et al., 1997). Furthermore, this double knockout mutation reveals a critical role for both p50 and p65 in the control of mature B cell activation because although normal hemopoietic cells can rescue the development of *Nfkb1*$^{-/-}$ *Rela*$^{-/-}$ mutants, they do not divide in response to mitogenic signals (Horwitz et al., 1999). Overall, the study of double knockout mice reveals a certain degree of redundancy in NF-κB subunits in the control of inflammation and immune development.

2.2 IKK Complex

2.2.1 IKKα

As previously mentioned, the IKK complex is an essential activator of NF-κB. In the classical pathway of NF-κB activation, IKKβ appears to be the most important kinase activity while IKKα alone is required for the activation of NF-κB by the noncanonical pathway (Scheidereit, 2006). Initial studies demonstrated that mutant mice lacking IKKα (*Ikka*$^{-/-}$) exhibit many skin and skeletal defects that are not seen in mice lacking one or more NF-κB subunits and revealed a kinase-independent role for IKKα (Cao et al., 2001). These developmental defects are not observed in mice with a catalytically inactive allele of IKKα (*Ikka*$^{AA/AA}$). The importance of IKKα in the activation of the noncanonical pathway is very obvious in *Ikka*$^{AA/AA}$ mice which are phenotypically similar to *Nfkb2*$^{-/-}$, *ltβ*$^{-/-}$, and *Nik*$^{-/-}$ mice in that they exhibit defects in the development of secondary lymphoid organs and do not produce mature fDCs (Bonizzi et al., 2004; Gerondakis et al., 2006). Somewhat counter intuitively, *Ikka*$^{AA/AA}$ mice also display a heightened inflammatory response to LPS stimulation due to increased proinflammatory cytokine expression by *Ikka*$^{AA/AA}$ macrophages (Lawrence et al., 2005). The exact mechanism behind this hyper-responsiveness is unclear and it has been proposed IKKα phosphorylation of c-Rel and p65 promotes their degradation (Lawrence et al., 2005). However, it has also been suggested that in the absence of IKKα kinase activity there is a decreased poststimulus induction of IκBα due to increased IKKβ activity, indicating a role for IKKα in the regulation of the IKK complex (Li et al., 2005). This is further supported by studies from IKKα-null bone marrow chimeras that demonstrate immune system abnormalities, including altered NF-κB activation in macrophages and defective B-cell function (Senftleben et al., 2001a). In humans, autosomal recessive loss-of-function mutations in IKKα result in fetal lethality, whereas IKKβ deficiency causes severe combined immunodeficiency (SCID) (Pannicke et al., 2013).

2.2.2 IKKβ

The importance of IKKβ in the classical pathway of NF-κB activation is apparent from the phenotype of *Ikkb*$^{-/-}$ mice which broadly encompasses the phenotypes of *Nfkb1*$^{-/-}$, *c-Rel*$^{-/-}$, and *Rela*$^{-/-}$ mice. Similar to *Rela*$^{-/-}$ mice, *Ikkb*$^{-/-}$ mice die as embryos due to TNFα-induced hepatocyte apoptosis (Li et al., 1999). This has meant that much of our understanding of IKKβ function in vivo has come from studies performed using tissue- or cell-specific deletion of IKKβ. This has identified important cell-specific roles for IKKβ in regulating inflammation. IKKβ expression in macrophages is required for the production of proinflammatory cytokines (Lawrence et al., 2005) and there has been a link established between IKKβ and the inflammation observed during obesity-induced insulin resistance (Arkan et al., 2005). Hepatocyte-specific inactivation of IKKβ led to the liver still being responsive to insulin, whereas muscle and fat tissue became insulin-resistant. However, mice with IKKβ deleted in myeloid cells retained systemic sensitivity to insulin, suggesting that the production of proinflammatory mediators by myeloid cells as regulated by NF-κB is key to the development of systemic insulin resistance that is the hallmark of type II diabetes. In keratinocytes, the opposite finding was observed, wherein inactivating IKKβ resulted in inflammation (Pasparakis et al., 2002). Mice with epidermis-specific deletion of IKKβ developed severe, TNFα-dependent inflammation in the skin shortly after birth. These studies suggest that IKKβ in epidermal keratinocytes has a critical function in maintaining immune homeostasis of the skin. This highlights the complex role IKKβ plays in the regulation of inflammation whereby for monocytes IKKβ is essential for inflammatory responses, whereas in skin cells, its inactivation leads to an inflammatory response (Pasparakis et al., 2002). In human studies, patients that lack IKKβ are born without any obvious defects but develop SCID after birth that is in contrast to *Ikkb*$^{-/-}$ mice, which die as embryos, and highlights a major physiological difference in the role of IKKβ in mice and humans. Cells from patients with a homozygous mutation in *IKBKB* had impaired responses to stimulation through T-cell receptors, B-cell receptors, toll-like receptors, inflammatory cytokine receptors, and mitogens (Zhang et al., 2017). In the absence of IKKβ, the IKK complex consists of IKKα homodimers thus altering the response to some stimuli; for example, phosphorylation and degradation of IκBα and induction of IL-6 are eliminated in response to TLR5 stimulation, but was only marginally affected following IL-1β stimulation (Pannicke et al., 2013). These studies demonstrate that IKKα homodimers can mediate classical pathway signaling for some stimuli, suggesting that IKKα and IKKβ may serve different receptors (Solt et al., 2007; Zhang et al., 2017).

2.2.3 NEMO

NEMO is a regulatory component of the IKK complex that is essential for the activation of the classical NF-κB pathway. NEMO$^{-/-}$ (*Ikbkg*$^{-/-}$) mice, similar to *Rela*$^{-/-}$ and *Ikkb*$^{-/-}$ mice, die during embryogenesis due to hepatocyte apoptosis induced by TNFα (Kim et al., 2003). The *Ikbkg* gene is located on the X chromosome so heterozygous null females are viable but exhibit an inflammatory skin disease soon after birth that is similar to incontinentia pigmenti, a human genetic condition caused by NEMO mutations (Makris et al., 2000). Indeed, in humans complete loss of function or amorphic mutations cause prenatal lethality in males (Zhang et al., 2017). Hypomorphic mutations in the gene encoding NEMO cause anhidrotic ectodermal dysplasia with immunodeficiency (Doffinger et al., 2001). Cells lacking NEMO also lack activation of the classical pathway in response to a number of immune mediators such as IL-1β, TNFα, and LPS (Hubeau et al., 2011). TNFα activates both apoptotic and NF-κB pathways and so the immune activation of cells requires NF-κB to induce genes that inhibit apoptosis (Karin and Lin, 2002). When NEMO is disrupted, NF-κB does not induce the expression of these inhibitory genes leading to TNFα-induced apoptosis and dysregulated immunity. Mice with the conditional deletion of NEMO in intestinal epithelial cells (NEMO$^{\Delta IEC}$) revealed a key role for NF-κB in maintaining epithelial barrier function and immune homeostasis in the gut (Nenci et al., 2007). As a result, NEMO$^{\Delta IEC}$ mice develop spontaneous colitis due to defective epithelial barrier function. Of note, this phenotype is observed in mice with the conditional deletion of both *Ikkb* and *Ikka* in intestinal epithelial cells, but not in mice with a conditional deletion of either *Ikkb* or *Ikka* alone (Nenci et al., 2007). These findings provide further proof that IKKα and IKKβ share some functional redundancy. T cell-specific deletion of NEMO or the replacement of IKKβ with a dominant-negative mutant prevents the development of mature peripheral T cells demonstrating that mature T cells require NF-κB activity (Schmidt-Supprian et al., 2003). Hepatocyte-restricted NEMO deletion leads to spontaneous chronic hepatitis in adult mice, which ultimately progresses to hepatocellular carcinoma through compensatory hepatocyte proliferation (Luedde et al., 2007). Indeed, it was these findings which provided the first major indication that the pharmacological inhibition of the IKK complex may not be a viable therapeutic strategy (Herrington et al., 2016).

2.2.4 NF-κB-Independent Functions of IKKs

The IKKs were long considered kinases specific for IκB proteins; however, a significant number of additional substrates have now been identified which reveal important NF-κB-independent roles for these kinases. The NF-κB-independent functions of IKKs have been reviewed in depth elsewhere (Chariot, 2009), and so here we will briefly discuss only those aspects most relevant to the regulation of inflammation. IKK-mediated phosphorylation and activation of the related TBK1 and IKKε kinases promotes the TLR-induced transcription of interferons via activation of interferon response factors (IRF) (Clark et al., 2011). In addition, IKKα-mediated phosphorylation of IRF7 promotes TLR7- and TLR9-induced interferon production (Hoshino et al., 2006), while IKKα-mediated phosphorylation of IRF5 appears to inhibit its activity (Balkhi et al., 2010). IKKβ plays a key role in TLR- and TNFR-induced activation of the MAPK pathway by regulating the activation of the MAP3K Tpl-2 (Beinke and Ley, 2004). In resting cells, Tpl-2 is inactive and bound by NF-κB p105. TLR activation induces IKKβ phosphorylation of p105, which triggers its proteasomal degradation, and the release of active Tpl-2 which initiates the MAPK cascade by phosphorylating and activating MEK1. IKK phosphorylation of histone H3 (Anest et al., 2003; Yamamoto et al., 2003) and a number of nuclear coactivators such as SMRT (Hoberg et al., 2004) and CBP (Huang et al., 2007) promotes inducible gene transcription. In mast cells, IKKβ phosphorylation of SNAP-23 is an important step in IgE-induced degranulation and subsequent allergic reactions (Suzuki and Verma, 2008). IKKs also participate in the termination of inflammatory responses through the phosphorylation of a variety of factors including PIAS1 (Liu et al., 2007), Dok1 (Lee et al., 2004), NEMO (Prajapati and Gaynor, 2002), and NF-κB p65 (Lawrence et al., 2005). Additional IKK substrates, such as p53 (Xia et al., 2009), cyclin D1 (Kwak et al., 2005), and the proapoptotic protein BAD (Yan et al., 2013), indicate an important role in the regulation of cell proliferation and survival. These NF-κB-independent roles of IKK confound the original view of these kinases as dedicated gatekeepers to inflammation and are likely important factors in the failure of IKK-centered therapeutic approaches for the treatment of inflammatory disease. These studies also suggest caution in interpreting some of the studies performed using transgenic mice where the contribution of the NF-κB-independent roles of IKKs cannot be ruled out.

2.3 IκB Proteins

2.3.1 Prototypical IκB Proteins

As previously described, the primary means for controlling NF-κB transcriptional activity are the IκB proteins. The prototypical IκB proteins are IκBα, IκBβ, and IκBε all of which undergo signal-induced IKK-mediated phosphorylation and proteasomal degradation. IκBα, IκBβ, and IκBε have distinct but overlapping selectivity in their interaction with NF-κB dimers; IκBα preferentially bind dimers containing p50, p65, and c-Rel; IκBβ preferentially binds p65 and c-Rel-containing heterodimers; and IκBε preferentially binds p65 and c-Rel but may also bind p65:p50 dimers (Hinz et al., 2012). Importantly, the degradation and stimulus-induced expression profile of each IκB subunit is distinct and appears linked to the specific role of each protein in regulating inflammation. The rapid degradation and resynthesis of IκBα allow for the rapid inhibition of NF-κB activity while the slower kinetics of IκBβ and IκBε degradation act to dampen long-term oscillations of the NF-κB response (Hoffmann et al., 2002). The nonredundant functions of the prototypical IκB proteins were revealed by the generation of knockout mice. IκBα-deficient (*Nfkbia*$^{-/-}$) mice die shortly after birth with severe granulopoeisis and inflammatory dermatitis due to uncontrolled and sustained activation of NF-κB (Beg et al., 1995). In addition, *Nfkbia*$^{-/-}$ mice show defects in secondary lymphoid organ structure and NF-κB activity in B cells. IκBα is essential in regulating the expression of proinflammatory genes as controlled by NF-κB, therefore, the inflammatory dermatitis observed in *Nfkbia*$^{-/-}$ mice is driven by increased levels of TNFα, G-CSF, MIP-2, and eotaxin (Klement et al., 1996). Mice lacking IκBβ (*Nfkbib*$^{-/-}$) or IκBε (*Nfkbie*$^{-/-}$) are viable and display only minor immune defects such as increased expression of specific cytokines (Memet et al., 1999). Remarkably, *Nfkbib*$^{-/-}$ mice were found to be resistant to LPS-induced septic shock and collagen-induced arthritis (Rao et al., 2010). Subsequent analysis revealed that following resynthesis, IκBβ is retained in the nucleus where it complexes with DNA bound p65 and c-Rel to prolong the NF-κB-dependent transcription of certain genes such as TNFα and IL1β (Scheibel et al., 2010). Thus, IκBβ serves to both inhibit and enhance the inflammatory response. In humans, heterozygous gain-of-function mutations of *NFKBIA* cause autosomal dominant conditions that resemble those seen in NEMO and IKKβ deficiencies (Schimke et al., 2013). Most mutations of *NFKBIA* are either substitutions of the phosphorylation sites at residues S32 and S36, or N-terminal truncations that remove these sites altogether which results in super-repressor forms of IκBα.

2.3.2 Atypical IκB Proteins

In contrast to the prototypical IκB proteins, which function predominantly as inhibitors of NF-κB activation, the atypical IκB proteins Bcl-3, IκBζ, IκBNS, and IκBη function largely to modulate the transcriptional activity of NF-κB in the nucleus. This modulation is achieved through a number of mechanisms including NF-κB stabilization, cofactor recruitment, and the enhancement or inhibition of NF-κB transactivational activity (Hinz et al., 2012).

2.3.3 Bcl-3

Bcl-3 is an atypical IκB protein that was initially identified at the t(14;19) chromosomal breakpoint found in some patients with B cell chronic lymphocytic leukemia (Ohno et al., 1990). Bcl-3 contains a central seven ANK that mediates the selective interaction with homodimers of the p50 and p52 subunits (Nolan et al., 1993). In addition, Bcl-3 has a well-defined TAD, which may enhance the expression of certain NF-κB target genes (Ghosh and Hayden, 2008). *Bcl-3*$^{-/-}$ mice have defective secondary lymphoid organ development including a lack of splenic germinal centers, reduced Peyers' patch numbers, and disrupted lymph node architecture (Franzoso et al., 1997; Gerondakis et al., 2006; Schwarz et al., 1997). *Bcl-3*$^{-/-}$ mice also show reduced numbers of follicular B cells but increased numbers of marginal zone B cells. Conversely, conditional overexpression of Bcl-3 in

B cells leads to splenomegaly, the accumulation of follicular B cells, and a lack of marginal zone B cells (Hovelmeyer et al., 2014; Zhang et al., 2013). *Bcl-3*$^{-/-}$ mice are sensitive to infection with *L. monocytogenes, S. pneumonia, Klebsiella pneumonia,* and *T. gondii* resulting from an inability to effectively clear bacteria (Pene et al., 2011; Schwarz et al., 1997). The hypersensitivity of *Bcl-3*$^{-/-}$ mice to infection results from a hyperresponsiveness to activation of TLRs that leads to an exaggerated proinflammatory response and an imbalanced cytokine and chemokine response. Bcl-3 limits proinflammatory cytokine expression through the inhibition of p50 homodimer ubiquitination which results in a stable DNA bound Bcl-3:p50 repressor complex (Carmody et al., 2007b). In the absence of Bcl-3, p50 homodimers undergo polyubiquitination and proteasomal degradation, leading to increased transcription of proinflammatory cytokines by NF-κB dimers containing p65 or c-Rel subunits (Carmody et al., 2007b). Thus, *Bcl-3*$^{-/-}$ macrophages, DCs and B cells are hyperresponsive to TLR stimulation as measured by increased expression of cytokines including TNFα and IL-6. Furthermore, as discussed later, *Bcl-3*$^{-/-}$ mice and macrophage are defective in TLR tolerance (Carmody et al., 2007b). Interaction with p50 is required for the antiinflammatory function of Bcl-3 (Collins et al., 2014) and cell permeable Bcl-3-derived peptides exhibit antiinflammatory activity in vitro and in vivo (Collins et al., 2015).

Mice deficient in either p100/p52 or Bcl-3 do not show overt signs of autoimmunity; however, *Nfkb2* and *Bcl-3* double knockout mice die within a month of birth from severe multiorgan inflammation (Zhang et al., 2007). This is the result of defective central tolerance in the thymus, which leads to the generation of autoreactive T cells. The lack of central tolerance results from defective medullary thymic epithelial cells and abnormal thymic development. In addition to a lack of central tolerance, Bcl-3 deficiency leads to impaired Th2 T cell differentiation resulting from reduced expression of GATA3 a key transcription factor required for Th2 differentiation (Corn et al., 2005). Consequently, *Bcl-3*$^{-/-}$ Th2 cells secrete reduced levels of IL-4, IL5, and IL-13. Th1 differentiation appears normal in the absence of Bcl-3 but Bcl-3 appears to inhibit the differentiation of Th17 cells from IFNγ producing Th1 cells in vivo (Tang et al., 2014). A study employing overexpression and knockdown of Bcl-3 in mouse CD4+ T cells identified a role for Bcl-3 in T_{FH} development (Meguro et al., 2015). Here, Bcl-3 appears to enhance the expression of Bcl-6 in order to promote T_{FH} differentiation. Furthermore, overexpression of Bcl-3 promotes T cell survival (Mitchell et al., 2001), while activated *Bcl-3*$^{-/-}$ T cells are more likely to undergo apoptosis than wild type counterparts (Bauer et al., 2006). In early rheumatoid arthritis patients, the expression of Bcl-3 in CD4+ T cells is predictive for disease and appears to be induced by elevated levels of circulating IL-6 (Pratt et al., 2012). The consequences of increased expression of Bcl-3 in the context of rheumatoid arthritis are not clear, but a link between IL-6-induced expression of Bcl-3 and enhanced T_{FH} cell development has been proposed (Meguro et al., 2015; Pratt et al., 2016).

Bcl-3$^{-/-}$ mice are more sensitive to autoimmune diabetes, a finding that correlated with increased IL-17A expression (Ruan et al., 2010). In a mouse model of DSS-induced colitis, *Bcl-3*$^{-/-}$ mice showed an unexpected resistance to disease due to the enhanced regeneration of intestinal epithelial cell and restoration of barrier function (O'Carroll et al., 2013). The conditional deletion of Bcl-3 in intestinal epithelial cells yielded similar results and demonstrated that Bcl-3 also suppresses inflammation-associated tumourigenesis of intestinal epithelial cells through a TNFα-dependent mechanism (Tang et al., 2016).

2.3.4 IκBζ

IκBζ (INAP/MAIL) selectively interacts with the p50 and p65 subunits of NF-κB (Yamazaki et al., 2001). Its expression is induced by a number of cytokines, TLRs and the B cell receptor (Annemann et al., 2016). The majority of IκBζ knockout mice (*Nfkbiz*$^{-/-}$) die during embryogenesis and those surviving develop atopic dermatitis and conjunctivitis (Yamamoto et al., 2004). *Nfkbiz*$^{-/-}$ mice have defective Th17 development and are resistant to experimental autoimmune encephalomyelitis (Okamoto et al., 2010). In T cells, IκBζ cooperates with RORγT and RORα to promote the expression of IL-17a and promotes the expression of IL-21, IL-22, and IL-23R. In macrophages, IκBζ promotes the expression of IL-6 and IL-12p40 through interaction with p50 on gene promoters (Motoyama et al., 2005). In addition, IκBζ is also required for the expression of CCL2, GM-CSF, and G-CSF but inhibits the expression of TNFα in macrophages (Kayama et al., 2008; Motoyama et al., 2005).

2.3.5 IκBNS

IκBNS preferentially interacts with p50 homodimers and selectively inhibits the expression of TLR-induced IL-6, IL-12p40, and IL-18 (Kuwata et al., 2006). Deletion of IκBNS leads to intestinal inflammation and increased sensitivity to LPS-induced septic shock, most likely because of increased cytokine production (Kuwata et al., 2006). IκBNS is required for T cell proliferation and IκBNS$^{-/-}$ T cells produce less IL-2 and IFNγ. Promoter analysis suggests that IκBNS is recruited to the IL-2 promoter and may act to enhance NF-κB-dependent transcription of IL-2 (Touma et al., 2007).

2.3.6 *IκBη*

IκBη is the most recently identified member of the atypical IκB family. It interacts selectively with the p50 subunit of NF-κB and appears to act as a coactivator of NF-κB-dependent transcription of IL-6, IL-1β, and G-CSF in macrophages (Yamauchi et al., 2010). Studies on IκBη have to date been performed using siRNA-mediated knockdown and the physiological role of IκBη is regulating immune development and inflammation is not yet known.

3. POSTTRANSLATIONAL MODIFICATION AND THE REGULATION OF NF-κB TRANSCRIPTIONAL ACTIVITY

The proinflammatory nature of the transcriptional targets of NF-κB and their inherent potential for damage to host tissue necessitates tight control of NF-κB activation and transcriptional activity. The consequences of uncontrolled, inappropriate, or dysregulated inflammation are manifest in a range of diseases including rheumatoid arthritis, autoimmunity, inflammatory bowel disease, as well as atherosclerosis and cancer. As we have described, the primary point of control of NF-κB transcriptional activity is through the prototypical IκB proteins that control the subcellular localization of NF-κB. Indeed, it is clear from the genetic studies performed to date that the combination of NF-κB subunits and the interplay with IκB proteins provides a complex regulatory network for the control of inflammation at the transcriptional level. However, it is also clear that the degradation of the IκB proteins and the translocation of NF-κB dimers to the nucleus, although necessary, are not sufficient for the optimal NF-κB-directed transcriptional response to inflammatory stimuli (Oeckinghaus and Ghosh, 2009). Many studies over the past 20 years have identified additional mechanisms, mostly involving posttranslational modification of the NF-κB subunits, that regulate NF-κB-mediated transcriptional responses (Christian et al., 2016). An array of posttranslational modifications have been identified including phosphorylation, ubiquitination, acetylation, glycosylation, and nitrosylation all of which directly impact of NF-κB transcriptional activity. A clear emerging theme from studies on NF-κB modification is that many modifications regulate NF-κB transcriptional activity in a gene-specific manner. These studies demonstrate the further complexity of NF-κB regulatory networks that incorporate the local environment of promoter and enhancer regions. A detailed description of NF-κB posttranslational modifications is beyond the scope of this review and a number of in depth reviews on the phosphorylation (Christian et al., 2016) and ubiquitination (Collins et al., 2016) of NF-κB are already available. Instead, we will highlight a number of key studies that demonstrate the importance of NF-κB modifications to the regulation of inflammation (Fig. 4).

3.1 Phosphorylation

Site-specific phosphorylation has been described for all NF-κB subunits, but the role of phosphorylation in regulating NF-κB transcriptional activity is best studied for the p65 subunit (Christian et al., 2016). p65 is phosphorylated on at least 24 sites by a range of kinases (Lanucara et al., 2016), with some sites targeted by multiple kinases (Christian et al., 2016). Of the identified phosphorylation sites of p65, phosphorylation of S536, S468, and S276 are the best characterized with respect to the impact on NF-κB transcriptional activity.

The phosphorylation of p65 at S536, which is contained within the TAD, enhances transactivation by increasing the binding of CBP/p300 (Chen et al., 2005b). S536 is targeted for phosphorylation by a number of kinases including IKKα, IKKβ, IKKε, RSK1, and TBK1 (Christian et al., 2016). Studies using IKKα^{AA} mice and more recently RelA^{S536A} knockin mice demonstrated that phosphorylation of S536 increases proteasomal degradation and reduces the half-life of activated p65 (Lawrence et al., 2005; Pradere et al., 2016). This increased instability of phosphorylated p65 is likely mediated by an E3 ligase complex composed of SOCS1, COMMD1, and Cullin2 which recognizes p65 phosphorylated at S536 (Wang et al., 2014). As a result, RelA^{S536A} mice are more sensitive to LPS-induced septic shock and display elevated serum levels of TNFα and IL-1β (Pradere et al., 2016). Of note, the impact of S536 phosphorylation on the transcription of NF-κB target genes appears to be both stimulus specific and gene specific (Pradere et al., 2016). What determines the selectivity of transcriptional control by S536 phosphorylation is not clear but early studies suggested that phosphorylation of p65 at S536 may determine the ability of resynthesized IκBα to inhibit transcription in a gene selective manner (Sasaki et al., 2005).

Similar to S536, S276 of p65 can also be phosphorylated by a number of kinases including PKA, PKC, MSK1/2, RSK1, and Pim1 (Christian et al.,

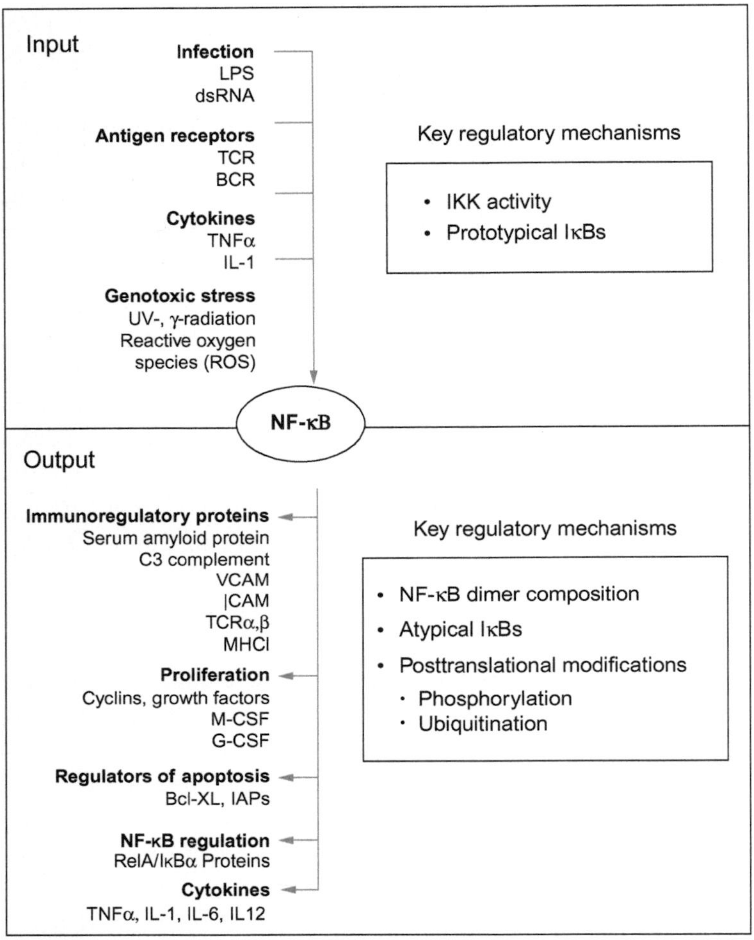

Fig. 4 NF-κB is activated by a diverse range of stimuli (Input) including microbes, antigens, cytokines, as well as environmental stress factors. The key regulators controlling activation of NF-κB are IKK kinase activity and the prototypical IκB proteins. Once activated NF-κB can direct a significant number of transcriptional outputs including the expression of immunoregulatory proteins, cytokines, and factors that regulated cell proliferation and apoptosis. The transcriptional output of activated NF-κB is primarily regulated by NF-κB dimer-specific roles, the atypical IκB proteins, and the posttranslational modification of NF-κB subunits, including phosphorylation and ubiquitination.

2016). As with S536, phosphorylation of S276 has a selective effect on the transcription of NF-κB target genes. Phosphorylation of S276 selectively promotes the expression of a subset of NF-κB target genes by promoting the recruitment of a cyclin-dependent kinase 9/cyclin1 complex to gene promoters (Nowak et al., 2008). The S276A mutation is embryonic lethal in mice due to variegated developmental abnormalities (Dong et al., 2008). These developmental abnormalities result from the recruitment of histone deacetylases to chromatin by nonphosphorylated p65 that leads to repression of proximal chromatin and appears to be unrelated to p65 transcriptional activity (Dong et al., 2008).

The phosphorylation of p65 at S468 promotes the binding of an E3 ubiquitin ligase complex composed of GCN5, COMMD1, Cullin2, and SOCS1 which leads to the ubiquitination and proteasomal degradation of p65 (Mao et al., 2009). Importantly, the impact of S468 phosphorylation on transcription is also gene specific and limits the expression of only a subset of NF-KB target genes; for example, S468A mutation in fibroblasts inhibits the IL-1β and TNFα inducible expression of CCL5 but not CCL2 (Geng et al., 2009; Mao et al., 2009).

While the regulation of NF-κB transcriptional activity by phosphorylation extends far beyond these three examples of p65 phosphorylation, they are exemplars of how phosphorylation at specific sites of NF-κB subunits may regulate transcription in a gene selective manner. The most recent phospho-proteomic analysis suggests that the phospho-regulatory network of NF-κB is extremely complex, particularly when considering the potential combinatorial outcomes of phosphorylation at different sites (Lanucara et al., 2016). Moreover, how the phosphorylation of one subunit influences the function of a partner subunit in NF-κB heterodimers is as of yet unknown.

3.2 Ubiquitination

Recent studies have demonstrated that not only does ubiquitination play an important role in the events leading to the activation of NF-κB, but also in the regulation of the NF-κB transcriptional response. The ubiquitination and subsequent proteasomal degradation of NF-κB subunits in the nucleus is a key determinant of transcriptional activity and is a major rate limiting factor for the expression of proinflammatory genes (Collins et al., 2016). The role of ubiquitination in regulating NF-κB transcriptional activity was initially discovered in IκBα$^{-/-}$ cells that, despite lacking the critical negative feedback loop, still terminate NF-κB transcriptional activity following stimulation (Saccani et al., 2004). The IκBα-independent termination of NF-κB transcriptional activity is dependent on the ubiquitination and proteasomal degradation of p65, which in turn is dependent on the binding of p65 to DNA (Saccani et al., 2004). The full extent of activation-induced p65 ubiquitination and its importance in regulating inflammatory responses was revealed following the identification of USP7 as a p65 deubiquitinase (DUB) (Colleran et al., 2013). USP7 is a nuclear DUB that is recruited to NF-κB target promoters bound by p65. USP7 removes polyubiquitin chains from p65 and promotes transcription by increasing the stability and half-life of transcriptionally active p65. Blockade of USP7 activity increases p65 ubiquitination and degradation, and inhibits TLR- and TNFα-induced cytokine and chemokine expression in macrophages (Colleran et al., 2013).

The predominant form of p65 ubiquitination is K48-linked polyubiquitin chains which target p65 for degradation by the proteasome (Collins et al., 2016). p65 polyubiquitination with other, nondegradative ubiquitin linkages has been identified, but the functional role of these types of chains is not yet known (Li et al., 2012). The monoubiquitination of p65 has also been demonstrated, which negatively regulates NF-κB transcriptional activity through a proteasome-independent mechanism (Hochrainer et al., 2012). A number of E3 ubiquitin ligases for NF-κB have been identified; the E3 ligases SOCS1 (Ryo et al., 2003), ING4 (Hou et al., 2014), PDLIM2 (Tanaka et al., 2007), and PPARγ (Hou et al., 2012) target p65 for ubiquitination, while Peli1 targets c-Rel for ubiquitination (Chang et al., 2011). To date no E3 ligase has been described for the p50 and RelB subunits. It is not yet clear whether there is redundancy in the activity of the E3 ligases on NF-κB subunits, but studies to date suggest that there may be gene selective effects of the E3 ligases for p65. TLR-stimulated *Pdlim2*$^{-/-}$ DCs show reduced p65 ubiquitination and degradation and increased expression of IL-6 and IL-12, and *Pdlim2*$^{-/-}$ mice are more sensitive to LPS-induced septic shock (Tanaka et al., 2007). *Ing4*$^{-/-}$ mice are also hypersensitive to LPS-induced septic shock, and LPS-stimulated macrophages show increased levels of nuclear p65 protein associated with increased expression of a subset of NF-κB genes including IL-6, CCL2, and CXCL10 (Coles et al., 2010). Interestingly, LPS-stimulated ING4$^{-/-}$ macrophages express reduced levels of TNFα as well as reduced levels of IκBα and IκBβ, demonstrating the selective regulation of NF-κB target genes (Coles et al., 2010). The ubiquitination of NF-κB is dependent on DNA binding but may also be regulated by site-specific phosphorylation. Phosphorylation of p65 at S536 promotes proteasomal degradation (Wang et al., 2014) but is also required for interaction with USP7 (Colleran et al., 2013). Phosphorylation of p65 at S468 promotes the recognition of p65 by an E3 ligase complex composed of Cullin2, SOCS1, COMMD1, and GCN5 and leads to polyubiquitination and degradation (Geng et al., 2009; Mao et al., 2009). As previously discussed, S468 phosphorylation selectively regulates the transcription of a subset of NF-κB target genes which may also be linked to the selective effects of this E3 ligase complex on the transcription NF-κB target genes (Geng et al., 2009).

The ubiquitination and proteasomal degradation of p50 homodimers is a critical determinant of inflammatory responses in vitro and in vivo (Carmody et al., 2007b). The p50 subunit of NF-κB lacks a TAD and so in the homodimeric form has the potential to act as a repressor of other transcriptionally active NF-κB dimers. As with p65, the ubiquitination of p50 is dependent on DNA binding and, as previously described; p50 ubiquitination is inhibited through interaction with the atypical IκB protein Bcl-3 (Carmody et al., 2007b). Indeed, the antiinflammatory properties of Bcl-3 are dependent on its interaction with, and stabilization of, DNA bound p50 homodimers (Collins et al., 2014). The stability of p50 homodimers by Bcl-3 is not only important for limiting the transcriptional response to acute TLR stimulation but also for establishing an antiinflammatory state in macrophages termed endotoxin or TLR tolerance (Carmody et al., 2007b; Collins and Carmody, 2015).

Although the study of NF-κB subunit ubiquitination has mostly focused on the p50 and p65 subunits, the ubiquitination of RelB and c-Rel subunits has also been reported. In contrast to p65 and p50, RelB ubiquitination does not require DNA binding and appears to promote rather than inhibit transcriptional activity (Leidner et al., 2008). The basis for these differences is not fully clear but may be explained by nondegradative polyubiquitin linkages. c-Rel undergoes ubiquitination following T cell stimulation and is dependent on the E3 ligase Peli1 (Chang et al., 2011). c-Rel ubiquitination promotes proteasomal degradation and appears to be important in limiting c-Rel-dependent T cell responses following activation.

3.3 TLR Tolerance

TLR tolerance (also termed endotoxin tolerance) is an altered responsiveness of cells, mostly macrophages, to repeated or chronic exposure to TLR ligands. TLR tolerance is typically characterized by the reduced expression of proinflammatory cytokines such as TNFα and IL-6 which serves to limit the potential for damage to host tissue during injury or infection (Collins and Carmody, 2015). A comprehensive transcriptomic analysis of TLR tolerance in mouse macrophages revealed that TLR tolerance represents a global transcriptional shift from a proinflammatory response to an antiinflammatory or proresolution response (Foster et al., 2007). This investigation categorized the two classes of genes that are induced by LPS stimulation as tolerisable genes that are repressed during tolerance and are proinflammatory in nature, and nontolerisable genes that are unchanged or induced at even greater levels during tolerance and which are antimicrobial, antiinflammatory, and proresolution in nature. TLR tolerance is a transient state that can be largely reversed both in vivo and in vitro (O'Carroll et al., 2014).

An analysis of tolerisable and nontolerisable genes revealed that almost 70% contain NF-κB binding sites for in their promoter regions (Yan et al., 2012). Subsequent analysis revealed that the presence of a promoter NF-κB binding site determined whether a gene was repressed during TLR tolerance. Indeed, the addition of a NF-κB binding site to the promoter of a nontolerisable gene was sufficient to convert the gene to a tolerisable one, and vice versa. NF-κB binding sites enforce tolerance by recruiting repressive p50 homodimers, stabilized by interaction with Bcl-3 (Yan et al., 2012). Thus, NF-κB binding sites in the promoter regions of proinflammatory genes are not only important for their rapid and robust expression following TLR activation, but also for their repression under tolerising conditions. Stably bound p50 homodimers also recruit histone deacetylases to the promoters of tolerised genes to enforce transcriptional repression further (Yan et al., 2012). The importance of p50 and Bcl-3 in tolerance is demonstrated further by the lack of tolerance in *Nfkb1*$^{-/-}$ and *Bcl3*$^{-/-}$ mice and macrophages (Bohuslav et al., 1998; Carmody et al., 2007b; Ziegler-Heitbrock, 2001).

3.4 NF-κB as a Therapeutic Target

Considering the ubiquity of NF-κB in the control of hundreds of biologically important genes and its critical role in the regulation of inflammation and immune development, it is a highly attractive pharmaceutical target. If its activity can be attenuated this may provide potential treatments for chronic inflammatory diseases and cancer. Significant attention has been given to the potential development and use of modulators of the NF-κB signaling pathway as therapeutics in the treatment of autoimmune disease. Initial efforts focused on targeting the IKK kinases in order to inhibit the activation of NF-κB with the expectation that this would have significant antiinflammatory effects. Although preclinical trials of a number of IKK inhibitors were promising, very few agents made it to clinical trials despite enormous investment by the pharmaceutical industry

(Herrington et al., 2016). The toxicity associated with IKK inhibitors is perhaps to be expected, considering the abnormalities observed in genetic models, and in retrospect a selective targeting of NF-κB transcriptional activity rather than the broad blockade of NF-κB activation may be a better therapeutic strategy. Indeed, the selective effects of many of the posttranslational modification of NF-κB subunits identified to date, in particular phosphorylation and ubiquitination, suggest that these may be more effective therapeutic targets. TLR tolerance provides a model for how selective regulation of gene transcription can lead to beneficial functional changes in the activity of immune cells. Our understanding of the molecular mechanisms underpinning this process identifies additional routes for targeting NF-κB; for example, mimicking the selective ability of Bcl-3 to inhibit proinflammatory gene expression has been demonstrated to be effective at inhibiting inflammation in vivo (Collins et al., 2015).

4. CONCLUDING REMARKS

Over the 30 years since NF-κB was first discovered there has been an exponential increase in our understanding of its physiological roles and the molecular processes that regulate its activity. The available experimental data describe a complex system for the interpretation of inflammatory stimuli and the generation of meaningful and potent inflammatory responses that are tightly controlled. The importance of the transcriptional programmes regulated by NF-κB is evident from the consequences of its dysregulation manifest in the available genetic models and in numerous inflammatory diseases, although clear differences in functionality have become evident between the mouse and human genetic studies. Despite the advances made to date, the ability to therapeutically target NF-κB for the control of inflammatory disease remains elusive. However, the potential reward to be gained from achieving such therapeutic control of NF-κB ensures its continued investigation.

ACKNOWLEDGMENTS

The authors wish to acknowledge the work of colleagues that could not be included due to space limitations. This work was supported by the Arthritis Research UK Rheumatoid Arthritis Pathogenesis Centre for Excellence (RACE) Grant 20298, Medical Research Council Grant MR/M010694/1, and Biotechnology and Biological Research Sciences Research Council Grant BB/M003671/1.

REFERENCES

Alcamo, E., Mizgerd, J.P., Horwitz, B.H., Bronson, R., Beg, A.A., Scott, M., Doerschuk, C.M., Hynes, R.O., Baltimore, D., 2001. Targeted mutation of TNF receptor I rescues the RelA-deficient mouse and reveals a critical role for NF-kappa B in leukocyte recruitment. J. Immunol. 167, 1592–1600.

Alcamo, E., Hacohen, N., Schulte, L.C., Rennert, P.D., Hynes, R.O., Baltimore, D., 2002. Requirement for the NF-kappaB family member RelA in the development of secondary lymphoid organs. J. Exp. Med. 195, 233–244.

Anest, V., Hanson, J.L., Cogswell, P.C., Steinbrecher, K.A., Strahl, B.D., Baldwin, A.S., 2003. A nucleosomal function for IkappaB kinase-alpha in NF-kappaB-dependent gene expression. Nature 423, 659–663.

Annemann, M., Plaza-Sirvent, C., Schuster, M., Katsoulis-Dimitriou, K., Kliche, S., Schraven, B., Schmitz, I., 2016. Atypical IkappaB proteins in immune cell differentiation and function. Immunol. Lett. 171, 26–35.

Arkan, M.C., Hevener, A.L., Greten, F.R., Maeda, S., Li, Z.W., Long, J.M., Wynshaw-Boris, A., Poli, G., Olefsky, J., Karin, M., 2005. IKK-beta links inflammation to obesity-induced insulin resistance. Nat. Med. 11, 191–198.

Artis, D., Kane, C.M., Fiore, J., Zaph, C., Shapira, S., Joyce, K., Macdonald, A., Hunter, C., Scott, P., Pearce, E.J., 2005. Dendritic cell-intrinsic expression of NF-kappa B1 is required to promote optimal Th2 cell differentiation. J. Immunol. 174, 7154–7159.

Ashall, L., Horton, C.A., Nelson, D.E., Paszek, P., Harper, C.V., Sillitoe, K., Ryan, S., Spiller, D.G., Unitt, J.F., Broomhead, D.S., Kell, D.B., Rand, D.A., See, V., White, M.R., 2009. Pulsatile stimulation determines timing and specificity of NF-kappaB-dependent transcription. Science 324, 242–246.

Baeuerle, P.A., Baltimore, D., 1988. Activation of DNA-binding activity in an apparently cytoplasmic precursor of the NF-kappa B transcription factor. Cell 53, 211–217.

Baker, R., Ghosh, S., 2010. Direct activation of protein kinases by ubiquitin. J. Mol. Cell Biol. 2, 20–22.

Balkhi, M.Y., Fitzgerald, K.A., Pitha, P.M., 2010. IKKalpha negatively regulates IRF-5 function in a MyD88-TRAF6 pathway. Cell. Signal. 22, 117–127.

Baltimore, D., 2011. NF-kappaB is 25. Nat. Immunol. 12, 683–685.

Banerjee, D., Liou, H.C., Sen, R., 2005. c-Rel-dependent priming of naive T cells by inflammatory cytokines. Immunity 23, 445–458.

Banerjee, A., Gugasyan, R., McMahon, M., Gerondakis, S., 2006. Diverse toll-like receptors utilize Tpl2 to activate extracellular signal-regulated kinase (ERK) in hemopoietic cells. Proc. Natl. Acad. Sci. U. S. A. 103, 3274–3279.

Bauer, A., Villunger, A., Labi, V., Fischer, S.F., Strasser, A., Wagner, H., Schmid, R.M., Hacker, G., 2006. The NF-kappaB regulator Bcl-3 and the BH3-only proteins Bim and Puma control the death of activated T cells. Proc. Natl. Acad. Sci. U. S. A. 103, 10979–10984.

Beg, A.A., Baltimore, D., 1996. An essential role for NF-kappaB in preventing TNF-alpha-induced cell death. Science 274, 782–784.

Beg, A.A., Sha, W.C., Bronson, R.T., Baltimore, D., 1995. Constitutive NF-kappa B activation, enhanced granulopoiesis, and neonatal lethality in I kappa B alpha-deficient mice. Genes Dev. 9, 2736–2746.

Beinke, S., Ley, S.C., 2004. Functions of NF-kappaB1 and NF-kappaB2 in immune cell biology. Biochem. J. 382, 393–409.

Bohuslav, J., Kravchenko, V.V., Parry, G.C., Erlich, J.H., Gerondakis, S., Mackman, N., Ulevitch, R.J., 1998. Regulation of an essential innate immune response by the p50 subunit of NF-kappaB. J. Clin. Invest. 102, 1645–1652.

Bonizzi, G., Karin, M., 2004. The two NF-κB activation pathways and their role in innate and adaptive immunity. Cell 25, 280–288.

Bonizzi, G., Bebien, M., Otero, D.C., Johnson-Vroom, K.E., Cao, Y., Vu, D., Jegga, A.G., Aronow, B.J., Ghosh, G., Rickert, R.C., Karin, M., 2004. Activation of IKKalpha target genes depends on recognition of specific kappaB binding sites by RelB:p52 dimers. EMBO J. 23, 4202–4210.

Boztug, H., Hirschmugl, T., Holter, W., Lakatos, K., Kager, L., Trapin, D., Pickl, W., Forster-Waldl, E., Boztug, K., 2016. NF-kappaB1 haploinsufficiency causing immunodeficiency and EBV-driven lymphoproliferation. J. Clin. Immunol. 36, 533–540.

Burkly, L., Hession, C., Ogata, L., Reilly, C., Marconi, L.A., Olson, D., Tizard, R., Cate, R., Lo, D., 1995. Expression of relB is required for the development of thymic medulla and dendritic cells. Nature 373, 531–536.

Caamano, J.H., Rizzo, C.A., Durham, S.K., Barton, D.S., Raventos-Suarez, C., Snapper, C.M., Bravo, R., 1998. Nuclear factor (NF)-kappa B2 (p100/p52) is required for normal splenic microarchitecture and B cell-mediated immune responses. J. Exp. Med. 187, 185–196.

Caamano, J., Tato, C., Cai, G., Villegas, E.N., Speirs, K., Craig, L., Alexander, J., Hunter, C.A., 2000. Identification of a role for NF-kappa B2 in the regulation of apoptosis and in maintenance of T cell-mediated immunity to toxoplasma gondii. J. Immunol. 165, 5720–5728.

Cao, Y., Bonizzi, G., Seagroves, T.N., Greten, F.R., Johnson, R., Schmidt, E.V., Karin, M., 2001. IKKalpha provides an essential link between RANK signaling and cyclin D1 expression during mammary gland development. Cell 107, 763–775.

Cariappa, A., Liou, H.C., Horwitz, B.H., Pillai, S., 2000. Nuclear factor kappa B is required for the development of marginal zone B lymphocytes. J. Exp. Med. 192, 1175–1182.

Carmody, R.J., Chen, Y.H., 2007. Nuclear factor-kappaB: activation and regulation during toll-like receptor signaling. Cell. Mol. Immunol. 4, 31–41.

Carmody, R.J., Ruan, Q., Liou, H.C., Chen, Y.H., 2007a. Essential roles of c-Rel in TLR-induced IL-23 p19 gene expression in dendritic cells. J. Immunol. 178, 186–191.

Carmody, R.J., Ruan, Q., Palmer, S., Hilliard, B., Chen, Y.H., 2007b. Negative regulation of toll-like receptor signaling by NF-kappaB p50 ubiquitination blockade. Science 317, 675–678.

Chang, M., Jin, W., Chang, J.H., Xiao, Y., Brittain, G.C., Yu, J., Zhou, X., Wang, Y.H., Cheng, X., Li, P., Rabinovich, B.A., Hwu, P., Sun, S.C., 2011. The ubiquitin ligase Peli1 negatively regulates T cell activation and prevents autoimmunity. Nat. Immunol. 12, 1002–1009.

Chariot, A., 2009. The NF-kappaB-independent functions of IKK subunits in immunity and cancer. Trends Cell Biol. 19, 404–413.

Chen, X., Wang, J., Woltring, D., Gerondakis, S., Shannon, M.F., 2005a. Histone dynamics on the interleukin-2 gene in response to T-cell activation. Mol. Cell. Biol. 25, 3209–3219.

Chen, L.F., Williams, S.A., Mu, Y., Nakano, H., Duerr, J.M., Buckbinder, L., Greene, W.C., 2005b. NF-kappaB RelA phosphorylation regulates RelA acetylation. Mol. Cell. Biol. 25, 7966–7975.

Chen, K., Coonrod, E.M., Kumanovics, A., Franks, Z.F., Durtschi, J.D., Margraf, R.L., Wu, W., Heikal, N.M., Augustine, N.H., Ridge, P.G., Hill, H.R., Jorde, L.B., Weyrich, A.S., Zimmerman, G.A., Gundlapalli, A.V., Bohnsack, J.F., Voelkerding, K.V., 2013. Germline mutations in NFKB2 implicate the noncanonical NF-kappaB pathway in the pathogenesis of common variable immunodeficiency. Am. J. Hum. Genet. 93, 812–824.

Cheng, S., Hsia, C.Y., Leone, G., Liou, H.C., 2003. Cyclin E and Bcl-xL cooperatively induce cell cycle progression in c-Rel−/− B cells. Oncogene 22, 8472–8486.

Christian, F., Smith, E.L., Carmody, R.J., 2016. The regulation of NF-kappaB subunits by phosphorylation. Cells 5, 12.

Clark, K., Peggie, M., Plater, L., Sorcek, R.J., Young, E.R., Madwed, J.B., Hough, J., McIver, E.G., Cohen, P., 2011. Novel cross-talk within the IKK family controls innate immunity. Biochem. J. 434, 93–104.
Claudio, E., Brown, K., Park, S., Wang, H., Siebenlist, U., 2002. BAFF-induced NEMO-independent processing of NF-kappa B2 in maturing B cells. Nat. Immunol. 3, 958–965.
Coles, A.H., Gannon, H., Cerny, A., Kurt-Jones, E., Jones, S.N., 2010. Inhibitor of growth-4 promotes IkappaB promoter activation to suppress NF-kappaB signaling and innate immunity. Proc. Natl. Acad. Sci. U. S. A. 107, 11423–11428.
Colleran, A., Collins, P.E., O'Carroll, C., Ahmed, A., Mao, X., McManus, B., Kiely, P.A., Burstein, E., Carmody, R.J., 2013. Deubiquitination of NF-kappaB by ubiquitin-specific protease-7 promotes transcription. Proc. Natl. Acad. Sci. U. S. A. 110, 618–623.
Collins, P.E., Carmody, R.J., 2015. The regulation of endotoxin tolerance and its impact on macrophage activation. Crit. Rev. Immunol. 35, 293–323.
Collins, P.E., Kiely, P.A., Carmody, R.J., 2014. Inhibition of transcription by B cell leukemia 3 (Bcl-3) protein requires interaction with nuclear factor kappaB (NF-kappaB) p50. J. Biol. Chem. 289, 7059–7067.
Collins, P.E., Grassia, G., Colleran, A., Kiely, P.A., Ialenti, A., Maffia, P., Carmody, R.J., 2015. Mapping the interaction of B cell leukemia 3 (BCL-3) and nuclear factor kappaB (NF-kappaB) p50 identifies a BCL-3-mimetic anti-inflammatory peptide. J. Biol. Chem. 290, 15687–15696.
Collins, P.E., Mitxitorena, I., Carmody, R.J., 2016. The ubiquitination of NF-kappaB subunits in the control of transcription. Cells 5, 23.
Corn, R.A., Hunter, C., Liou, H.C., Siebenlist, U., Boothby, M.R., 2005. Opposing roles for RelB and Bcl-3 in regulation of T-box expressed in T cells, GATA-3, and Th effector differentiation. J. Immunol. 175, 2102–2110.
Courtois, G., Gilmore, T.D., 2006. Mutations in the NF-kappaB signaling pathway: implications for human disease. Oncogene 25, 6831–6843.
Das, J., Chen, C.H., Yang, L., Cohn, L., Ray, P., Ray, A., 2001. A critical role for NF-kappa B in GATA3 expression and TH2 differentiation in allergic airway inflammation. Nat. Immunol. 2, 45–50.
Dennis, A., Kudo, T., Kruidenier, L., Girard, F., Crepin, V.F., MacDonald, T.T., Frankel, G., Wiles, S., 2008. The p50 subunit of NF-kappaB is critical for in vivo clearance of the noninvasive enteric pathogen Citrobacter rodentium. Infect. Immun. 76, 4978–4988.
Doffinger, R., Smahi, A., Bessia, C., Geissmann, F., Feinberg, J., Durandy, A., Bodemer, C., Kenwrick, S., Dupuis-Girod, S., Blanche, S., Wood, P., Rabia, S.H., Headon, D.J., Overbeek, P.A., Le Deist, F., Holland, S.M., Belani, K., Kumararatne, D.S., Fischer, A., Shapiro, R., Conley, M.E., Reimund, E., Kalhoff, H., Abinun, M., Munnich, A., Israel, A., Courtois, G., Casanova, J.L., 2001. X-linked anhidrotic ectodermal dysplasia with immunodeficiency is caused by impaired NF-kappaB signaling. Nat. Genet. 27, 277–285.
Doi, T.S., Marino, M.W., Takahashi, T., Yoshida, T., Sakakura, T., Old, L.J., Obata, Y., 1999. Absence of tumor necrosis factor rescues RelA-deficient mice from embryonic lethality. Proc. Natl. Acad. Sci. U. S. A. 96, 2994–2999.
Dong, J., Jimi, E., Zhong, H., Hayden, M.S., Ghosh, S., 2008. Repression of gene expression by unphosphorylated NF-kappa B p65 through epigenetic mechanisms. Genes Dev. 22, 1159–1173.
Dutta, J., Fan, Y., Gupta, N., Fan, G., Gelinas, C., 2006. Current insights into the regulation of programmed cell death by NF-kappaB. Oncogene 25, 6800–6816.
Feldman, N., Rotter-Maskowitz, A., Okun, E., 2015. DAMPs as mediators of sterile inflammation in aging-related pathologies. Ageing Res. Rev. 24, 29–39.

Fliegauf, M., Bryant, V.L., Frede, N., Slade, C., Woon, S.T., Lehnert, K., Winzer, S., Bulashevska, A., Scerri, T., Leung, E., Jordan, A., Keller, B., de Vries, E., Cao, H., Yang, F., Schaffer, A.A., Warnatz, K., Browett, P., Douglass, J., Ameratunga, R.V., van der Meer, J.W., Grimbacher, B., 2015. Haploinsufficiency of the NF-kappaB1 subunit p50 in common variable immunodeficiency. Am. J. Hum. Genet. 97, 389–403.

Foster, S.L., Hargreaves, D.C., Medzhitov, R., 2007. Gene-specific control of inflammation by TLR-induced chromatin modifications. Nature 447, 972–978.

Franzoso, G., Carlson, L., Xing, L., Poljak, L., Shores, E.W., Brown, K.D., Leonardi, A., Tran, T., Boyce, B.F., Siebenlist, U., 1997. Requirement for NF-kappaB in osteoclast and B-cell development. Genes Dev. 11, 3482–3496.

Franzoso, G., Carlson, L., Poljak, L., Shores, E.W., Epstein, S., Leonardi, A., Grinberg, A., Tran, T., Scharton-Kersten, T., Anver, M., Love, P., Brown, K., Siebenlist, U., 1998. Mice deficient in nuclear factor (NF)-kappa B/p52 present with defects in humoral responses, germinal center reactions, and splenic microarchitecture. J. Exp. Med. 187, 147–159.

Gantke, T., Sriskantharajah, S., Sadowski, M., Ley, S.C., 2012. IkappaB kinase regulation of the TPL-2/ERK MAPK pathway. Immunol. Rev. 246, 168–182.

Geng, H., Wittwer, T., Dittrich-Breiholz, O., Kracht, M., Schmitz, M.L., 2009. Phosphorylation of NF-kappa B p65 at Ser 468 controls its COMMD1-dependent ubiquitination and target gene-specific proteasomal elimination. EMBO Rep. 10, 381–386.

Gerondakis, S., Strasser, A., 2003. The role of Rel/NF-kappaB transcription factors in B lymphocyte survival. Semin. Immunol. 15, 159–166.

Gerondakis, S., Strasser, A., Metcalf, D., Grigoriadis, G., Scheerlinck, J.Y., Grumont, R.J., 1996. Rel-deficient T cells exhibit defects in production of interleukin 3 and granulocyte-macrophage colony-stimulating factor. Proc. Natl. Acad. Sci. U. S. A. 93, 3405–3409.

Gerondakis, S., Grumont, R., Gugasyan, R., Wong, L., Isomura, I., Ho, W., Banerjee, A., 2006. Unravelling the complexities of the NF-kappaB signalling pathway using mouse knockout and transgenic models. Oncogene 25, 6781–6799.

Ghosh, S., Hayden, M.S., 2008. New regulators of NF-kappaB in inflammation. Nat. Rev. Immunol. 8, 837–848.

Grumont, R.J., Rourke, I.J., O'Reilly, L.A., Strasser, A., Miyake, K., Sha, W., Gerondakis, S., 1998. B lymphocytes differentially use the Rel and nuclear factor kappaB1 (NF-kappaB1) transcription factors to regulate cell cycle progression and apoptosis in quiescent and mitogen-activated cells. J. Exp. Med. 187, 663–674.

Grumont, R., Hochrein, H., O'Keeffe, M., Gugasyan, R., White, C., Caminschi, I., Cook, W., Gerondakis, S., 2001. c-Rel regulates interleukin 12 p70 expression in CD8(+) dendritic cells by specifically inducing p35 gene transcription. J. Exp. Med. 194, 1021–1032.

Guerin, S., Baron, M.L., Valero, R., Herrant, M., Auberger, P., Naquet, P., 2002. RelB reduces thymocyte apoptosis and regulates terminal thymocyte maturation. Eur. J. Immunol. 32, 1–9.

Hatada, E.N., Do, R.K., Orlofsky, A., Liou, H.C., Prystowsky, M., MacLennan, I.C., Caamano, J., Chen-Kiang, S., 2003. NF-kappa B1 p50 is required for BLyS attenuation of apoptosis but dispensable for processing of NF-kappa B2 p100 to p52 in quiescent mature B cells. J. Immunol. 171, 761–768.

Hayden, M.S., Ghosh, S., 2008. Shared principles in NF-kappaB signaling. Cell 132, 344–362.

Hayden, M.S., Ghosh, S., 2014. Regulation of NF-kappaB by TNF family cytokines. Semin. Immunol. 26, 253–266.

Herrington, F.D., Carmody, R.J., Goodyear, C.S., 2016. Modulation of NF-kappaB signaling as a therapeutic target in autoimmunity. J. Biomol. Screen. 21, 223–242.

Hilliard, B.A., Mason, N., Xu, L., Sun, J., Lamhamedi-Cherradi, S.E., Liou, H.C., Hunter, C., Chen, Y.H., 2002. Critical roles of c-Rel in autoimmune inflammation and helper T cell differentiation. J. Clin. Invest. 110, 843–850.

Hinz, M., Arslan, S.C., Scheidereit, C., 2012. It takes two to tango: IkappaBs, the multifunctional partners of NF-kappaB. Immunol. Rev. 246, 59–76.

Hoberg, J.E., Yeung, F., Mayo, M.W., 2004. SMRT derepression by the IkappaB kinase alpha: a prerequisite to NF-kappaB transcription and survival. Mol. Cell 16, 245–255.

Hochrainer, K., Racchumi, G., Zhang, S., Iadecola, C., Anrather, J., 2012. Monoubiquitination of nuclear RelA negatively regulates NF-kappaB activity independent of proteasomal degradation. Cell. Mol. Life Sci. 69, 2057–2073.

Hoffmann, A., Levchenko, A., Scott, M.L., Baltimore, D., 2002. The IkappaB-NF-kappaB signaling module: temporal control and selective gene activation. Science 298, 1241–1245.

Horwitz, B.H., Scott, M.L., Cherry, S.R., Bronson, R.T., Baltimore, D., 1997. Failure of lymphopoiesis after adoptive transfer of NF-kappaB-deficient fetal liver cells. Immunity 6, 765–772.

Horwitz, B.H., Zelazowski, P., Shen, Y., Wolcott, K.M., Scott, M.L., Baltimore, D., Snapper, C.M., 1999. The p65 subunit of NF-kappa B is redundant with p50 during B cell proliferative responses, and is required for germline CH transcription and class switching to IgG3. J. Immunol. 162, 1941–1946.

Hoshino, K., Sugiyama, T., Matsumoto, M., Tanaka, T., Saito, M., Hemmi, H., Ohara, O., Akira, S., Kaisho, T., 2006. IkappaB kinase-alpha is critical for interferon-alpha production induced by toll-like receptors 7 and 9. Nature 440, 949–953.

Hou, Y., Moreau, F., Chadee, K., 2012. PPARgamma is an E3 ligase that induces the degradation of NFkappaB/p65. Nat. Commun. 3, 1300.

Hou, Y., Zhang, Z., Xu, Q., Wang, H., Xu, Y., Chen, K., 2014. Inhibitor of growth 4 induces NFkappaB/p65 ubiquitin-dependent degradation. Oncogene 33, 1997–2003.

Hovelmeyer, N., Worns, M.A., Reissig, S., Adams-Quack, P., Leclaire, J., Hahn, M., Wortge, S., Nikolaev, A., Galle, P.R., Waisman, A., 2014. Overexpression of Bcl-3 inhibits the development of marginal zone B cells. Eur. J. Immunol. 44, 545–552.

Huang, W.C., Ju, T.K., Hung, M.C., Chen, C.C., 2007. Phosphorylation of CBP by IKKalpha promotes cell growth by switching the binding preference of CBP from p53 to NF-kappaB. Mol. Cell 26, 75–87.

Hubeau, M., Ngadjeua, F., Puel, A., Israel, L., Feinberg, J., Chrabieh, M., Belani, K., Bodemer, C., Fabre, I., Plebani, A., Boisson-Dupuis, S., Picard, C., Fischer, A., Israel, A., Abel, L., Veron, M., Casanova, J.L., Agou, F., Bustamante, J., 2011. New mechanism of X-linked anhidrotic ectodermal dysplasia with immunodeficiency: impairment of ubiquitin binding despite normal folding of NEMO protein. Blood 118, 926–935.

Ishikawa, H., Carrasco, D., Claudio, E., Ryseck, R.P., Bravo, R., 1997. Gastric hyperplasia and increased proliferative responses of lymphocytes in mice lacking the COOH-terminal ankyrin domain of NF-kappaB2. J. Exp. Med. 186, 999–1014.

Ishikawa, H., Claudio, E., Dambach, D., Raventos-Suarez, C., Ryan, C., Bravo, R., 1998. Chronic inflammation and susceptibility to bacterial infections in mice lacking the polypeptide (p)105 precursor (NF-kappaB1) but expressing p50. J. Exp. Med. 187, 985–996.

Karin, M., Lin, A., 2002. NF-kappaB at the crossroads of life and death. Nat. Immunol. 3, 221–227.

Kayama, H., Ramirez-Carrozzi, V.R., Yamamoto, M., Mizutani, T., Kuwata, H., Iba, H., Matsumoto, M., Honda, K., Smale, S.T., Takeda, K., 2008. Class-specific regulation of pro-inflammatory genes by MyD88 pathways and IkappaBzeta. J. Biol. Chem. 283, 12468–12477.

Kim, S., La Motte-Mohs, R.N., Rudolph, D., Zuniga-Pflucker, J.C., Mak, T.W., 2003. The role of nuclear factor-kappaB essential modulator (NEMO) in B cell development and survival. Proc. Natl. Acad. Sci. U. S. A. 100, 1203–1208.

Klement, J.F., Rice, N.R., Car, B.D., Abbondanzo, S.J., Powers, G.D., Bhatt, P.H., Chen, C.H., Rosen, C.A., Stewart, C.L., 1996. IkappaBalpha deficiency results in a sustained NF-kappaB response and severe widespread dermatitis in mice. Mol. Cell. Biol. 16, 2341–2349.

Kuwata, H., Matsumoto, M., Atarashi, K., Morishita, H., Hirotani, T., Koga, R., Takeda, K., 2006. IkappaBNS inhibits induction of a subset of toll-like receptor-dependent genes and limits inflammation. Immunity 24, 41–51.

Kwak, Y.T., Li, R., Becerra, C.R., Tripathy, D., Frenkel, E.P., Verma, U.N., 2005. IkappaB kinase alpha regulates subcellular distribution and turnover of cyclin D1 by phosphorylation. J. Biol. Chem. 280, 33945–33952.

Lanucara, F., Lam, C., Mann, J., Monie, T.P., Colombo, S.A., Holman, S.W., Boyd, J., Dange, M.C., Mann, D.A., White, M.R., Eyers, C.E., 2016. Dynamic phosphorylation of RelA on Ser42 and Ser45 in response to TNFalpha stimulation regulates DNA binding and transcription. Open Biol. 6.

Lawrence, T., Bebien, M., Liu, G.Y., Nizet, V., Karin, M., 2005. IKKalpha limits macrophage NF-kappaB activation and contributes to the resolution of inflammation. Nature 434, 1138–1143.

Lee, S., Andrieu, C., Saltel, F., Destaing, O., Auclair, J., Pouchkine, V., Michelon, J., Salaun, B., Kobayashi, R., Jurdic, P., Kieff, E.D., Sylla, B.S., 2004. IkappaB kinase beta phosphorylates Dok1 serines in response to TNF, IL-1, or gamma radiation. Proc. Natl. Acad. Sci. U. S. A. 101, 17416–17421.

Leidner, J., Palkowitsch, L., Marienfeld, U., Fischer, D., Marienfeld, R., 2008. Identification of lysine residues critical for the transcriptional activity and polyubiquitination of the NF-kappaB family member RelB. Biochem. J. 416, 117–127.

Li, Q., Van Antwerp, D., Mercurio, F., Lee, K.F., Verma, I.M., 1999. Severe liver degeneration in mice lacking the IkappaB kinase 2 gene. Science 284, 321–325.

Li, M., Shillinglaw, W., Henzel, W.J., Beg, A.A., 2001. The Rela(p65) subunit of NF-kappaB is essential for inhibiting double-stranded RNA-induced cytotoxicity. J. Biol. Chem. 276, 1185–1194.

Li, Q., Lu, Q., Bottero, V., Estepa, G., Morrison, L., Mercurio, F., Verma, I.M., 2005. Enhanced NF-kappaB activation and cellular function in macrophages lacking IkappaB kinase 1 (IKK1). Proc. Natl. Acad. Sci. U. S. A. 102, 12425–12430.

Li, H., Wittwer, T., Weber, A., Schneider, H., Moreno, R., Maine, G.N., Kracht, M., Schmitz, M.L., Burstein, E., 2012. Regulation of NF-kappaB activity by competition between RelA acetylation and ubiquitination. Oncogene 31, 611–623.

Liu, B., Yang, Y., Chernishof, V., Loo, R.R., Jang, H., Tahk, S., Yang, R., Mink, S., Shultz, D., Bellone, C.J., Loo, J.A., Shuai, K., 2007. Proinflammatory stimuli induce IKKalpha-mediated phosphorylation of PIAS1 to restrict inflammation and immunity. Cell 129, 903–914.

Lo, J.C., Basak, S., James, E.S., Quiambo, R.S., Kinsella, M.C., Alegre, M.L., Weih, F., Franzoso, G., Hoffmann, A., Fu, Y.X., 2006. Coordination between NF-kappaB family members p50 and p52 is essential for mediating LTbetaR signals in the development and organization of secondary lymphoid tissues. Blood 107, 1048–1055.

Luedde, T., Beraza, N., Kotsikoris, V., van Loo, G., Nenci, A., De Vos, R., Roskams, T., Trautwein, C., Pasparakis, M., 2007. Deletion of NEMO/IKKgamma in liver parenchymal cells causes steatohepatitis and hepatocellular carcinoma. Cancer Cell 11, 119–132.

Makris, C., Godfrey, V.L., Krahn-Senftleben, G., Takahashi, T., Roberts, J.L., Schwarz, T., Feng, L., Johnson, R.S., Karin, M., 2000. Female mice heterozygous for IKK

gamma/NEMO deficiencies develop a dermatopathy similar to the human X-linked disorder incontinentia pigmenti. Mol. Cell 5, 969–979.
Mao, X.C., Gluck, N., Li, D., Maine, G.N., Li, H.Y., Zaidi, I.W., Repaka, A., Mayo, M.W., Burstein, E., 2009. GCN5 is a required cofactor for a ubiquitin ligase that targets NF-kappa B/RelA. Genes Dev. 23, 849–861.
Mason, N.J., Liou, H.C., Hunter, C.A., 2004. T cell-intrinsic expression of c-Rel regulates Th1 cell responses essential for resistance to toxoplasma gondii. J. Immunol. 172, 3704–3711.
Meguro, K., Suzuki, K., Hosokawa, J., Sanayama, Y., Tanaka, S., Furuta, S., Ikeda, K., Takatori, H., Suto, A., Sakamoto, A., Ohara, O., Nakajima, H., 2015. Role of Bcl-3 in the development of follicular helper T cells and in the pathogenesis of rheumatoid arthritis. Arthritis Rheumatol. 67, 2651–2660.
Memet, S., Laouini, D., Epinat, J.C., Whiteside, S.T., Goudeau, B., Philpott, D., Kayal, S., Sansonetti, P.J., Berche, P., Kanellopoulos, J., Israel, A., 1999. IkappaBepsilon-deficient mice: reduction of one T cell precursor subspecies and enhanced Ig isotype switching and cytokine synthesis. J. Immunol. 163, 5994–6005.
Mintern, J.D., Belz, G., Gerondakis, S., Carbone, F.R., Heath, W.R., 2002. The cross-priming APC requires a Rel-dependent signal to induce CTL. J. Immunol. 168, 3283–3287.
Mitchell, T.C., Hildeman, D., Kedl, R.M., Teague, T.K., Schaefer, B.C., White, J., Zhu, Y., Kappler, J., Marrack, P., 2001. Immunological adjuvants promote activated T cell survival via induction of Bcl-3. Nat. Immunol. 2, 397–402.
Motoyama, M., Yamazaki, S., Eto-Kimura, A., Takeshige, K., Muta, T., 2005. Positive and negative regulation of nuclear factor-kappaB-mediated transcription by IkappaB-zeta, an inducible nuclear protein. J. Biol. Chem. 280, 7444–7451.
Nelson, D.E., Ihekwaba, A.E., Elliott, M., Johnson, J.R., Gibney, C.A., Foreman, B.E., Nelson, G., See, V., Horton, C.A., Spiller, D.G., Edwards, S.W., McDowell, H.P., Unitt, J.F., Sullivan, E., Grimley, R., Benson, N., Broomhead, D., Kell, D.B., White, M.R., 2004. Oscillations in NF-kappaB signaling control the dynamics of gene expression. Science 306, 704–708.
Nenci, A., Becker, C., Wullaert, A., Gareus, R., van Loo, G., Danese, S., Huth, M., Nikolaev, A., Neufert, C., Madison, B., Gumucio, D., Neurath, M.F., Pasparakis, M., 2007. Epithelial NEMO links innate immunity to chronic intestinal inflammation. Nature 446, 557–561.
Nolan, G.P., Fujita, T., Bhatia, K., Huppi, C., Liou, H.C., Scott, M.L., Baltimore, D., 1993. The bcl-3 proto-oncogene encodes a nuclear I kappa B-like molecule that preferentially interacts with NF-kappa B p50 and p52 in a phosphorylation-dependent manner. Mol. Cell. Biol. 13, 3557–3566.
Nowak, D.E., Tian, B., Jamaluddin, M., Boldogh, I., Vergara, L.A., Choudhary, S., Brasier, A.R., 2008. RelA Ser(276) phosphorylation is required for activation of a subset of NF-kappa B-dependent genes by recruiting cyclin-dependent kinase 9/cyclin T1 complexes. Mol. Cell. Biol. 28, 3623–3638.
O'Carroll, C., Moloney, G., Hurley, G., Melgar, S., Brint, E., Nally, K., Nibbs, R.J., Shanahan, F., Carmody, R.J., 2013. Bcl-3 deficiency protects against dextran-sodium sulphate-induced colitis in the mouse. Clin. Exp. Immunol. 173, 332–342.
O'Carroll, C., Fagan, A., Shanahan, F., Carmody, R.J., 2014. Identification of a unique hybrid macrophage-polarization state following recovery from lipopolysaccharide tolerance. J. Immunol. 192, 427–436.
Oeckinghaus, A., Ghosh, S., 2009. The NF-kappaB family of transcription factors and its regulation. Cold Spring Harb. Perspect. Biol. 1, a000034.
Ohno, H., Takimoto, G., McKeithan, T.W., 1990. The candidate proto-oncogene bcl-3 is related to genes implicated in cell lineage determination and cell cycle control. Cell 60, 991–997.

Okamoto, K., Iwai, Y., Oh-Hora, M., Yamamoto, M., Morio, T., Aoki, K., Ohya, K., Jetten, A.M., Akira, S., Muta, T., Takayanagi, H., 2010. IkappaBzeta regulates T(H)17 development by cooperating with ROR nuclear receptors. Nature 464, 1381–1385.

O'Keeffe, M., Grumont, R.J., Hochrein, H., Fuchsberger, M., Gugasyan, R., Vremec, D., Shortman, K., Gerondakis, S., 2005. Distinct roles for the NF-kappaB1 and c-Rel transcription factors in the differentiation and survival of plasmacytoid and conventional dendritic cells activated by TLR-9 signals. Blood 106, 3457–3464.

Pannicke, U., Baumann, B., Fuchs, S., Henneke, P., Rensing-Ehl, A., Rizzi, M., Janda, A., Hese, K., Schlesier, M., Holzmann, K., Borte, S., Laux, C., Rump, E.M., Rosenberg, A., Zelinski, T., Schrezenmeier, H., Wirth, T., Ehl, S., Schroeder, M.L., Schwarz, K., 2013. Deficiency of innate and acquired immunity caused by an IKBKB mutation. N. Engl. J. Med. 369, 2504–2514.

Pasparakis, M., Courtois, G., Hafner, M., Schmidt-Supprian, M., Nenci, A., Toksoy, A., Krampert, M., Goebeler, M., Gillitzer, R., Israel, A., Krieg, T., Rajewsky, K., Haase, I., 2002. TNF-mediated inflammatory skin disease in mice with epidermis-specific deletion of IKK2. Nature 417, 861–866.

Pasparakis, M., Luedde, T., Schmidt-Supprian, M., 2006. Dissection of the NF-kappaB signalling cascade in transgenic and knockout mice. Cell Death Differ. 13, 861–872.

Pene, F., Paun, A., Sonder, S.U., Rikhi, N., Wang, H., Claudio, E., Siebenlist, U., 2011. The IkappaB family member Bcl-3 coordinates the pulmonary defense against *Klebsiella pneumoniae* infection. J. Immunol. 186, 2412–2421.

Pohl, T., Gugasyan, R., Grumont, R.J., Strasser, A., Metcalf, D., Tarlinton, D., Sha, W., Baltimore, D., Gerondakis, S., 2002. The combined absence of NF-kappa B1 and c-Rel reveals that overlapping roles for these transcription factors in the B cell lineage are restricted to the activation and function of mature cells. Proc. Natl. Acad. Sci. U. S. A. 99, 4514–4519.

Pradere, J.P., Hernandez, C., Koppe, C., Friedman, R.A., Luedde, T., Schwabe, R.F., 2016. Negative regulation of NF-kappaB p65 activity by serine 536 phosphorylation. Sci. Signal. 9, ra85.

Prajapati, S., Gaynor, R.B., 2002. Regulation of Ikappa B kinase (IKK)gamma/NEMO function by IKKbeta -mediated phosphorylation. J. Biol. Chem. 277, 24331–24339.

Pratt, A.G., Swan, D.C., Richardson, S., Wilson, G., Hilkens, C.M., Young, D.A., Isaacs, J.D., 2012. A CD4 T cell gene signature for early rheumatoid arthritis implicates interleukin 6-mediated STAT3 signalling, particularly in anti-citrullinated peptide antibody-negative disease. Ann. Rheum. Dis. 71, 1374–1381.

Pratt, A.G., Anderson, A.E., Carmody, R.J., Isaacs, J.D., 2016. Bcl-3 in CD4+ T cell-mediated rheumatoid arthritis pathogenesis: comment on the article by Meguro et al. Arthritis Rheumatol. 68, 770–771.

Prendes, M., Zheng, Y., Beg, A.A., 2003. Regulation of developing B cell survival by RelA-containing NF-kappa B complexes. J. Immunol. 171, 3963–3969.

Rao, P., Hayden, M.S., Long, M., Scott, M.L., West, A.P., Zhang, D., Oeckinghaus, A., Lynch, C., Hoffmann, A., Baltimore, D., Ghosh, S., 2010. IkappaBbeta acts to inhibit and activate gene expression during the inflammatory response. Nature 466, 1115–1119.

Roget, K., Ben-Addi, A., Mambole-Dema, A., Gantke, T., Yang, H.T., Janzen, J., Morrice, N., Abbott, D., Ley, S.C., 2012. IkappaB kinase 2 regulates TPL-2 activation of extracellular signal-regulated kinases 1 and 2 by direct phosphorylation of TPL-2 serine 400. Mol. Cell. Biol. 32, 4684–4690.

Ruan, Q., Zheng, S.J., Palmer, S., Carmody, R.J., Chen, Y.H., 2010. Roles of Bcl-3 in the pathogenesis of murine type 1 diabetes. Diabetes 59, 2549–2557.

Ryo, A., Suizu, F., Yoshida, Y., Perrem, K., Liou, Y.C., Wulf, G., Rottapel, R., Yamaoka, S., Lu, K.P., 2003. Regulation of NF-kappaB signaling by pin

1-dependent prolyl isomerization and ubiquitin-mediated proteolysis of p65/RelA. Mol. Cell 12, 1413–1426.
Saccani, S., Marazzi, I., Beg, A.A., Natoli, G., 2004. Degradation of promoter-bound p65/RelA is essential for the prompt termination of the nuclear factor kappaB response. J. Exp. Med. 200, 107–113.
Sanjabi, S., Hoffmann, A., Liou, H.C., Baltimore, D., Smale, S.T., 2000. Selective requirement for c-Rel during IL-12 P40 gene induction in macrophages. Proc. Natl. Acad. Sci. U. S. A. 97, 12705–12710.
Sasaki, C.Y., Barberi, T.J., Ghosh, P., Longo, D.L., 2005. Phosphorylation of RelA/p65 on serine 536 defines an I{kappa}B{alpha}-independent NF-{kappa}B pathway. J. Biol. Chem. 280, 34538–34547.
Scheibel, M., Klein, B., Merkle, H., Schulz, M., Fritsch, R., Greten, F.R., Arkan, M.C., Schneider, G., Schmid, R.M., 2010. IkappaBbeta is an essential co-activator for LPS-induced IL-1beta transcription in vivo. J. Exp. Med. 207, 2621–2630.
Scheidereit, C., 2006. IkB kinase complexes: gateways to NF-jB activation and transcription. Oncogene 25, 6685–6705.
Schimke, L.F., Rieber, N., Rylaarsdam, S., Cabral-Marques, O., Hubbard, N., Puel, A., Kallmann, L., Sombke, S.A., Notheis, G., Schwarz, H.P., Kammer, B., Hokfelt, T., Repp, R., Picard, C., Casanova, J.L., Belohradsky, B.H., Albert, M.H., Ochs, H.D., Renner, E.D., Torgerson, T.R., 2013. A novel gain-of-function IKBA mutation underlies ectodermal dysplasia with immunodeficiency and polyendocrinopathy. J. Clin. Immunol. 33, 1088–1099.
Schmidt-Supprian, M., Courtois, G., Tian, J., Coyle, A.J., Israel, A., Rajewsky, K., Pasparakis, M., 2003. Mature T cells depend on signaling through the IKK complex. Immunity 19, 377–389.
Schwarz, E.M., Krimpenfort, P., Berns, A., Verma, I.M., 1997. Immunological defects in mice with a targeted disruption in Bcl-3. Genes Dev. 11, 187–197.
Sen, R., Baltimore, D., 1986. Inducibility of kappa immunoglobulin enhancer-binding protein NF-kappa B by a posttranslational mechanism. Cell 47, 921–928.
Senftleben, U., Cao, Y., Xiao, G., Greten, F.R., Krahn, G., Bonizzi, G., Chen, Y., Hu, Y., Fong, A., Sun, S.C., Karin, M., 2001a. Activation by IKKalpha of a second, evolutionary conserved, NF-kappa B signaling pathway. Science 293, 1495–1499.
Senftleben, U., Li, Z.W., Baud, V., Karin, M., 2001b. IKKbeta is essential for protecting T cells from TNFalpha-induced apoptosis. Immunity 14, 217–230.
Sha, W.C., Liou, H.C., Tuomanen, E.I., Baltimore, D., 1995. Targeted disruption of the p50 subunit of NF-kappa B leads to multifocal defects in immune responses. Cell 80, 321–330.
Sharfe, N., Merico, D., Karanxha, A., Macdonald, C., Dadi, H., Ngan, B., Herbrick, J.A., Roifman, C.M., 2015. The effects of RelB deficiency on lymphocyte development and function. J. Autoimmun. 65, 90–100.
Smale, S.T., 2012. Dimer-specific regulatory mechanisms within the NF-kappaB family of transcription factors. Immunol. Rev. 246, 193–204.
Solt, L.A., Madge, L.A., Orange, J.S., May, M.J., 2007. Interleukin-1-induced NF-kappaB activation is NEMO-dependent but does not require IKKbeta. J. Biol. Chem. 282, 8724–8733.
Speirs, K., Caamano, J., Goldschmidt, M.H., Hunter, C.A., Scott, P., 2002. NF-kappa B2 is required for optimal CD40-induced IL-12 production but dispensable for Th1 cell differentiation. J. Immunol. 168, 4406–4413.
Strasser, A., Grumont, R.J., Stanley, M.L., Gerondakis, S., 1999. The transcriptional regulator Rel is essential for antigen receptor-mediated stimulation of mature T cells but dispensable for positive and negative selection of thymocytes and T cell apoptosis. Eur. J. Immunol. 29, 928–935.

Sun, S.C., Ganchi, P.A., Ballard, D.W., Greene, W.C., 1993. NF-kappa B controls expression of inhibitor I kappa B alpha: evidence for an inducible autoregulatory pathway. Science 259, 1912–1915.

Suzuki, K., Verma, I.M., 2008. Phosphorylation of SNAP-23 by IkappaB kinase 2 regulates mast cell degranulation. Cell 134, 485–495.

Tam, W.F., Lee, L.H., Davis, L., Sen, R., 2000. Cytoplasmic sequestration ofvrel proteins by IκBα requires CRM1-dependent nuclear export. Mol. Cell. Biol. 20, 2269–2284.

Tanaka, T., Grusby, M.J., Kaisho, T., 2007. PDLIM2-mediated termination of transcription factor NF-kappaB activation by intranuclear sequestration and degradation of the p65 subunit. Nat. Immunol. 8, 584–591.

Tang, W., Wang, H., Claudio, E., Tassi, I., Ha, H.L., Saret, S., Siebenlist, U., 2014. The oncoprotein and transcriptional regulator Bcl-3 governs plasticity and pathogenicity of autoimmune T cells. Immunity 41, 555–566.

Tang, W., Wang, H., Ha, H.L., Tassi, I., Bhardwaj, R., Claudio, E., Siebenlist, U., 2016. The B-cell tumor promoter Bcl-3 suppresses inflammation-associated colon tumorigenesis in epithelial cells. Oncogene 35, 6203–6211.

Tato, C.M., Mason, N., Artis, D., Shapira, S., Caamano, J.C., Bream, J.H., Liou, H.C., Hunter, C.A., 2006. Opposing roles of NF-kappaB family members in the regulation of NK cell proliferation and production of IFN-gamma. Int. Immunol. 18, 505–513.

Touma, M., Antonini, V., Kumar, M., Osborn, S.L., Bobenchik, A.M., Keskin, D.B., Connolly, J.E., Grusby, M.J., Reinherz, E.L., Clayton, L.K., 2007. Functional role for I kappa BNS in T cell cytokine regulation as revealed by targeted gene disruption. J. Immunol. 179, 1681–1692.

Turner, D.A., Paszek, P., Woodcock, D.J., Nelson, D.E., Horton, C.A., Wang, Y., Spiller, D.G., Rand, D.A., White, M.R., Harper, C.V., 2010. Physiological levels of TNFalpha stimulation induce stochastic dynamics of NF-kappaB responses in single living cells. J. Cell Sci. 123, 2834–2843.

Wang, Y., Hu, L., Tong, X., Ye, X., 2014. Casein kinase 1gamma1 inhibits the RIG-I/TLR signaling pathway through phosphorylating p65 and promoting its degradation. J. Immunol. 192, 1855–1861.

Weih, F., Caamano, J., 2003. Regulation of secondary lymphoid organ development by the nuclear factor-kappaB signal transduction pathway. Immunol. Rev. 195, 91–105.

Weih, F., Durham, S.K., Barton, D.S., Sha, W.C., Baltimore, D., Bravo, R., 1997. p50–NF-kappaB complexes partially compensate for the absence of RelB: severely increased pathology in p50(−/−)relB(−/−) double-knockout mice. J. Exp. Med. 185, 1359–1370.

Wertz, I.E., Dixit, V.M., 2010. Signaling to NF-kappaB: regulation by ubiquitination. Cold Spring Harb. Perspect. Biol. 2, a003350.

Wu, L., D'Amico, A., Winkel, K.D., Suter, M., Lo, D., Shortman, K., 1998. RelB is essential for the development of myeloid-related CD8alpha- dendritic cells but not of lymphoid-related CD8alpha+ dendritic cells. Immunity 9, 839–847.

Xia, Y., Padre, R.C., De Mendoza, T.H., Bottero, V., Tergaonkar, V.B., Verma, I.M., 2009. Phosphorylation of p53 by IkappaB kinase 2 promotes its degradation by beta-TrCP. Proc. Natl. Acad. Sci. U. S. A. 106, 2629–2634.

Xiao, Y., Jin, J., Chang, M., Nakaya, M., Hu, H., Zou, Q., Zhou, X., Brittain, G.C., Cheng, X., Sun, S.C., 2014. TPL2 mediates autoimmune inflammation through activation of the TAK1 axis of IL-17 signaling. J. Exp. Med. 211, 1689–1702.

Yamamoto, Y., Verma, U.N., Prajapati, S., Kwak, Y.T., Gaynor, R.B., 2003. Histone H3 phosphorylation by IKK-alpha is critical for cytokine-induced gene expression. Nature 423, 655–659.

Yamamoto, M., Yamazaki, S., Uematsu, S., Sato, S., Hemmi, H., Hoshino, K., Kaisho, T., Kuwata, H., Takeuchi, O., Takeshige, K., Saitoh, T., Yamaoka, S., Yamamoto, N., Yamamoto, S., Muta, T., Takeda, K., Akira, S., 2004. Regulation of toll/IL-1-receptor-mediated gene expression by the inducible nuclear protein IkappaBzeta. Nature 430, 218–222.

Yamauchi, S., Ito, H., Miyajima, A., 2010. IkappaBeta, a nuclear IkappaB protein, positively regulates the NF-kappaB-mediated expression of proinflammatory cytokines. Proc. Natl. Acad. Sci. U. S. A. 107, 11924–11929.

Yamazaki, S., Muta, T., Takeshige, K., 2001. A novel IkappaB protein, IkappaB-zeta, induced by proinflammatory stimuli, negatively regulates nuclear factor-kappaB in the nuclei. J. Biol. Chem. 276, 27657–27662.

Yan, Q., Carmody, R.J., Qu, Z., Ruan, Q., Jager, J., Mullican, S.E., Lazar, M.A., Chen, Y.H., 2012. Nuclear factor-kappaB binding motifs specify toll-like receptor-induced gene repression through an inducible repressosome. Proc. Natl. Acad. Sci. U. S. A. 109, 14140–14145.

Yan, J., Xiang, J., Lin, Y., Ma, J., Zhang, J., Zhang, H., Sun, J., Danial, N.N., Liu, J., Lin, A., 2013. Inactivation of BAD by IKK inhibits TNFalpha-induced apoptosis independently of NF-kappaB activation. Cell 152, 304–315.

Yang, H.T., Papoutsopoulou, S., Belich, M., Brender, C., Janzen, J., Gantke, T., Handley, M., Ley, S.C., 2012. Coordinate regulation of TPL-2 and NF-kappaB signaling in macrophages by NF-kappaB1 p105. Mol. Cell. Biol. 32, 3438–3451.

Yilmaz, Z.B., Weih, D.S., Sivakumar, V., Weih, F., 2003. RelB is required for Peyer's patch development: differential regulation of p52-RelB by lymphotoxin and TNF. EMBO J. 22, 121–130.

Zhang, X., Wang, H., Claudio, E., Brown, K., Siebenlist, U., 2007. A role for the IkappaB family member Bcl-3 in the control of central immunologic tolerance. Immunity 27, 438–452.

Zhang, X., Paun, A., Claudio, E., Wang, H., Siebenlist, U., 2013. The tumor promoter and NF-kappaB modulator Bcl-3 regulates splenic B cell development. J. Immunol. 191, 5984–5992.

Zhang, Q., Lenardo, M.J., Baltimore, D., 2017. 30 years of NF-κB: a blossoming of relevance to human pathobiology. Cell 168, 37–57.

Ziegler-Heitbrock, L., 2001. The p50-homodimer mechanism in tolerance to LPS. J. Endotoxin Res. 7, 219–222.

CHAPTER THREE

Alternative Splicing in Genetic Diseases: Improved Diagnosis and Novel Treatment Options

Atze J. Bergsma*,†,‡, Erik van der Wal*,†,‡, Mike Broeders*,†,‡, Ans T. van der Ploeg†,‡, W.W.M. Pim Pijnappel*,†,‡,1

*Molecular Stem Cell Biology, Department of Clinical Genetics, Erasmus Medical Center, Rotterdam, The Netherlands
†Department of Pediatrics, Erasmus Medical Center, Rotterdam, The Netherlands
‡Center for Lysosomal and Metabolic Diseases, Erasmus Medical Center, Rotterdam, The Netherlands
[1]Corresponding author: e-mail address: w.pijnappel@erasmusmc.nl

Contents

International Review of Cell and Molecular Biology, Volume 335
ISSN 1937-6448
https://doi.org/10.1016/bs.ircmb.2017.07.008

Abstract

Alternative splicing is an important mechanism to regulate gene expression and to expand the repertoire of gene products in order to accommodate an increase in complexity of multicellular organisms. It needs to be precisely regulated, which is achieved via RNA structure, splicing factors, transcriptional regulation, and chromatin. Changes in any of these factors can lead to disease. These may include the core spliceosome, splicing enhancer/repressor sequences and their interacting proteins, the speed of transcription by RNA polymerase II, and histone modifications. While the basic principle of splicing is well understood, it is still very difficult to predict splicing outcome, due to the multiple levels of regulation. Current molecular diagnostics mainly uses Sanger sequencing of exons, or next-generation sequencing of gene panels or the whole exome. Functional analysis of potential splicing variants is scarce, and intronic variants are often not considered. This likely results in underestimation of the percentage of splicing variants. Understanding how sequence variants may affect splicing is not only crucial for confirmation of diagnosis and for genetic counseling, but also for the development of novel treatment options. These include small molecules, transsplicing, antisense oligonucleotides, and gene therapy. Here we review the current state of molecular mechanisms of splicing regulation and how deregulation can lead to human disease, diagnostics to detect splicing variants, and novel treatment options based on splicing correction.

LIST OF ABBREVIATIONS

2′O-mePS 2′O-methyl phosphorothioate
AON antisense oligonucleotide
BMD Becker muscular dystrophy
CADD combined annotation-dependent depletion
cDNA complementary DNA
CLIP crosslinking and immunoprecipitation
DMD Duchenne muscular dystrophy
ESC embryonic stem cell
ESE exonic splicing enhancer
ESS exonic splicing silencer
GAA acid α-glucosidase
HDR homology directed repair
HGMD human gene mutation database
hnRNPs heterogeneous nuclear ribonucleoproteins
iCLIP individual-nucleotide resolution UV CLIP
ISS intronic splicing silencer
MOE 2′O-methoxyethyl
mRNA messenger RNA
NGD No-go decay
NGS next-generation sequencing
NHEJ nonhomologous end joining

NMD nonsense-mediated decay
NSD nonstop decay
PAR-CLIP photoactivatable-ribonucleoside enhanced CLIP
PMO phosphorodiamidate morpholino oligomer
rAAVs recombinant adeno-associated viruses
RRM RNA recognition motif
SF3B splicing factor 3B
SMA spinal muscular atrophy
SNP single-nucleotide polymorphism
snRNA small nuclear RNA
snRNP small nuclear ribonuclear protein
sQTL splicing quantitative trait locus
SR serine/arginine-rich
SREs splicing regulatory elements
SRSF SR splicing factor
SSOs splice-switching oligonucleotides
WES whole exome sequencing
WGS whole genome sequencing

1. INTRODUCTION

Determination of the pathogenicity of sequence variants is important for prediction of disease severity and genetic counseling. Furthermore, understanding the basic mechanism underlying disease caused by pathogenic variants is critical for development of new therapies. To date, approximately 192,000 disease-related lesions have been described in the human gene mutation database (HGMD). These occur in more than 30% of human protein-coding genes (http://www.hgmd.cf.ac.uk). Of all variants that are annotated in HGMD, 17,439 entries are believed to alter canonical splicing (HGMD, December 2016). These account for 9.1% of all disease-associated variants in the database, although this likely represents an underestimation as most efforts are directed toward missense and nonsense variants and because splicing is difficult to predict. With the advent of new next-generation sequencing (NGS) techniques (Goodwin et al., 2016), many more variants and genes that are linked to disease are likely to be added to the HGMD, broadening our insight of genotype/phenotype relations. However, there is an enormous pace at which new potentially pathogenic variants are discovered, with associated challenges to determine their pathogenicity. This review highlights advances in diagnostics, our understanding of the different levels at which splicing is regulated in healthy individuals and in disease, and the development of potential novel treatment options based on splicing correction.

2. DIFFERENT OUTCOMES OF ALTERNATIVE SPLICING AND THEIR IMPACT ON PROTEIN EXPRESSION

Current estimates state that the human genome contains close to 21,000 protein-coding genes (Moraes and Goes, 2016). Alternative promoter usage, transcription termination, RNA processing, and posttranslational modifications further expand the repertoire for protein diversity. A striking example of diversity generated by alternative splicing is the *Dscam* gene in Drosophila, from which potentially 38,016 different transcripts can be expressed (Park and Graveley, 2007). Interestingly, almost all human gene transcripts undergo alternative splicing and are able to produce multiple types of transcripts (Barbosa-Morais et al., 2012; Pan et al., 2008; Wang et al., 2008). These potentially could expand the repertoire of human proteins by approximately 10-fold (Nilsen and Graveley, 2010). Indeed, the major fraction of alternatively spliced messenger RNA (mRNA) transcripts is engaged by ribosomes, suggesting these transcripts are translated into protein (Weatheritt et al., 2016). It will be important to confirm this hypothesis with proteomic approaches (Tress et al., 2017). Another important role for alternative splicing is the regulation of mRNA expression by generating out of frame products resulting in mRNA-decay, for example intron retention (Wong et al., 2016). One such case is the autoregulatory negative feedback loop that controls expression of Chtop (chromatin target of Prmt1). In this loop, Chtop binds its own mRNA and regulates intron 2 retention. Subsequently, a premature termination codon in the 5′ end of intron 2 leads to nonsense-mediated decay (NMD) of the mRNA (Izumikawa et al., 2016). Furthermore, the extent of alternative splicing differs per type of tissue, with relatively high levels in brain compared to other tissues (Barbosa-Morais et al., 2012). Alternative splicing is thought to be important for normal development and homeostasis of various tissues including the nervous system (Raj and Blencowe, 2015), the immune system (Carpenter et al., 2014), and muscle (Jangi et al., 2014).

Multiple types of alternative splicing are known. Many genes constitutively express one major isoform which contains all canonical exons (Fig. 1). During alternative splicing, canonical splice sites can be skipped and/or noncanonical splice sites can be utilized to form different transcripts.

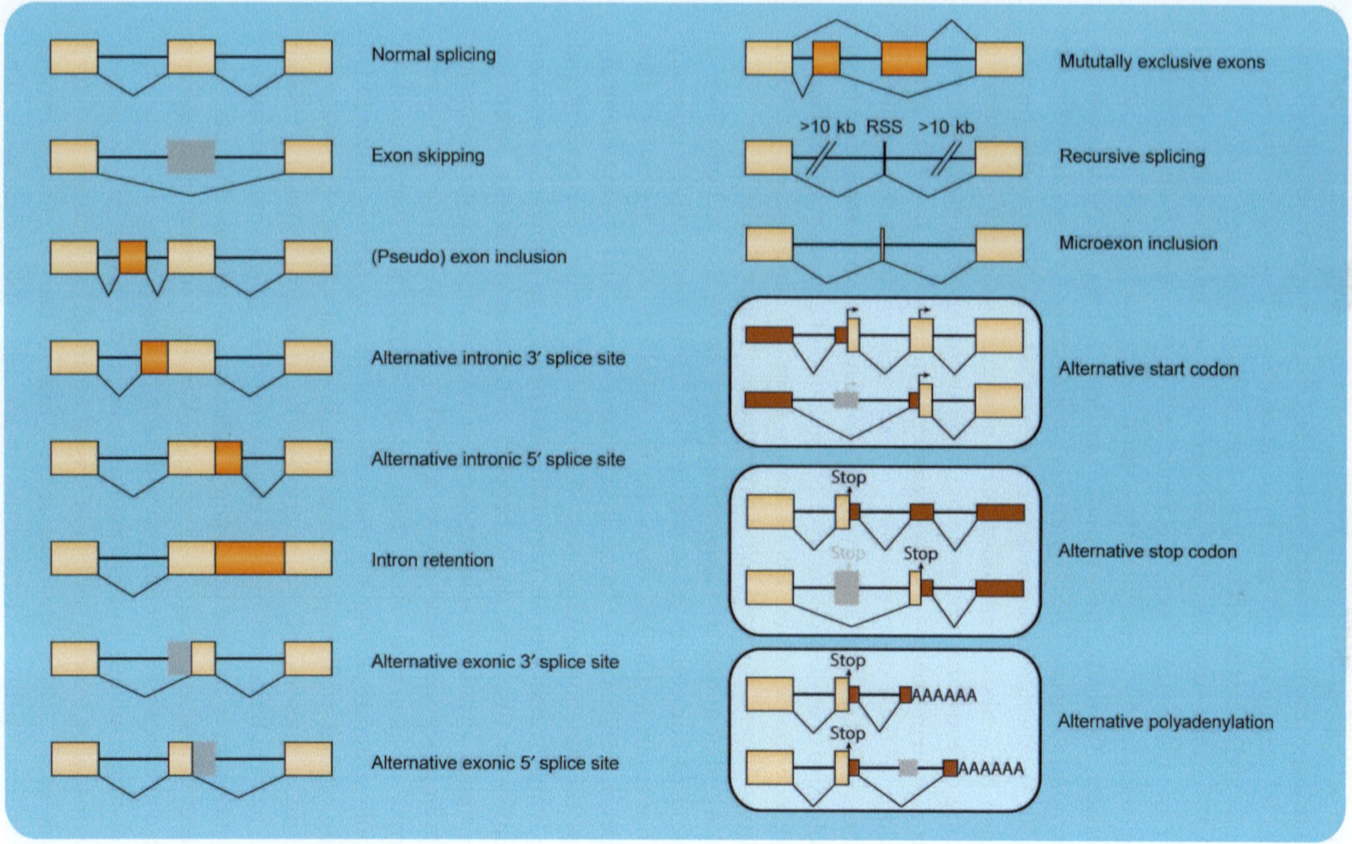

Fig. 1 Potential outcomes of alternative splicing. Constitutive exons are highlighted in *beige*, alternative exons and retained intronic sequences in *orange*, (partially) skipped exons in *gray*, 3′ and 5′ UTRs in *brown*, introns are depicted as *black lines*.

The consequences include exon skipping, exon inclusion, alternative 3′ and 5′ splice site usage, and intron retention (Fig. 1). Genetic variants can shift splicing patterns by altering the recognition of splice sites and/or splicing regulatory elements (SREs) by the spliceosome. This can have multiple outcomes.

(1) When the new length of the coding region in the mature mRNA cannot be divided by three, the result will be a reading frame shift and the generation of a premature stop-codon or the abolishment of a natural stop-codon. Normally, these products are degraded via the NMD pathway or the nonstop decay (NSD) pathway, respectively (reviewed by Simms et al., 2017). Aberrant proteins generated by NSD–mRNA are subsequently degraded by a ribosome-associated E3 ubiquitin ligase that is capable of detecting defective nascent proteins, targeting them for degradation. This acts as a protein quality control measure (Bengtson and Joazeiro, 2010). Activation of these decay pathways is often seen in human disease (Miller and Pearce, 2014). Interestingly, they also play important roles in normal physiology to provide an extra layer of regulation of gene expression. Examples include the determination of human stem cell fate (Lou et al., 2016), inflammation, and myeloid cell differentiation (Saul et al., 2016).

(2) Another outcome of alternative splicing can be that inclusion and/or exclusion of nucleotides leaves the reading frame intact, which generally does not result in mRNA decay via the NMD or NSD pathways. This includes the utilization of mutually exclusive exons, and it can potentially reshape the function of the encoded protein. In normal development and homeostasis, it provides an important mechanism to generate alternative proteins with tissue-specific expression patterns and distinct functions. For example, alternative splicing can alter protein localization via changes in signaling peptides (Balthazar et al., 2017), protein–protein interaction networks in neuronal tissue (Ellis et al., 2012), and organization of T-tubule structure during the maturation of muscle fibers via alternative splicing of four trafficking proteins (Giudice et al., 2016). Furthermore, inclusion of microexons, which are 3–50 bp in length, leaves the reading frame intact in 80%–90% of cases in neuronal tissue, suggesting that microexon inclusion is another important mechanism for protein diversification (Irimia et al., 2014; Li et al., 2015) (Fig. 1). Variations in the human genome can lead to alternative splicing with pathogenic consequences. In human disease, aberrant splicing can generate protein products with loss or gain of function and/or toxicity, such as in several forms of Tauopathies (reviewed in Park et al., 2016). In-frame splice products can potentially activate another decay pathway, termed the no-go decay (NGD) pathway. This

pathway is initiated by barriers that block translation elongation in the ribosome, such as the presence of RNA structures like hairpins and pseudoknots (Doma and Parker, 2006). NGD is initiated in a similar manner compared to NSD, with endonucleolytic cleavage of stalled mRNAs close to the stall site, leaving an NSD-like substrate. This cleavage is stimulated by the DOM34: HBS1 complex both in NSD and in NGD (Tsuboi et al., 2012).

(3) Besides changes within the coding region of pre-mRNA, alternative splicing can also induce the usage of alternative translation start/stop sites and of different polyadenylation signals, which in turn causes changes in the 5′ and 3′ UTRs, respectively (Fig. 1). These variations can result in differential binding of RNA-binding proteins (RBPs), leading to the use of different polyadenylation signals or inclusion/exclusion of miRNA binding sites. One example is the upregulation of expression of functional *DUX4* mRNA. Upon contraction of the D4Z4 repeat at the telomeric end of chromosome 10, *DUX4* expression is induced. However, only in the context of a particular genetic background in which *DUX4* splicing leads to inclusion of a polyA signal in the mRNA, functional DUX4 protein is produced (Daxinger et al., 2015). DUX4 protein is toxic to skeletal muscle cells, and these events result in facioscapulohumeral muscular dystrophy.

3. MECHANISMS OF SPLICING REGULATION IN HOMEOSTASIS AND DISEASE

3.1 Spliceosome

Pre-mRNA splicing is regulated at several layers. The first is at the level of the core splicing machinery, which consists of a large RNA/protein complex termed the spliceosome. It exists in two forms: the major and the minor spliceosomes, of which the major spliceosome accounts for splicing of more than 99% of all introns (Turunen et al., 2013). The major spliceosome is composed of the small nuclear ribonuclear proteins (snRNPs) U1, U2, U4, U5, and U6, which assemble in a specific order (Fig. 2). The mechanism of action has been extensively reviewed by Lee and Rio (2015). The minor spliceosome has a similar mechanism as the major spliceosome. It recognizes slightly different consensus intronic sequences and utilizes the functional analogs U11, U12, U4atac, and U6atac of the U1, U2, U4, and U6 snRNPs from the major spliceosome, respectively. U5 remains shared between both complexes (Turunen et al., 2013). The spliceosome identifies target introns by recognition of conserved splice elements that are present in the native pre-mRNA sequence including the 5′ and 3′ splice sites, the branch point site

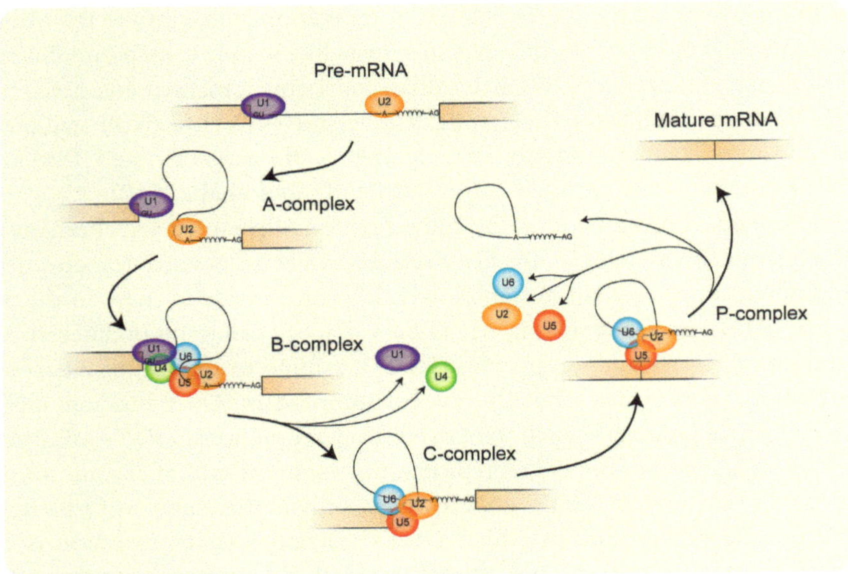

Fig. 2 Schematic representation of splicing performed by the major spliceosome. Pre-mRNA splicing starts with recognition of the 5′ and 3′ splice sites. Looping occurs to form the A-complex, generating a platform for the binding of the U4–U5–U6 tri-snRNP to form the B-complex. After activation of the B-complex, two transesterification steps follow to cut out the intron lariat via C- and P-complexes formation and ligation of the remaining exons, resulting in completion of the splicing process and formation of the mature mRNA.

and the polypyrimidine tract (Fig. 2). U1 recognizes the 5′ splice site, while U2 recognizes the branch point and the 3′ splice site to form the A-complex. In a next step, U4, U5, and U6 snRNPs join to form the B-complex. Two transesterification reactions follow in which the intron lariat is cut and exons are ligated during processing through the C- and P-complexes (reviewed in Moore and Sharp, 1993). It is well known that variations in any of the conserved splicing elements can result in aberrant splicing patterns. In fact, we calculated that 63% of splicing variants annotated in the HGMD are located within the two bases at the beginning and the end of introns (http://www.hgmd.cf.ac.uk/ac/hoho2.php). Pathogenic splicing variants that cause monogenic disorders have been identified in hundreds of disease-causing genes. These variants are often located at well-defined splice sites.

3.2 The Spliceosome in Human Disease

Pathogenic variants in genes encoding for subunits of the major spliceosome occur in various cancers and in retinitis pigmentosa (see review in Singh and

Cooper, 2012). Variants causing retinitis pigmentosa were identified in U4/U5/U6 tri-snRNP-specific proteins, but these only affected splicing of retina-related genes (Ruzickova and Stanek, 2016). The question is why only the retina is affected by defects in the major spliceosome. It has been hypothesized that splicing activity is only partially decreased in retinitis pigmentosa, and that only weakly spliced introns present in retina-specific genes are affected (Linder et al., 2011).

Additional evidence for a role of the major spliceosome in human development and disease was reported recently for germ cell development and variants that may cause male infertility (Wu et al., 2016). A genome-wide association study, in which single-nucleotide polymorphisms (SNPs) are compared, identified significant differences in the genomes of 981 patients with nonobstructive azoospermia (NOA) disorder, compared to the genomes of 1,657 healthy controls. This demonstrated linkage with the locus containing the *SNRPA1* gene, which encodes for the U2A subunit of the major spliceosome. Missense variants in the human *SNRPA1* expressed in *Drosophila* resulted in deregulation of the major spliceosome and disrupted spermatogenesis in *Drosophila*. In addition, knockdown of other core subunits of the major spliceosome (U1, U2, U4, and U5) in Drosophila resulted in male infertility. These results suggest that the major spliceosome is involved in spermatogenesis, and that its dysfunction could explain infertility in patients with NOA. Variants in U4atac, U11/U12–65k, and U12 of the minor spliceosome cause Taybi–Linder syndrome, isolated growth hormone deficiency, and early onset cerebellar ataxia, respectively. This is caused by intron retention of transcripts that are processed by the minor spliceosome (Argente et al., 2014; Edery et al., 2011; Elsaid et al., 2017).

The examples above show that variants in components of the core spliceosomal machinery can cause various human disorders that affect very distinct cell types (Fig. 3). While this seems counterintuitive, it may indicate that splicing outcome is a trade-off between many regulatory pathways, and that the requirements for splicing regulatory factors to ensure correct splicing can be highly cell type specific.

3.3 Splicing Enhancers and Repressors

Besides the spliceosome, additional layers for the regulation of splicing exist. SREs are encoded in pre-mRNA transcripts. These are called exonic splicing enhancers (ESEs), intronic splicing enhancers, exonic splicing silencers (ESSs) or intronic splicing silencers (ISSs), and can be located at any position along the pre-mRNA (Fig. 3). Over 300 splicing regulatory proteins exist

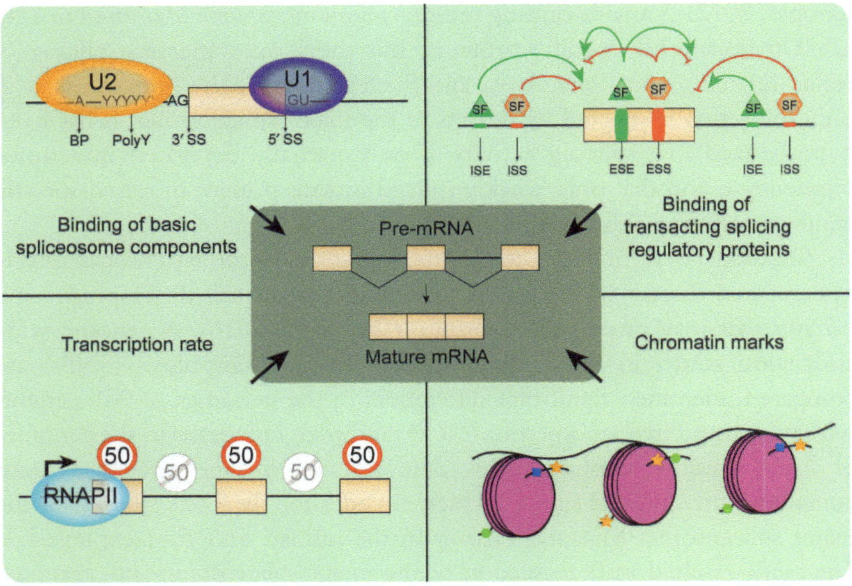

Fig. 3 Regulation of splicing. Splicing regulatory elements are utilized to guide binding of constitutive or tissue-specific splicing factors to modulate splicing. Transcription rate and changes in chromatin marks can also have a profound effect on splicing outcome.

that are thought to act via these elements (Hoskins and Moore, 2012). Following their binding to splice enhancer or silencer sequences, these proteins confer a positive or negative signal, influencing recognition of nearby splice sites by the spliceosome, thereby influencing splicing outcome (Fig. 3). These signals are important for the recognition of splice sites and have been shown to be able to function over distances >300 bp (Schultz et al., 2016). There are several different types of splicing regulatory protein groups, including heterogeneous nuclear ribonucleoproteins (hnRNPs), serine/argenine-rich (SR) proteins, and tissue-specific RBPs. Each of these families can have both positive and negative effects on splicing.

The hnRNP family of splicing regulatory proteins is a group consisting of at least 20 proteins that exert different functions, dictated by their domain organization and cellular localization (reviewed in Geuens et al., 2016). hnRNPs are involved in multiple RNA-related processes during the regulation of transcription and translation (Park et al., 2015), such as RNA trafficking and localization (Gao et al., 2008; Shan et al., 2003), loading of mRNA into exosomes (Villarroya-Beltri et al., 2013), telomere maintenance (Pont et al., 2012), and splicing. hnRNPs can influence exon definition either positively or negatively to modulate splicing. For example, hnRNP A1 primarily binds to splicing silencers from which it negatively influences exon definition in various ways (Blanchette and Chabot, 1999; Krecic and Swanson, 1999; Okunola and Krainer, 2009). hnRNP A1 has also been implicated in stimulation of exon inclusion (Martinez-Contreras et al., 2006; Oh et al., 2013). Similarly, hnRNP L can bind to multiple SREs to influence exon definition either positively or negatively (Motta-Mena et al., 2010). Because of the heterogeneous functions of members of the hnRNP family, it remains difficult to predict splicing-outcome when variants disrupt binding of these proteins. However, it is thought that hnRNPs tend to have more negative than positive effects on exon definition (Geuens et al., 2016).

hnRNPs have also been implicated in disease. An example is hnRNP C. It is one of the first family members that have been shown to be essential for splicing (Choi et al., 1986). hnRNP C plays a role in packaging of native RNA transcripts (Choi et al., 1986; Dreyfuss et al., 1993) and suppression of inclusion of alternative exons (Konig et al., 2010), especially of cryptic exons generated by transposition of Alu-elements (Zarnack et al., 2013). Dysregulation of hnRNP C can lead to altered expression of cancer-related genes such as *BRCA* (Anantha et al., 2013), indicating its potential role in certain malignancies. Furthermore, hnRNPs have been shown to play a major role in the regulation of splicing in neuronal cell homeostasis and have been implicated in diseases including ALS/FTLD, fragile X syndrome, Alzheimer's disease, and spinal muscular atrophy (SMA) (reviewed in Geuens et al., 2016).

SR proteins contain an N-terminal RNA recognition motif (RRM) and a C-terminal SR domain (Long and Caceres, 2009). Like hnRNPs, SR proteins can influence splice site definition in both positive and negative ways (Pandit et al., 2013; Zhou and Fu, 2013). Among the first identified SR proteins are SR splicing factor 1 (SRSF1) (also known as SF2/ASF) and SRSF2 (also known as SC35) (Fu et al., 1992; Krainer et al., 1990; Manley and Krainer, 2010). So far, 12 SRSF factors have been described, all of which function in splicing regulation (reviewed in Howard and Sanford, 2015).

Pathogenic variants that disrupt SREs can result in altered splicing patterns and disease. An example can be found in patients with SMA. SMA is caused by pathogenic variants in the *SMN1* gene. Its sister gene *SMN2* is inactivated by a point variant in exon 7, which disrupts an ESE that is normally recognized by SRSF1 (Cartegni et al., 2006), and/or generates an ESS that is bound by hnRNP A1 and A2, resulting in exon skipping (Kashima et al., 2007). Antisense therapies are now being evaluated that block an ISS located in *SMN2* intron 7 to promote canonical splicing (Singh et al., 2006; see Section 5.3). Another example is pyruvate dehydrogenase complex deficiency, which causes Leigh's encephalomyelopathies and lactic acidosis, and is predominantly caused by pathogenic variants in the *E1α* gene. One variant has been identified that causes missplicing of *E1α* exon 7 due to an increase in the strength of an SRSF2 binding site in intron 7, which results in activation of a cryptic splice donor site (Gabut et al., 2005).

3.4 Tissue-Specific Master Splicing Factors

Tissue-specific splicing proteins play important roles in cellular differentiation, cell-fate determination, and regulation of cellular homeostasis. A number of master splicing factors have been identified that regulate downstream splicing networks in a tissue-specific manner (Jangi and Sharp, 2014). Dysregulation or pathogenic variants in these factors can lead to several diseases. Examples are found within the RNA binding motif (RBM) family, the Rbfox family, and the muscleblind family.

3.5 RBM Family

RBM24 and its close paralog RBM38 have been found to regulate alternative splicing and to play an important role in myogenesis. Jin et al. showed that RBM24 is involved in the posttranscriptional regulation of myogenin and p21. RBM24 regulates the mRNA levels of myogenin and p21, in part, through a myogenin-dependent posttranscriptional regulatory pathway and promotes myogenic differentiation (Jiang et al., 2014; Jin et al., 2010). Knockdown of RBM24 resulted in a decreased half-life of myogenin, suppression of differentiation-associated cell cycle arrest, and a delayed myogenic differentiation (Jin et al., 2010; Miyamoto et al., 2009). Knockdown of RBM38 had similar effects on cell cycle arrest and myogenic differentiation (Miyamoto et al., 2009). Overexpression of either RBM24 or RBM38 had the opposite effect. It induced cell cycle arrest, and promoted myogenic differentiation (Miyamoto et al., 2009). Yang et al. identified RBM24 as a crucial muscle-specific splicing factor for skeletal muscle and cardiac development. In vitro splicing assays showed that muscle-specific exon inclusion in nuclear extracts of nonmuscle cells was promoted by recombinant RBM24, and that RBM24 is crucial for at least 68 splicing events, mainly exon inclusions (Yang et al., 2014).

RBM24 and RBM38 have been suggested to act as tumor suppressors. RBM38 is repressed in acute myeloid leukemia (AML) patients, and low levels of *RBM38* mRNA and downstream effector proteins may contribute to the differentiation block in AML (Wampfler et al., 2016). RBM24 is thought to act as a tumor suppressor through the upregulation of miR-25 in nasopharyngeal carcinoma (Hua et al., 2016). RBM24 has also been linked to Hirschsprung disease where it binds to *MIR143HG* and accelerates its transcript degradation (Du et al., 2016).

RBM20 directly regulates splicing of Titin through binding to certain regions of the pre-mRNA where it represses splicing (Li et al., 2013). In the remaining regions, where there is no binding of RBM20, Titin pre-mRNA is normally spliced, which generates a partially processed Titin pre-mRNA. Instead of being exported from the nucleus, these partially processed Titin pre-mRNAs are stored in the nucleus awaiting further processing. The 5′ splice site and 3′ splice site of the exons flanking the RBM20-repressed regions are then utilized to remove the repressed internal exons and introns (Li et al., 2013).

RBM20 has also been related to human heart disease. Pathogenic variants in *RBM20* lead to several cardiomyopathies like sudden cardiac death and dilated cardiomyopathy (Brauch et al., 2009). A deletion of the *RBM20*-RRM in a mouse model and knockout of *RBM20* in rat showed more expression of the cardiomyopathy-related compliant Titin isoforms and recapitulated the human RBM20-associated cardiomyopathy (Guo et al., 2012; Methawasin et al., 2014). Hinze et al. used a Titin N2B knockout mouse ($Titin^{\Delta N2B/\Delta N2B}$), which shows a complex cardiac phenotype including cardiac atrophy and dysfunction. Altering differential splicing of RBM20 substrates showed positive effects on cardiac function in this model, which suggested RBM20 as a therapeutic target (Hinze et al., 2016). In double-deficient mice ($Titin^{\Delta N2B/\Delta N2B}$ $RBM20^{\Delta RRM/WT}$), splicing of the compliant Titin isoforms was reduced and this was accompanied by positive effects on mRNA levels of genes related to the cAMP response, cardiac dimensions, and oxidative phosphorylation. In this splice-rescue mouse model, a reduction of 50% of RMB20 function was sufficient to restore the diastolic dysfunction completely and reverted the cardiac atrophy, while no obvious extracardiac phenotypes with respect to weight, fertility, and grooming behavior were observed (Hinze et al., 2016).

3.6 Rbfox Family

The Rbfox family regulates alternative splicing in multiple cell types, including embryonic stem cells (ESCs), and cells from neuronal and myogenic lineages (Gehman et al., 2012; Jin et al., 2003; Zhang et al., 2008). The Rbfox1 and Rbfox3 family members are expressed in muscle and neuronal tissues, whereas Rbfox2 regulates expression of more than 70 RBPs in mouse ESCs, thereby inducing alternative splicing-mediated NMD. This results in differentiation toward the mesodermal lineage (Jangi et al., 2014).

Rbfox1 has been linked to several disorders, one of which is autism spectrum disorder (ASD) (Martin et al., 2007; Sebat et al., 2007). In patients with ASD, downregulation of Rbfox1 in the brain was associated with altered splicing patterns of 48 ASD-susceptibility genes (Weyn-Vanhentenryck et al., 2014), and correlated with altered splicing of its predicted target exons (Bhalla et al., 2004; Voineagu et al., 2011). Hamada et al. showed that a specific isoform of Rbfox, Rbfox-isoform1 (iso1), plays an important role in synapse network formation and neuronal migration and that this may relate to ASD. Rbfox-iso1 knockdown in vitro resulted in a reduction of mature spine number and spine density, while in vivo knockdown showed defects in the radial migration and terminal translocation of cortical neurons. Involvement of Rbfox1-iso1 in the formation or maintenance of the synapse network was proposed (Hamada et al., 2016).

Gao et al. have shown that Rbfox1 is also an important regulator of alternative RNA splicing during heart failure, and proposed Rbfox1-mediated pre-mRNA splicing as a new potential therapeutic target (Gao et al., 2016). In a mouse model, Rbfox1 deficiency promoted pressure overload-induced heart failure and cardiac hypertrophy, whereas cardiac reexpression of Rbfox1 rescued this pathology. The same study showed that Rbfox1 expression is decreased in human failing hearts, and that Rbfox1 expression is essential for normal heart development in the developing zebrafish (Gao et al., 2016). The proposed mechanism involves the MEF2 transcription factor family. Rbfox regulates the switch from MEF2 isoform α1 to α2, which have differential effects on expression of cardiac hypertrophy genes. Similar Rbfox1 binding motifs are found near the targeted *MEF2 α2* exons across zebrafish, mice, rats, and humans, suggesting a conserved Rbfox1/*MEF2* mechanism.

In a similar fashion, dysregulation of Rbfox2 has been linked to other cardiac pathologies like hypoplastic left heart syndrome (HLHS) and cardiac pathology in diabetes. A study by Verma et al. showed that Rbfox2 is a major contributor to transcriptome changes in HLHS patients. In these patients, Rbfox2 carries pathogenic variants that alter its function in pre-mRNA splicing. In particular, Rbfox2 binding sites were found in 936 differentially expressed transcripts in the right ventricle of infant HLHS patients (Verma et al., 2016). Nutter et al. showed that in diabetic hearts, 73% of the mis-spliced transcripts have an Rbfox2 binding site. A dominant-negative mechanism has been proposed for Rbfox2, in which an Rbfox2 isoform binds to wild-type Rbfox2 and subsequently inhibits Rbfox2-mediated splicing (Nutter et al., 2016). Rbfox3 has been less well studied in relation to disease, although a recent study linked variants in *RBFOX3* with sleep latency (Amin et al., 2016).

3.7 Muscleblind Family

Another well-studied splicing network is controlled by the muscleblind family. Members of this family play a role in cell differentiation, inferred from an improved efficiency of cellular reprogramming to induced pluripotent stem cells after knockdown of muscleblind-like splicing regulator 1 (MBNL1) and the MBNL-associated changes in splicing patterns upon mesodermal differentiation (Han et al., 2013; Venables et al., 2013). A variety of alternative splicing events involving cassette exons are directly regulated by MBNL1 and MBNL2 and these differ between ESCs and other cell types. In ESCs, overexpression of MBNL proteins shifted the splicing pattern toward a differentiation-like splicing pattern, whereas MBNL knockdown in differentiated cells promoted an ESC-like splicing pattern (Han et al., 2013). Another group showed that during cardiomyocyte maturation MBNL1-dependent splicing changes were enriched for spliceosomal genes as well as genes relevant to vesicular trafficking and membrane organization (Giudice et al., 2014). The function of MBNL and its effect on splicing regulation in myotonic dystrophy (DM) has been extensively reviewed in Cardani et al. (2014). A recent study showed that splicing events are responsive to the dose of MBNL1 protein. Splicing events were evaluated upon a range of MBNL1 protein levels, and each splicing event presented a unique dose–response curve. It has been suggested that the alternative splicing events can be used as a biomarker to accurately estimate the functional levels of MBNL1 protein in DM (Wagner et al., 2016).

The above tissue-specific regulators of splicing have many downstream targets and can be considered important master regulators. It will be interesting to further define the hierarchy of tissue-specific splicing regulation. This will be important to understand the consequences of misregulation of these factors for human development and disease.

3.8 Chromatin and Splicing Regulation

The cotranscriptional nature of pre-mRNA imposes transcription and chromatin structure as potential levels at which splicing may be regulated (Fig. 3). Chromatin structure differentiates intronic from exonic sequences (Schwartz et al., 2009b). One of the factors influencing this is GC content. In general, GC content tends to be higher in exonic regions. GC-rich regions are more likely to result in DNA methylation, as this occurs at CG dinucleotides (Gelfman and Ast, 2013). When DNA is methylated, it is more densely packed in chromatin, thereby slowing down passage of RNA-polymerase II (RNAPII) during transcription. This leads to a lower transcription rate, resulting in increased accumulation of RNAPII over intron–exon boundaries and exon bodies, and ultimately increases the chance of exon inclusion during cotranscriptional splicing as it allows more time for splicing (Adelman and Lis, 2012; Jonkers et al., 2014; Kwak and Lis, 2013). Furthermore, levels of specific histone modifications are different in exonic regions compared to intronic regions. One example is histone 3 lysine 36 trimethylation (H3K36me3), which is associated with transcriptional elongation and tends to coaccumulate with RNAPII at canonical splice sites and exonic regions (Ye et al., 2014). H3K36me3 also seems to be more enriched at constitutive exons compared to alternative exons (Kolasinska-Zwierz et al., 2009). The histone deacetylase complex RPD3S is recruited to the H3K36 di- and trimethylation marks and removes acetylation marks from nearby nucleosomes. Acetylation of histones generally leads to a more open chromatin structure, and removal of these marks leads to a more closed structure, resulting in a lowered transcription rate at these sites and consequently more time for correct splicing. Variants in the *SETD2* gene, which encodes for the enzyme responsible for H3K36 trimethylation, have been linked to human kidney cancer via alterations in RNA processing of highly transcribed genes including intron retention and aberrant splicing (Simon et al., 2014). Furthermore, alteration of H3K36me3 levels result in increased risk of renal cell carcinoma-specific death via changes in RNA processing (Ho et al., 2016).

Other histone marks that have been associated with splicing include H3K4me3, H4K20me1, and H3K79 di- and trimethylation (Andersson et al., 2009; Luco et al., 2010; Shindo et al., 2013; Ye et al., 2014). All these marks have established roles during transcriptional regulation, and their effects on splicing may either be explained by indirect effects on the speed of transcription or by providing binding platforms for splicing proteins that modulate splicing cotranscriptionally. H3K4me3, which is associated with transcriptional activation, has been linked both with inclusion and exclusion of exons in several studies (Luco et al., 2010; Shindo et al., 2013). Enrichment in H3K79me1, which is involved in various roles of chromatin remodeling, correlates with exon inclusion in human embryonic and fetal lung fibroblast cell lines in a similar way as H3K36me3 (Shindo et al., 2013). H4K20me1 is enriched in exons and has been coupled to transcription elongation (Schwartz et al., 2009b). Global cassette exon inclusion levels positively correlate with global H4K20me1 levels of nucleosomes at the position of cassette exons in B-lymphocytes and chronic myelogenous leukemia cell lines (Liu et al., 2014).

Recruitment of splicing regulatory proteins can occur via chromatin-binding factors, as is the case with MRG15, which is part of several histone-modifying complexes. MRG15 is recruited to the H3K36me3 histone mark. By binding to the PTB splicing regulatory factor, MRG15 can form an active cotranscriptional complex that acts as a scaffold for splicing machinery components (Llorian et al., 2010; Luco et al., 2010). Another interaction between chromatin-binding proteins and splicing regulatory proteins can be found in the interaction between PSIP1, a H3K36me3 reader protein, with SRSF1 (Pradeepa et al., 2012). Furthermore CHD1, and ATP-dependent chromatin remodeling factor, is known to interact with the U2 snRNP (Sims et al., 2007).

Chromatin-binding proteins can act as scaffolds to recruit elements of the splicing machinery toward actively transcribed genes, specifically toward intronic regions. These and other findings strongly support the "exon definition" model, which states that exonic marks are recognized by the splicing machinery. This model is indeed favored in vertebrate species, where intronic lengths are considerably longer than exonic lengths (Hollander et al., 2016). Variants that lead to changes in the distribution of histone marks that are associated with exon definition could potentially lead to altered splicing.

3.9 Recursive Splicing

In very large introns, the three reactive sites in splicing, i.e., the 5′ splice site, the branch point, and the 3′ splice site, are separated by long stretches of RNA sequence (Fig. 4A and B). How these sites come together in a three-dimensional space is hard to explain with a conventional splicing mechanism. In humans, over 3400 introns are longer than 50 kb, and over

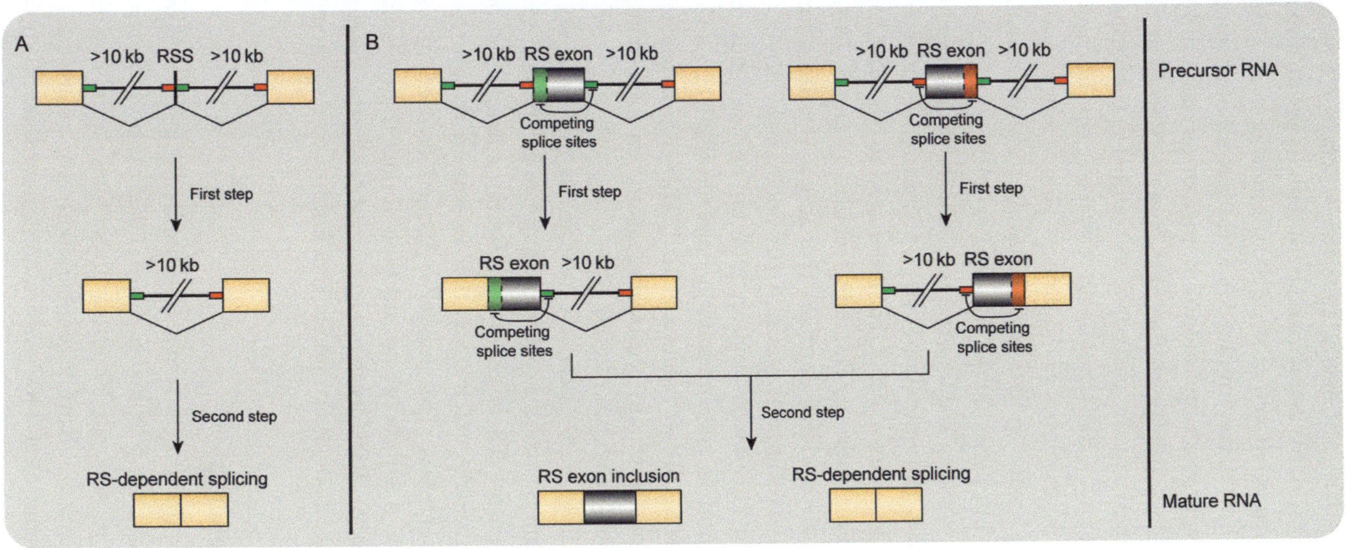

Fig. 4 Schematic representation of recursive splicing (RS). (A) Recursive splice sites deep in the intron are used to remove long intron sequences in a stepwise process. (B) RS utilizing a RS exon. Precursor RNAs are spliced using noncanonical splice sites to from a RS exon. In the first step, one part of the intron is spliced out creating the RS exon. This RS exon can be retained or removed in the second step, depending on which of two competing splice sites is used. 5′-Splice motif depicted in *green*, 3′ splice motif in *red*.

1200 introns are longer than 100 kb (Bradnam and Korf, 2008). A refined mechanism for splicing of large introns was proposed and termed recursive splicing (RS). RS is the removal of large introns using noncanonical splice sites deep in the intron to splice out the intron in a stepwise manner (Fig. 4A and B). This phenomenon was already described in 1998 when Hatton et al. reported the stepwise removal of an intron of the Ultrabithorax gene in *Drosophila melanogaster* (Hatton et al., 1998). Only recently has it become clear that RS is widespread in *Drosophila* and has been conserved between species (Duff et al., 2015; Kelly et al., 2015; Sibley et al., 2015).

A recent study showed a sequential and principally obligatory process for genes that have recursive splice sites in *Drosophila* (Duff et al., 2015). Using deep-sequencing methods, 197 functional recursive splice sites were identified that were highly conserved between *Drosophila* strains. One-hundred and fifteen *Drosophila* genes were reported to undergo RS. Sibley et al. reported that nine genes in the human brain utilized RS to remove an intron (Sibley et al., 2015). A study using human primary endothelial cells detected intermediate splicing products from eight recursive splice sites in a 134 kb intron of the *SAMD4A* gene that appeared and disappeared during transcript production (Kelly et al., 2015). In contrast to RS seen in *Drosophila*, in which the recursive introns are completely spliced out (Fig. 4A), humans harbor a "recursive splicing exon" that seems crucial for removing the long intron and can be retained (Fig. 4B). Two functions of the RS exon in human have been proposed. First, to serve as quality control. RS exons are completely spliced out in normal mRNA, but can be retained when the exon is generated by a faulty splicing event or from an aberrant promotor sequence. A retained RS exon in an mRNA transcript usually leads to NMD due to premature stop codons. Second, the recognition of recursive splice sites is improved through the process of exon definition, a mechanism that defines splice sites on either side of the exon by recruiting splice promoting proteins (Sibley et al., 2015).

To the best of our knowledge, no disease has been linked to RS yet, however, mutating three recursive splice sites in *SAMD4A* using CRISPR/Cas9 resulted in a 35%–50% reduction of mRNA levels (Kelly et al., 2015). Furthermore, an antisense oligonucleotide (AON) that blocked the recursive splice site in the zebrafish *cadm2a* gene resulted in a twofold reduction of mRNA levels in vivo (Sibley et al., 2015). These results highlight the importance of RS for gene expression, and they open the possibility that human disorders may exist that are caused by disruption of this process, for example by deep intronic variants that are usually missed in standard diagnostics.

Interestingly, a recent study by Gazzoli et al. reported that RS occurs in many introns of the dystrophin gene, and that current therapies for Duchenne muscular dystrophy (DMD) based on skipping of exons 45–55 might be affected by RS (Gazzoli et al., 2016).

4. IDENTIFICATION OF PATHOGENIC VARIANTS THAT AFFECT SPLICING

4.1 Molecular Diagnostics in the NGS Era

Massively parallel sequencing, or NGS, has the major benefit that it can query multiple genomic regions in the same sample. Regions of interest can be specifically enriched using different techniques including solid and liquid hybridization capture techniques and multiplexed PCR-based methods. This technique can be beneficial for single gene screening of variants, particularly in large genes like dystrophin (Wang et al., 2014), and the breast cancer genes *BRCA1* and *BRCA2* (Neveling et al., 2017), for which robust screening methods have been developed.

Enrichment and subsequent sequencing of genes that are known to be linked to a specific phenotype is known as gene panel sequencing. This technique is currently the most common NGS technique used in a diagnostic setting. Examples of gene panels include those for inherited retinal disease (Carrigan et al., 2016), early onset epileptic encephalopathy (Gokben et al., 2016), specific types of cancer (Stanislaw et al., 2016), eye disorders (Consugar et al., 2015), hearing disorders (Tekin et al., 2016), neurodegenerative disorders (Kruger et al., 2016), cardiac disorders (Proost et al., 2017), and many others. Gene panel sequencing is a popular tool in clinical genetic laboratories because costs for sequencing and data analysis are significantly lower than for whole genome sequencing (WGS) techniques.

Whole exome sequencing (WES), in which the protein-coding part of the genome is analyzed by high-throughput sequencing, is becoming the next standard in DNA sequencing of patients that lack a diagnosis. WES enrichment is performed in the same way as gene panel sequencing, in fact many gene panels are generated by filtering WES data. In WES, several enrichment strategies can be used to enrich for all coding regions of the entire genome, including in-solution hybridization, selective circularization PCR, and high-multiplex PCR techniques (reviewed in Ballester et al., 2016). In this way, the analysis is performed in a less biased manner and novel potentially disease-causing genes may be identified. Many genes have been identified by WES that have a previously unrecognized link to disease

(Beaulieu et al., 2014; Chong et al., 2015; Wright et al., 2015). The wealth of variants detected with WES that could potentially be pathogenic necessitates extensive filtering. These filtering steps can include discarding variants that are above a certain threshold for minor allele frequency, as well as synonymous variants and intronic variants. However, this filtering increases the risk of discarding pathogenic variants for which the mechanism of pathogenicity is difficult to predict, for example, variants that affect splicing.

All NGS techniques mentioned earlier have one major limitation, which is that targeted enrichment is performed before sequence analysis. This generates an inherent bias of these approaches, and likely leaves causative variants undetected. Indeed, a recent survey has shown that NGS-based screening results in a genetic diagnosis in only 23%–26% of cases (Sawyer et al., 2016; Yang et al., 2013).

The optimal NGS technique for implementation in DNA diagnostics would be WGS, provided that costs and analysis time can be reduced in future improvements. This technique analyzes the complete genome, removing the bias of targeted enrichment seen in WES or gene panel sequencing. It should be noted however that technical bias can still occur, for instance GC-rich regions are difficult to sequence and repeat-containing sequences are excluded. WGS datasets take a relatively long time and large computer power to analyze because of their size. Furthermore, due to inclusion of all regions of the DNA, including noncoding parts, many more variants are detected, and the task to identify the truly pathogenic variant is even more daunting. Often datasets are extensively filtered for variants that are likely nonsignificant using algorithms that predict variant pathogenicity. The implementation of WGS in diagnostics has led to discovery of new disease genes. An example is the study from Nishiguchi et al., which resulted in the identification of causative variants in patients with autosomal recessive retinitis pigmentosa, including variants in the NEK2 gene. These variants have not been linked previously to this disease (Nishiguchi et al., 2013).

4.2 In Silico Prediction of Functional Effects That Variants Have on Splicing

Various algorithms have been developed for splicing prediction. Currently, splicing prediction is routinely performed by diagnostics laboratories (Fig. 5). Many different online tools have been made available for prediction of splicing-related elements and/or to anticipate the effects that variants have on splicing (Table 1). Below, some of these are described in more detail.

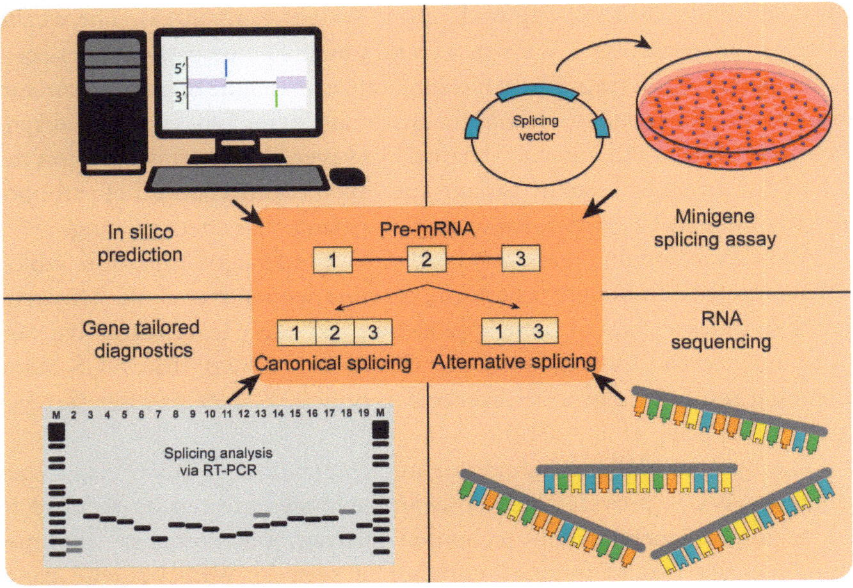

Fig. 5 Assays for testing alternative splicing.

A plethora of algorithms can be used to predict the presence of 3′ and 5′ splice sites (reviewed in Jian et al., 2014) as well as splicing enhancer and splicing silencer elements (Sironi et al., 2004; Soukarieh et al., 2016; Yeo et al., 2004). Human splicing finder (HSF) and SROOGLE are two widely used prediction tools that integrate many of different algorithms in one interface (Desmet et al., 2009; Schwartz et al., 2009a). However, interpretation of the output in these programs can be challenging (Van der Wal et al., 2017a). Recently, the Automated Splice Site and Exon Definition Analyses (ASSEDA) tool has become available online. It predicts splice site utilization based on predicted splice site strengths (Mucaki et al., 2013).

Although these algorithms can be used to predict splicing outcome, caution should be taken when interpreting results from such predictions, as it does not always reflect the true molecular outcome. This was exemplified by Schleit et al. who functionally investigated splicing outcome of *COL1A1* mRNA in 40 individuals suspected of osteogenesis imperfecta (Schleit et al., 2015). Subsequently, variants present within the *COL1A1* gene of these individuals were analyzed with three different splicing prediction algorithms. The outcome of these predictions was correct in less than 75% of instances, compared to functional tests.

Table 1 Splice Prediction Tools Accessible Online

Prediction Tool	Type of Motif Detected	Original Reference	Site
Splice site analyzer tool	Splice sites, branch site. SRE elements	Based on (Shapiro and Senapathy, 1987)	http://ibis.tau.ac.il/ssat/SpliceSiteFrame.htm
NetGene2	Splice sites	Hebsgaard et al. (1996)	http://www.cbs.dtu.dk/services/NetGene2/
NNSplice	Splice sites	Reese et al. (1997)	https://omictools.com/nnsplice-tool
GENSCAN	Splice sites	Burge and Karlin (1997)	http://genes.mit.edu/GENSCAN.html
SpliceView	Splice sites	Rogozin and Milanesi (1997)	http://bioinfo.itb.cnr.it/cgi-bin/oriel/wwwspliceview.pl
MaxEntScan	Splice sites	Yeo and Burge (2004)	http://genes.mit.edu/burgelab/maxent/Xmaxentscan_scoreseq.html
SplicePredictor	Splice sites	Brendel and Kleffe (1998)	http://sites.usd.edu/bioinformatics/projects/splicepredictor
Spliceport	Splice sites	Dogan et al. (2007)	http://spliceport.cbcb.umd.edu/
Human splicing finder	Splice sites, branch site. SRE elements	Desmet et al. (2009)	http://www.umd.be/HSF/
CRYP-SKIP	Splice sites	Divina et al. (2009)	http://cryp-skip.img.cas.cz/
EX-SKIP	Exonic splicing regulatory elements	Raponi et al. (2011)	http://ex-skip.img.cas.cz/
SROOGLE	Splice sites, branch site. SRE elements	Schwartz et al. (2009a)	http://sroogle.tau.ac.il/
GeneSplicer	Splice sites	Pertea et al. (2001)	https://ccb.jhu.edu/software/genesplicer/

Continued

Table 1 Splice Prediction Tools Accessible Online—cont'd

Prediction Tool	Type of Motif Detected	Original Reference	Site
AVISPA	Splice sites	Barash et al. (2013)	http://avispa.biociphers.org
ESEFinder	Exonic splicing enhancers	Cartegni et al. (2003)	http://krainer01.cshl.edu/cgi-bin/tools/ESE3/esefinder.cgi?process=home
[RESCUE]-ESE	Exonic splicing regulatory elements	Fairbrother et al. (2002)	http://genes.mit.edu/burgelab/rescue-ese/
ASPic	Splice sites	Martelli et al. (2011)	https://algolab.eu/research/aspic/
RegRNA2	Big range of motifs, including user defined	Chang et al. (2013)	http://regrna2.mbc.nctu.edu.tw/
Mutation Taster2	Protein and RNA changes	Schwarz et al. (2014)	http://www.mutationtaster.org/
Imhotep	Many different aspects (incl. DNA, RNA, and protein level)	Knecht et al. (2017)	http://www.uni-kiel.de/medinfo/cgi-bin/predictor/
AASsites	Classifies probable change in splicing	Faber et al. (2011)	http://genius.embnet.dkfz-heidelberg.de/menu/biounit/open-husar
SNPlice	Effect of variants on intron retention	Mudvari et al. (2015)	https://code.google.com/archive/p/snplice/
Spliceman	Effect of variants on splicing	Lim and Fairbrother (2012)	http://fairbrother.biomed.brown.edu/spliceman/
ASSEDA	Effect of variants on splicing	Mucaki et al. (2013)	http://splice.uwo.ca/

Most algorithms have been generated based on different datasets and use different calculations, which explains why splicing predictions can vary substantially between different algorithms. Alamut® is a commercial platform that uses five different algorithms to predict the strengths of both 3′ and 5′ splice sites, as well as algorithms that detect potential binding sites of multiple splicing factors. The program helps the user by visualizing the potential presence of these sites, an example of which can be found in Bergsma et al. (2016). Certain programs focus not only on variants that affect splicing, but also on changes to protein sequence and stability. One example is Mutationtaster, which integrates information from different biomedical databases to test evolutionary conservation, splice-site changes, loss of protein features, and changes that might affect the amount of mRNA (Schwarz et al., 2010). In the era of NGS, high-throughput data interpretation is becoming more and more essential. A new version, Mutationtaster2, has been generated to analyze multiple variants in a short period of time from deep-sequencing data (Schwarz et al., 2014). Another program that is well suited for multiple predictions of results from NGS datasets is combined annotation-dependent depletion (CADD), which incorporates multiple factors and gives a single pathogenicity score to variants (Kircher et al., 2014). More examples of splicing prediction programs include IntSplice for prediction of the splicing consequences of intronic single-nucleotide variations, and IMHOTEP, a composite score integrating popular tools for predicting the functional consequences of nonsynonymous sequence variants (Knecht et al., 2017; Shibata et al., 2016). Interestingly, a comparison of IMHOTEP against a variety of other prediction tools, including CADD, showed that it was superior in predicting variants that affect splicing (Knecht et al., 2017). Finally, to generate more accurate predictions, splice prediction programs have been generated based on functional data (Di Giacomo et al., 2013; Soukarieh et al., 2016). Although algorithms based on functional data will likely give a better representation of splicing in vivo, these two algorithms have been based on a relatively small dataset. With current RNA sequencing techniques, algorithms based on larger datasets can be generated, which will likely lead to better splicing prediction algorithms in the near future.

The programs and algorithms described above each have their benefits and drawbacks. Algorithms and prediction tools for pathogenicity will become increasingly important with implementation of NGS techniques in diagnostics. However, functional data that confirm these predictions are often lacking. In practice, predictions always should be verified using a functional assay to confirm a particular diagnosis.

4.3 Functional Assays

A widely used approach to evaluate functional effects that variants may have on splicing is to introduce part of the genomic region of the gene of interest in an artificial minigene system (Fig. 5). This technique can be used to study splicing of genes that have extremely low or no detectable expression in patient material or if the material is not available. The system is utilized to evaluate splicing of well-described genes associated with disease, such as dystrophin, *CFTR*, *BRCA1*, and *BRCA2* (Bonnet et al., 2008; Giorgi et al., 2015; Wang et al., 2016). These assays may utilize a dedicated exon trapping vector in which only the exon of interest and short (30–150 bp) flanking intronic regions are included. This system represents a quick and easy way to analyze the effect of selected variants on pre-mRNA splicing. Several recent studies have utilized minigene assays to validate the effect of variants on alternative splicing, including variants affecting splicing of *SLC26A4* mRNA causing autoimmune thyroid diseases (Bouattour et al., 2017) and the *PKD2* gene causing polycystic kidney disease (Gonzalez-Paredes et al., 2016). A disadvantage is that the assay is particularly biased, because the exon is placed outside its in vivo context, with altered transcription, transcript stability, and flanking intronic sequences. An improvement is to insert the exon of interest including neighboring introns and exons in an expression vector, provided that the total insert is not too long. Some of the above-mentioned disadvantages also apply to this option. Minigenes are useful tools to test potential pathogenic effects of variants, but care should be taken with translating results to regulation of splicing in vivo, as alternative splicing can differ between cell/tissue types.

With current NGS techniques, it is now possible to perform genome-wide mRNA expression analyses for diagnostic purposes to directly detect aberrant splicing in mRNA isolated from patient-derived cells (Fig. 5). Short sequence reads can be generated, from which splicing products need to be reassembled in silico, which can be a challenge. Numerous mapping protocols, programs, and pipelines have been generated to deal with this problem, and each of these have their own benefits and drawbacks (Engstrom et al., 2013). Recent advances have enabled long-read RNA sequencing, a third-generation sequencing technique, enabling sequencing of large stretches of contiguous mRNA avoiding in silico reconstruction of isoforms (reviewed in Rhoads and Au, 2015). This technique has led to the detection of many previously unknown splice isoforms (Sharon et al., 2013). Furthermore, it allows for the generation of full-length personal trancriptomes, highlighting the potential for clinical implementation (Tilgner et al., 2014).

Several challenges remain to be overcome before RNA-based NGS methods can be implemented in routine diagnostics (Byron et al., 2016). One reason for this is that it requires patient material, which can be impractical, or even impossible to obtain, depending on the tissue in which the gene of interest is expressed. In addition, expression levels of the gene of interest need to be high enough to detect aberrant splicing, which will affect the required depth of sequencing. However, for certain genes with ubiquitous expression patterns, gene-tailored diagnostics based on mRNA analysis should be possible. For example, it is feasible to perform unbiased PCR-based analyses of all exons from fibroblast cultures to detect splicing variants of the acid α-glucosidase (*GAA*) gene causing Pompe disease, a metabolic myopathy (Fig. 5). This approach has already revealed novel splicing variants (Bergsma et al., 2015, 2016). In principle, it should be possible to develop diagnostic kits tailored for the detection of splicing variants in genes that are expressed in easy-to-obtain human material such as blood, urine, and skin. At present, PCR-based methods are the cheapest and easiest way for gene-tailored mRNA analysis, but in the future it is likely that these will be replaced by NGS-based techniques.

Recently, genome-editing techniques such as CRISPR/Cas9 have opened new possibilities for investigating functional effects of alternative splicing. CRISPR/Cas9 can be used to induce splicing variants to test for pathogenic effects, as exemplified in Zhang et al. (2016). Another application is the generation of transgenic mice to modulate expression of specific alternative transcripts, in which inhibition of expression of a *Titin* isoform was used to study the effects on M-line formation in striated muscle (Charton et al., 2016). The relative ease of generating transgenic animal models will likely lead to more functional studies of alternative splicing in relation to disease.

4.4 Ultraviolet Crosslinking and Immunoprecipitation

RBPs are key players in posttranslational gene regulation and RNA processing (Gerstberger et al., 2014). To understand the mechanisms involved, systematic identification of endogenous protein–RNA interactions is crucial. In 2003, Ule et al. developed a method for the genome-wide identification of RBP-binding sites called ultraviolet (UV) crosslinking and immunoprecipitation (CLIP) (Ule et al., 2003). This method relies on crosslinking of the protein–RNA complexes, followed by purification and release of RNA fragments by protein digestion. The complementary DNA (cDNA) library generated from these RNA fragments can be subjected to deep sequencing. Bioinformatic analysis of these data can reveal the RBP–RNA interactions. A variety of bioinformatics pipelines for the handling of genome-wide CLIP data has been developed over the years, which has been extensively reviewed by Wang et al. (2015).

Since its introduction, the CLIP method has evolved into three distinct versions: (1) high-throughput sequencing of RNA isolated by CLIP (HITS-CLIP), the first CLIP platform developed for the genome-wide-identification of RBP binding sites (Licatalosi et al., 2008). (2) Photoactivatable-ribonucleoside enhanced CLIP (PAR-CLIP), which improved the signal-to-noise ratio by incorporating photoreactive ribonucleoside analogs into living cells via the culture system prior to UV treatment (Hafner et al., 2010). (3) Individual-nucleotide resolution UV CLIP (iCLIP), which is based on truncation of cDNAs at the crosslink site and identifies the positions of crosslink sites at single-nucleotide resolution (Konig et al., 2010). New variants such as the enhanced CLIP (eCLIP), which decreases PCR amplification by a 1000-fold (Van Nostrand et al., 2016), infrared-CLIP (irCLIP), which uses an infrared dye-conjugated and biotinylated ligation adaptor for visualization of UV-crosslinked protein–RNA complexes (Zarnegar et al., 2016), and bromodeoxyuridine CLIP (BrdU-CLIP), which incorporates BrdUTP into the cDNA during reverse transcription (Weyn-Vanhentenryck et al., 2014), have been developed more recently with undoubtedly many more variants to follow. With the introduction of CLIP and all its variants, the RNA-regulation field has rapidly developed and this has greatly increased our understanding of RBP–RNA interactions over the past few years.

This is well exemplified by recent studies on RBM10, a splicing regulatory protein. Wang et al. used PAR-CLIP to identify binding sites for RBM10 in HEK 293 cells. A large number of binding sites were identified and these were located in close proximity to splice sites. A role for RBM10 in splicing regulation was proposed, especially in alternative splicing resulting in exon skipping (Wang et al., 2013). Pathogenic variants in *RBM10* result in TARP (talipes equinovarus, atrial septal defect, robin sequence, and persistent left superior vena cava) syndrome. iCLIP was used to understand how loss of RBM10 function might result in TARP (Rodor et al., 2017). RNA targets of RBM10 in a mouse mandibular embryonic cell line were identified, and this revealed a clear enrichment of RBM10 binding in the proximity of the branch point and the 3′ splice site of protein-coding genes. In addition, RBM10 binding to small nuclear RNAs (snRNAs) of canonical and minor spliceosomes was shown. Together with full transcriptome analysis of gene expression and splicing changes in mouse ESCs with disrupted RBM10 expression and RBM10 knockout mouse mandibular cells, this revealed an important role of RBM10 in the regulation of alternative splicing. This suggests that TARP syndrome might be induced by loss-of-function variants in *RBM10* that cause dysregulation of alternative splicing and/or altered expression of other genes. One gene in which splicing changes were observed was OPA1, which is involved in optic atrophy that is also recognized in TARP syndrome (Rodor et al., 2017). In a different study iCLIP demonstrated the role of RBM10 in NUMB splicing regulation, a gene that is involved in the Notch signaling pathway and that is frequently altered in lung cancer. RBM10 variants were shown to promote cell growth by alternative splicing of NUMB pre-mRNA leading to lung adenocarcinoma progression (Bechara et al., 2013).

4.5 Are Splicing Variants Underestimated?

Of all variants that are listed in HGMD, 16,356 entries are known to alter splicing (HGMD, December 2016). These represent 9.1% of all disease-associated variants present in the database. There is increasing evidence that this percentage may be an underestimation for a number of reasons. In the early days of sequence analysis, functional analysis of reverse-transcribed mRNAs using PCR techniques was often employed for finding variants that alter the length of mRNA transcripts. The first dataset specific for pathogenic splice variants presented 101 variants, and these were thought to account for at least 15% of pathogenic variants (Krawczak et al., 1992). Sanger sequencing of genomic DNA is to date the golden standard in genetic diagnostics. Owing to the high number of common SNPs (minor allele frequency of >1%) that are present in noncoding intronic regions, exonic sequences are prioritized for detection of pathogenic variants, leading to an overrepresentation of variants other than splicing variants in current databases. Indeed, a recent survey shows that approximately 25% of missense and nonsense variants described in the HGM database may have an effect on splicing (Sterne-Weiler et al., 2011), suggesting that one-third of all pathogenic single-nucleotide variants in the database potentially affect splicing. Synonymous variants, that are usually disregarded as they do not change the amino acid sequence, can also have an effect on splicing. A study performed by Supek et al. indicates that synonymous variants occur more often in oncogenes where they might act as driver variants by changing SREs, ultimately resulting in altered splicing (Supek et al., 2014).

Functional investigation of all known variants present in a single exon in *BRCA2* and *MLH1* yielded a high percentage of variants that affect splicing

(42% and 77%, respectively) (Di Giacomo et al., 2013; Soukarieh et al., 2016). One mathematical model even suggests that up to 62% of all disease-causing variants affect splicing (Lopez-Bigas et al., 2005). These predictions should be further investigated at the functional level.

A more comprehensive understanding of the splicing code will be necessary to estimate the impact that variants have on splicing in general. A recent study showed that disease-causing variants located in deep intronic regions affect splicing nine times more often than SNPs (Xiong et al., 2015). Furthermore, if missense variants have a low predicted impact on protein function, the likelihood that these variants lead to aberrant splicing is five times higher.

As well as rare variants, polymorphisms can also impact splicing. It has been estimated that up to 45% of synonymous SNPs potentially alter pre-mRNA splicing patterns (Mueller et al., 2015). Furthermore, splicing quantitative trait loci (sQTLs), which are regions in the DNA that correlate with splicing patterns and are defined by the presence of SNPs, are thought to be important contributors to phenotypic variation (Li et al., 2016). In a recent study, Takata et al. analyzed tissue from the prefrontal cortex of 206 individuals and identified genome-wide alternative splicing events controlled by sQTLs (Takata et al., 2017). Investigation of sQTLs SNPs located in schizophrenia-associated loci linked to four potential candidate genes which underwent alternative splicing and could be implicated in schizophrenia.

The data highlighted earlier indicate that the effect of (pathogenic) variants on splicing of pre-mRNA may be considerably underestimated and warrants future investigation.

5. THERAPEUTIC APPROACHES TO MODULATE SPLICING

Human disease caused by aberrant splicing is amenable to modulation in several ways. In recent years, new developments have presented promising therapeutic strategies.

5.1 Small Molecule Modulators of Splicing

Small molecules that modulate alternative splicing or inhibit the major spliceosome have been reported for the treatment of cancer and monogenic disorders. The advantages of using small molecules include that these can be administered orally, and that these may cross the blood–brain barrier.

In cancer, splicing regulatory sequences in (proto-)oncogenes can be mutated, and/or splicing factors can have altered expression. This can cause alternative splicing and induce resistance to cancer drugs, or it can lead to escape of cancers cells from the immune system. This has been extensively reviewed in Lee and Abdel-Wahab (2016) and Salton and Misteli (2016) (Fig. 6). Small molecules have been developed that block early spliceosome formation via binding to the splicing factor 3B (SF3B) core spliceosome subunit, or that alter phosphorylation of splicing regulatory SR proteins. One of the earliest drugs to modulate splicing in cancer is the compound spliceostatin A, which targets the SF3B subunit of U2 snRNP, leading to cell cycle arrest in G1 and G2/M (Kaida et al., 2007). Unfortunately, a clinical trial with the SF3B inhibitor E7107 was suspended due to toxicity (Hong et al., 2014). Currently, one trial is recruiting patients for treatment with H3B-8800 (NCT02841540), another SF3B inhibitor.

Many small molecules have been reported that modulate aberrant splicing in monogenic disorders in vitro, but preclinical safety and efficacy for most of these compounds remain to be established (reviewed in Bates et al., 2017; Ohe and Hagiwara, 2015). It is worth noting that high-throughput screens with compounds that are already used in the clinic may identify other compounds that also modulate splicing. An example is

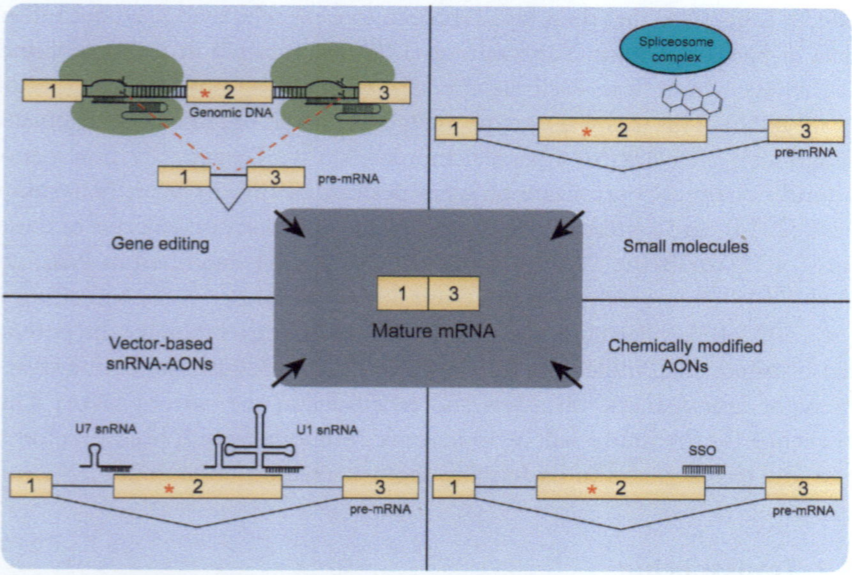

Fig. 6 Potential therapeutic strategies to interfere with splicing. *Red asterisks* indicate a hypothetical pathogenic variant. Potential therapeutic interventions include deletion of an exon from the DNA using gene-editing tools, small molecules that interfere with the activity of splicing proteins, vector-based snRNA-AONs, and chemically modified AONs.

provided by cardiotonic steroids. It was demonstrated that these regulate alternative splicing via depletion of SRSF3 and TRA2B, resulting in increased exon 10 inclusion in microtubule-associated protein tau (*MAPT*) in frontotemporal dementia in vitro (Anderson et al., 2012; Stoilov et al., 2008). Another example is DM type 1 (DM1), in which the master splicing factor MBNL1 is sequestered and depleted by CTG repeats in the myotonin protein kinase transcript, resulting in defects in alternative splicing. Metformin (Biguanides), which is commonly used for the treatment of type 2 diabetes, was shown to downregulate the master splicing factor RBM3 and alleviate some alternative splicing defects in DM1 (Laustriat et al., 2015). For SMA (discussed in more detail below), multiple splice-switching compounds were identified using large-scale drug screens. An example is Aclarubicin, which promotes inclusion of exon 7 of *SMN2* (Andreassi et al., 2001). More recently, additional molecules have been identified that can alter *SMN2* splicing to promote exon 7 inclusion using high-throughput screening. These include SMN-C1, SMN-C2, SMN-C3 (Naryshkin et al., 2014), and NVS-SM1 (Palacino et al., 2015). Both studies demonstrated enhanced survival of humanized SMN7delta mice after treatment with these compounds. Global gene expression analysis showed no major off-target effects, suggesting that these small molecules were selective. Lead optimization of SMN-C3 from Naryshkin et al. (2014) resulted in the compound RG7800, which was used in the clinical trial NCT02240355 (Ratni et al., 2016). This trial was unexpectedly terminated because of toxic effects found after 39 weeks in a parallel monkey study. To generate new compounds, chemical optimizations were performed on hits from Naryshkin et al. (2014), and a safety study was completed on a derivative of one of these hits (NCT02633709, Woll et al., 2016). NVS-SM1 described in Palacino et al. (2015) is currently used in a clinical trial to evaluate safety and efficacy (NCT02268552, Palacino et al., 2015). These findings highlight the potential of using small molecules for targeting splicing regulators for the treatment of monogenetic disorders and especially in the case of SMA. The challenge for the future will be to identify compounds with high selectivity and low toxicity, and to understand their mechanism of action.

5.2 Transsplicing

Another option for correcting aberrant splicing caused by the presence of splice altering variants is spliceosome-mediated RNA transsplicing (SMaRT) (reviewed in Berger et al., 2016). This technique is based on the property of the spliceosome to enable the combination of transcripts from two independent RNA molecules to form one mRNA transcript. Several applications of SMaRT have been reported. Examples include cystic fibrosis (Mansfield et al., 2000), Huntington's disease (Rindt et al., 2012), inherited tauopathy (Avale et al., 2013), and p53-defective cancers (He et al., 2015). The process of transsplicing rarely occurs in humans (Lei et al., 2016), and the challenge will be to increase the efficiency of this approach for clinical development.

5.3 Antisense Oligonucleotides

A specific way to influence gene expression can be accomplished with short (~15–30 nt) RNA-based molecules termed AONs that bind to RNA in a sequence-dependent manner. Several types of AONs can be distinguished. These include single-stranded and double-stranded AONs for mRNA degradation, and splice-switching oligonucleotides (SSOs). To induce mRNA degradation, either the RNA interference (RNAi) pathway (using dsRNA) or the RNase H pathway (using ssRNA) may be utilized. SSOs rely on the ability to bind to splicing regulator sequences at the pre-mRNA to modulate splicing outcome. AONs have chemically modified backbones to prevent degradation. The type of modification has consequences for the mechanism of action (e.g., (in)activity toward RNase H and RNAi pathways) and for biodistribution and pharmacokinetics. The difficulty to deliver RNA-based molecules to cells is an area of active research, including the development of conjugates and chemistries that improve cellular delivery.

In RNAi, double-stranded RNA is administered to cells, which is then processed and loaded on the RNA-induced silencing complex (RISC). The RNA binds to a target mRNA, which results in its degradation, or inhibition of protein translation (Elbashir et al., 2001). Although no RNAi drugs are on the market thus far, several are currently tested in clinical trials. The most clinically advanced example is RNAi-mediated treatment of diabetic macular edema, which is currently in Phase II trials (NCT01445899). RNAi has been used to modulate splicing of alternative exons by targeting sequences that are located in the vicinity of those exons (Allo et al., 2009). Although RNAi has been extensively investigated as a potential therapy, doubts have been raised on its clinical applicability (Krieg, 2011). Off-target effects are a concern (Jackson and Linsley, 2010). Moreover, the administration of double-stranded RNA can lead to an innate immune response via recognition of double-stranded RNA by Toll-like receptors (de Fougerolles et al., 2007). These drawbacks need to be overcome before RNAi can be applied to the clinic.

Single-stranded AONs can target RNA after which these are recognized by the endogenous RNA degradation pathways involving RNase H. The single-stranded AON is usually modified to prevent its degradation; however, these modifications often interfere with the RNase H pathway. To prevent this, modifications to the backbone of the RNA molecule are made at the nucleotides on the outside of the molecule, the result of which is termed a gapmer (Yu et al., 2013). In this way, exonucleases do not degrade the RNA target. RNA bases in the middle of the AON are not modified so that they are still recognized by RNAse H. One of the antisense gapmers under development is Mipomersen for intervention in patients with heterozygous familial hypercholesterolemia. This second-generation AON gapmer targets the splice acceptor of exon 21 in the *APOB* gene, causing both exon skipping of exon 21 and cleavage of the mRNA by RNAse H, both resulting in downregulation of expression of the apolipoprotein B enzyme (Kastelein et al., 2006). First indications of clinical trials using subcutaneous administration (3 × weekly) revealed that indeed ApoB production was lower in treated patients (Akdim et al., 2010; Akdim et al., 2011; Kastelein et al., 2006) (NCT00216463, NCT00231569). However, adverse events also occurred, including injection site reactions, increased liver fat percentage, and increased serum transaminase levels. Four Phase III clinical trials (NCT00607373, NCT00706849, NCT00794664, and NCT00694109) indicated that mipomersen treatment resulted in lower levels of LDL-C, non-HDL cholesterol and Lp(a). Furthermore, a dramatic reduction in cardiac events was seen, indicating that the drug has beneficial therapeutic effects (Duell et al., 2016). The drug was FDA approved in 2013 for homozygous hypercholesterolemia, however, it should be stated that it received a black box label for potential hepatotoxicity.

SSOs are designed to modulate pre-mRNA splicing (Fig. 6). These typically utilize chemically modified AONs that do not activate the RNase H or RNAi pathways. The first SSOs carried a 2′O-methylphosphorotioate (2′O-mePS) backbone and were targeted against cryptic splice sites in mRNAs derived from pathogenic variants in the *β-globin* gene (Dominski and Kole, 1993; Sierakowska et al., 1996). Currently, a host of different backbone structures are under investigation for application in the clinic. Frequently used backbone chemistries include 2′O-mePS, 2′O-methoxyethyl (MOE), phosphorodiamidate morpholino oligomer (PMO), and peptide nucleic acids (Geary et al., 2015). Recently, a new type of backbone chemistry, tricyclo-DNA, was suggested to enhance cellular uptake (Goyenvalle et al., 2015). Methods to improve uptake include the addition of cell-penetrating peptides (reviewed in Boisguerin et al., 2015) and delivery via nanoparticles (reviewed in Petrilli et al., 2014). Furthermore, coadministration of specific monosaccharides can also lead to improved uptake (Cao et al., 2016; Han et al., 2016). For many disorders, preclinical development of SSO-mediated therapy has been performed, including Hutchinson–Gilford progeria syndrome (Osorio et al., 2011), type I Usher syndrome (Lentz et al., 2013), cystic fibrosis (Igreja et al., 2016), Leber congenital amaurosis (Garanto et al., 2016), certain types of cancer (reviewed in Farooqi et al., 2014), and Pompe disease. In the latter, a substrate reduction strategy used SSO-mediated reduction of glycogen synthase expression (Clayton et al., 2014), while direct approaches used promotion of canonical *GAA* splicing via inhibition of cryptic splicing (Bergsma et al., 2016; Van der Wal et al., 2017a, b).

For two diseases, DMD and SMA, clinical development of AON therapy have also passed Phase III trials (ClinicIaTrials.gov identifiers NCT02255552 and NCT02193074, respectively) after extensive preclinical work (reviewed in Aartsma-Rus, 2010; Singh et al., 2015) and both drugs have been approved by the FDA. Eteplirsen, a PMO-based AON for DMD was approved by the FDA in September 2016 for IV administration (dose of 30 mg/kg weekly) (Niks and Aartsma-Rus, 2017). Nusinersen, an MOE phosphorothioate-modified AON for SMA, was approved by the FDA in December 2016 for intrathecal administration (three loading doses of 12 mg/kgL, fourth dose 30 days later, and every 3 months thereafter) (Chiriboga et al., 2017). New generation AONs are currently being developed to improve uptake and lower toxicity. For example, cellular uptake in heart, skeletal muscle, and the CNS is still inefficient following intravenous administration, and it can be anticipated that new generation SSOs with improved properties with respect to cellular uptake will be tested in future clinical trials. For SMA a new type of AON containing a morpholino backbone and a conjugated peptide showed improved uptake in vivo in a mouse model (Shabanpoor et al., 2017). Increasing amounts of reports now focus on improved AON therapeutic targeting strategies for previous and novel targets (d'Ydewalle et al., 2017; Osman et al., 2016).

5.4 Gene Therapy

5.4.1 U1/U7 snRNAs

Delivery of AONs into target tissues is challenging, and recent advances in viral-based gene therapy can overcome this limitation (Naldini, 2015). Naturally occurring snRNAs, which are normally involved in processing of pre-mRNAs, can be used for viral vector-based delivery of antisense sequences. U1 and U7 snRNAs have been used to modulate splicing (Imbert et al., 2017) (Fig. 6). The U7 snRNA is normally utilized in processing of histone pre-mRNAs. When the SM binding site is modified, it can function to target and block any desired sequence on a pre-mRNA of choice (Soldati and Schumperli, 1988 reviewed in Schumperli and Pillai, 2004). Natural U1 snRNA consists of four stem loops (sl) that bind the core spliceosomal components U1–70k (sl 1), U1-A (sl 2), U1-C (sl 3), and U2 snRNP (sl 4). It also contains an SM binding site for SM proteins. This assembly generates a functional U1 snRNP, which is involved in intron definition by binding to splice donor sites (Lee and Rio, 2015; Stark et al., 2001). Bases 3–10 of the U1 snRNA are involved in this binding and can be modified by insertion of up to 50 bases and then targeted to any desired sequence. For example, targeting to mutated (cryptic) splice donor sites can result in promotion of intron definition. It should be noted that intronic splice donor sites are often well conserved and therefore potential off-targets should be determined.

Evidence that the modified U1/U7 snRNA system is capable of targeting a heterologous pre-mRNA and to modulate its splicing was shown for beta thalassemia. The IVS2-705T>G variant in the pre-mRNA of *ß-globin* generates a cryptic splice site. Targeting this cryptic splice site with U1 or U7 snRNA in the pre-mRNA restored canonical splicing to up 56% and 65% of healthy individuals, respectively (Gorman et al., 1998, 2000). Many studies used the U7 snRNA system to modulate splicing in vitro. The targeting strategy is similar to that used with chemical stable AONs and the efficacy can be comparable. The U1 snRNA targeting implements a different strategy as it is semirestricted to the splice donor site. Proof of principle for a U1 snRNA-mediated therapy for SMA was reported by Rogalska et al., who demonstrated increased survival in a mouse model. Untreated mice died within 6 days, while insertion of a single germline copy of exon-specific U1 snRNA resulted in increased survival up to at least 250 days. Off-target effects were reduced by targeting a nonconserved intronic sequence upstream of the splice donor site, resulting in 12 genes with changed expression (Rogalska et al., 2016).

For in vivo targeting, the modified U1/U7 snRNA system can be inserted into different viruses including retroviruses, lentiviruses, adenoviruses, and recombinant adeno-associated viruses (rAAVs) (extensively reviewed in Imbert et al., 2017). rAAV is of particular interest because it can be targeted to a wide range of tissue types using different tropisms.

Furthermore, it is relative safe as it has low likelihood to integrate into the genome, and the transgene can be expressed over prolonged periods up to years. Disadvantages include antibody formation against the capsid (Kotterman and Schaffer, 2014), and, because rAAV is a noninsertional vector, decreased transgene expression over time due to cell division and cell loss. A long-term follow-up study of golden retriever muscular dystrophy dogs using rAAV-U7 snRNA for the skipping of exons 6 and 8, which was applied to skeletal muscle, demonstrated loss of rAAV copy number, and dystrophin positive fibers during a period of 5 years (Vulin et al., 2012). This may have been caused by turnover of transduced fibers, as the exon skipping strategy converted the DMD phenotype to a milder Becker muscular dystrophy (BMD) phenotype, which still has ongoing muscle wasting and regeneration.

5.4.2 Gene Editing

Permanent modulation of splicing has recently become possible with the development of gene-editing techniques, which use designer nucleases such as ZFNs, TALENs, or CRISPR/Cas9 (Fig. 6). These can generate double-strand breaks at predetermined locations in genomic DNA (Maggio and Goncalves, 2015). Repair of these breaks is performed by the host, either by using the nonhomologous end joining (NHEJ) pathway or by homology directed repair (HDR). The NHEJ pathway can be exploited to knockout genes (by disrupting the reading frame) or to generate deletions, while the HDR pathway requires a donor template with homology arms that can be used to repair or modify target genes. The NHEJ pathway is also active in nondividing cells, which allows gene editing of terminally differentiated cells (Lieber, 2010). The field is currently addressing the low efficiencies of HDR in nondividing cells, and a recent study demonstrated that transgene integration via homology-independent targeted integration (HITI) was possible via the NHEJ pathway (Suzuki et al., 2016).

Strategies that apply to splicing may include restoration of the reading frame by generation of a small nucleotide deletion or insertion, deletion of an exon to bypass variation hotspots and to restore the reading frame, or disruption of cryptic splice sites. To apply this strategy, designer nucleases have to be delivered to the target cells in vivo or ex vivo. In vitro studies demonstrated the potential of this approach by reframing dystrophin transcripts via removal of various exons to produce a shorter but functional protein, which is associated with a milder BMD phenotype (Iyombe-Engembe et al., 2016; Maggio et al., 2016; Young et al., 2016).

The first proof of principle study for in vivo targeting used either electroporation or adenoviral infection to deliver CRISPR/Cas9 in *mdx* mice. This successfully removed exon 23 in the dystrophin gene (Xu et al., 2016). Using the smaller SaCas9 (Ran et al., 2015) three studies demonstrated in vivo delivery of CRISPR/cas9 to muscle and heart tissue using systemic administration of rAAV8/9 (Long et al., 2016; Nelson et al., 2016; Tabebordbar et al., 2016). Muscle strength and force measurements showed improvements as compared with untreated *mdx* mice. Tabebordbar et al. also demonstrated that muscle stem cells were targeted, which is important for regeneration of myofibers with a reframed dystrophin protein (Tabebordbar et al., 2016). These studies presented a novel approach to modulate splicing using designer nucleases. Challenges for the future include the efficiency of in vivo gene editing, and the safety profile of such methods. In particular, undesired double-stranded DNA breaks generated by off-target effects would pose a threat for the development of tumorigenic cells. Ex vivo application of gene editing followed by cell engraftment is currently under clinical development for the treatment of HIV (Tebas et al., 2014), and has so far not reported serious adverse events.

6. FUTURE PERSPECTIVES

This review has highlighted the complexity of splicing and its crucial role in gene regulation in healthy and diseased individuals. Yet, our understanding of how sequence variants may affect splicing outcome and how this relates to human disease is at its infancy. Our ability to detect sequence variants at the DNA level have reached the previously unthinkable milestone of WGS at an affordable price and high speed. This has also exponentially increased the number of questions regarding pathogenicity of sequence variants. Currently, DNA NGS is mostly confined to gene panels or whole exomes, but in the near future it can be expected that WGS analysis will be increasingly applied, provided that the costs will further decrease. This will put an even larger pressure on our ability to assess which variants may be pathogenic. Improved in silico prediction programs would be important. Next-generation programs would be needed that take into account the different layers at which splicing is regulated, including expression and activity of splicing factors, presence, and strengths of cis acting elements, transcription rate, chromatin environment, activation of decay pathways, and other splicing mechanisms such as RS.

Functional assays to detect aberrant splicing will continue to be essential to validate prediction programs, but also to improve these. Single gene-based assays using PCR analysis of all exons are fast, simple, and feasible for a subset of genes with ubiquitous expression. In the future, it can be expected that PCR analysis will be replaced by NGS-based analysis. Upcoming is genome-wide analysis of mRNA transcripts using long-read RNAseq, which allow the analysis of isoforms without in silico reconstruction from short reads, and can even be analyzed on the single cell level (Karlsson and Linnarsson, 2017). In addition, CLIP-based technology to detect RNA–protein interactions will remain a crucial approach to understand splicing mechanisms, and optimization and diversifications of CLIP will expand our options for in-depth mechanistic studies.

Another challenge for the future is to link splicing outcome to disease. Is the variant dominant negative? Is there leaky wild-type splicing? How much normal product is needed to correct the disease? Why are certain organs/tissues particularly sensitive to variants in general splicing factors? Can we link splicing outcome to disease severity? The answers to these questions will be helpful to develop novel treatment options. While a number of options are under development and some have entered the clinic, therapeutic intervention with splicing still faces some important challenges. For AONs, cellular uptake and tolerability are key issues. The development of next-generation AONs with improved properties is promising and may play an important role in upcoming trials for a variety of disorders. Small molecule drugs that interfere with expression or activity of splicing factors face the challenge of specificity. Gene therapy is a promising, rapidly developing field, and potentially offers a long- or life-term treatment option. The challenges depend on the type of gene therapy. Safety with respect to genome integrity and immune responses to viral vectors, and long-term efficacy are important aspects that need to be addressed. Current AAV and LV vectors appear effective and safe in a number of clinical trials. The excellent progress that has been made in the past few years in splicing research is expected to result in improved DNA and RNA-based diagnosis and methods to predict or determine splicing outcome, a better understanding of splicing mechanisms, and clinical development of more treatment options for human diseases based on splicing correction.

ACKNOWLEDGMENTS

We thank Dr. Lies Hoefsloot for discussion. We thank Katherine MacKenzie and Pablo Herrero Hernández for critical reading of the manuscript. The authors are supported by grants from Sophia Children's Hospital Foundation (SSWO, grant S17-32), Stichting Metakids, The Prinses Beatrix Spierfonds/Stichting Spieren voor Spieren (grant W.OR13-21), Tex Net, the WE Foundation, ZonMW—the Netherlands Organization for Health Research and Development (project number 152001005), and by the Erasmus MC.

REFERENCES

Aartsma-Rus, A., 2010. Antisense-mediated modulation of splicing: therapeutic implications for Duchenne muscular dystrophy. RNA Biol. 7 (4), 453–461.

Adelman, K., Lis, J.T., 2012. Promoter-proximal pausing of RNA polymerase II: emerging roles in metazoans. Nat. Rev. Genet. 13 (10), 720–731.

Akdim, F., Stroes, E.S., Sijbrands, E.J., Tribble, D.L., Trip, M.D., Jukema, J.W., Flaim, J.D., Su, J., Yu, R., Baker, B.F., Wedel, M.K., Kastelein, J.J., 2010. Efficacy and safety of mipomersen, an antisense inhibitor of apolipoprotein B, in hypercholesterolemic subjects receiving stable statin therapy. J. Am. Coll. Cardiol. 55 (15), 1611–1618.

Akdim, F., Tribble, D.L., Flaim, J.D., Yu, R., Su, J., Geary, R.S., Baker, B.F., Fuhr, R., Wedel, M.K., Kastelein, J.J., 2011. Efficacy of apolipoprotein B synthesis inhibition in subjects with mild-to-moderate hyperlipidaemia. Eur. Heart J. 32 (21), 2650–2659.

Allo, M., Buggiano, V., Fededa, J.P., Petrillo, E., Schor, I., de la Mata, M., Agirre, E., Plass, M., Eyras, E., Elela, S.A., Klinck, R., Chabot, B., Kornblihtt, A.R., 2009. Control of alternative splicing through siRNA-mediated transcriptional gene silencing. Nat. Struct. Mol. Biol. 16 (7), 717–724.

Amin, N., Allebrandt, K.V., van der Spek, A., Muller-Myhsok, B., Hek, K., Teder-Laving, M., Hayward, C., Esko, T., van Mill, J.G., Mbarek, H., Watson, N.F., Melville, S.A., Del Greco, F.M., Byrne, E.M., Oole, E., Kolcic, I., Chen, T.H., Evans, D.S., Coresh, J., Vogelzangs, N., Karjalainen, J., Willemsen, G., Gharib, S.A., Zgaga, L., Mihailov, E., Stone, K.L., Campbell, H., Brouwer, R.W., Demirkan, A., Isaacs, A., Dogas, Z., Marciante, K.D., Campbell, S., Borovecki, F., Luik, A.I., Li, M., Hottenga, J.J., Huffman, J.E., van den Hout, M.C., Cummings, S.R., Aulchenko, Y.S., Gehrman, P.R., Uitterlinden, A.G., Wichmann, H.E., Muller-Nurasyid, M., Fehrmann, R.S., Montgomery, G.W., Hofman, A., Kao, W.H., Oostra, B.A., Wright, A.F., Vink, J.M., Wilson, J.F., Pramstaller, P.P., Hicks, A.A., Polasek, O., Punjabi, N.M., Redline, S., Psaty, B.M., Heath, A.C., Merrow, M., Tranah, G.J., Gottlieb, D.J., Boomsma, D.I., Martin, N.G., Rudan, I., Tiemeier, H., van, I.W.F., Penninx, B.W., Metspalu, A., Meitinger, T., Franke, L., Roenneberg, T., van Duijn, C.M., 2016. Genetic variants in RBFOX3 are associated with sleep latency. Eur. J. Hum. Genet. 24 (10), 1488–1495.

Anantha, R.W., Alcivar, A.L., Ma, J., Cai, H., Simhadri, S., Ule, J., Konig, J., Xia, B., 2013. Requirement of heterogeneous nuclear ribonucleoprotein C for BRCA gene expression and homologous recombination. PLoS One 8 (4), e61368.

Anderson, E.S., Lin, C.H., Xiao, X., Stoilov, P., Burge, C.B., Black, D.L., 2012. The cardiotonic steroid digitoxin regulates alternative splicing through depletion of the splicing factors SRSF3 and TRA2B. RNA 18 (5), 1041–1049.

Andersson, R., Enroth, S., Rada-Iglesias, A., Wadelius, C., Komorowski, J., 2009. Nucleosomes are well positioned in exons and carry characteristic histone modifications. Genome Res. 19 (10), 1732–1741.

Andreassi, C., Jarecki, J., Zhou, J., Coovert, D.D., Monani, U.R., Chen, X., Whitney, M., Pollok, B., Zhang, M., Androphy, E., Burghes, A.H., 2001. Aclarubicin treatment restores SMN levels to cells derived from type I spinal muscular atrophy patients. Hum. Mol. Genet. 10 (24), 2841–2849.

Argente, J., Flores, R., Gutierrez-Arumi, A., Verma, B., Martos-Moreno, G.A., Cusco, I., Oghabian, A., Chowen, J.A., Frilander, M.J., Perez-Jurado, L.A., 2014. Defective minor spliceosome mRNA processing results in isolated familial growth hormone deficiency. EMBO Mol. Med. 6 (3), 299–306.

Avale, M.E., Rodriguez-Martin, T., Gallo, J.M., 2013. Trans-splicing correction of tau isoform imbalance in a mouse model of tau mis-splicing. Hum. Mol. Genet. 22 (13), 2603–2611.

Ballester, L.Y., Luthra, R., Kanagal-Shamanna, R., Singh, R.R., 2016. Advances in clinical next-generation sequencing: target enrichment and sequencing technologies. Expert Rev. Mol. Diagn. 16 (3), 357–372.

Balthazar, P., Tucunduva, D.C., Luce, M.J., Scott, M.S., 2017. Widespread pre-translational regulation of the inclusion of signal peptides in human proteins. Genomics 109, 113–122.

Barash, Y., Vaquero-Garcia, J., Gonzalez-Vallinas, J., Xiong, H.Y., Gao, W., Lee, L.J., Frey, B.J., 2013. AVISPA: a web tool for the prediction and analysis of alternative splicing. Genome Biol. 14 (10), R114.

Barbosa-Morais, N.L., Irimia, M., Pan, Q., Xiong, H.Y., Gueroussov, S., Lee, L.J., Slobodeniuc, V., Kutter, C., Watt, S., Colak, R., Kim, T., Misquitta-Ali, C.M., Wilson, M.D., Kim, P.M., Odom, D.T., Frey, B.J., Blencowe, B.J., 2012. The evolutionary landscape of alternative splicing in vertebrate species. Science 338 (6114), 1587–1593.

Bates, D.O., Morris, J.C., Oltean, S., Donaldson, L.F., 2017. Pharmacology of modulators of alternative splicing. Pharmacol. Rev. 69 (1), 63–79.

Beaulieu, C.L., Majewski, J., Schwartzentruber, J., Samuels, M.E., Fernandez, B.A., Bernier, F.P., Brudno, M., Knoppers, B., Marcadier, J., Dyment, D., Adam, S., Bulman, D.E., Jones, S.J., Avard, D., Nguyen, M.T., Rousseau, F., Marshall, C., Wintle, R.F., Shen, Y., Scherer, S.W., Consortium, F.C., Friedman, J.M., Michaud, J.L., Boycott, K.M., 2014. FORGE Canada consortium: outcomes of a 2-year national rare-disease gene-discovery project. Am. J. Hum. Genet. 94 (6), 809–817.

Bechara, E.G., Sebestyen, E., Bernardis, I., Eyras, E., Valcarcel, J., 2013. RBM5, 6, and 10 differentially regulate NUMB alternative splicing to control cancer cell proliferation. Mol. Cell 52 (5), 720–733.

Bengtson, M.H., Joazeiro, C.A., 2010. Role of a ribosome-associated E3 ubiquitin ligase in protein quality control. Nature 467 (7314), 470–473.

Berger, A., Maire, S., Gaillard, M.C., Sahel, J.A., Hantraye, P., Bemelmans, A.P., 2016. mRNA trans-splicing in gene therapy for genetic diseases. Wiley Interdiscip. Rev. RNA 7 (4), 487–498.

Bergsma, A.J., Kroos, M., Hoogeveen-Westerveld, M., Halley, D., van der Ploeg, A.T., Pijnappel, W.W., 2015. Identification and characterization of aberrant GAA pre-mRNA splicing in Pompe disease using a generic approach. Hum. Mutat. 36 (1), 57–68.

Bergsma, A.J., In 't Groen, S.L., Verheijen, F.W., van der Ploeg, A.T., Pijnappel, W.P., 2016. From cryptic toward canonical pre-mRNA splicing in Pompe disease: a pipeline for the development of antisense oligonucleotides. Mol. Ther. Nucleic Acids 5, e361.

Bhalla, K., Phillips, H.A., Crawford, J., McKenzie, O.L., Mulley, J.C., Eyre, H., Gardner, A.E., Kremmidiotis, G., Callen, D.F., 2004. The de novo chromosome 16 translocations of two patients with abnormal phenotypes (mental retardation and epilepsy) disrupt the A2BP1 gene. J. Hum. Genet. 49 (6), 308–311.

Blanchette, M., Chabot, B., 1999. Modulation of exon skipping by high-affinity hnRNP A1-binding sites and by intron elements that repress splice site utilization. EMBO J. 18 (7), 1939–1952.

Boisguerin, P., O'Donovan, L., Gait, M.J., Lebleu, B., 2015. In vitro assays to assess exon skipping in Duchenne muscular dystrophy. Methods Mol. Biol. 1324, 317–329.

Bonnet, C., Krieger, S., Vezain, M., Rousselin, A., Tournier, I., Martins, A., Berthet, P., Chevrier, A., Dugast, C., Layet, V., Rossi, A., Lidereau, R., Frebourg, T.,

Hardouin, A., Tosi, M., 2008. Screening BRCA1 and BRCA2 unclassified variants for splicing mutations using reverse transcription PCR on patient RNA and an ex vivo assay based on a splicing reporter minigene. J. Med. Genet. 45 (7), 438–446.
Bouattour, R.K., Maalej, S.B., Zouari-Bradai, E., Mnif, M., Abid, M., Kacem, H.H., 2017. Intronic variants of SLC26A4 gene enhance splicing efficiency in hybrid minigene assay. Gene 620, 10–14.
Bradnam, K.R., Korf, I., 2008. Longer first introns are a general property of eukaryotic gene structure. PLoS One 3 (8), e3093.
Brauch, K.M., Karst, M.L., Herron, K.J., de Andrade, M., Pellikka, P.A., Rodeheffer, R.J., Michels, V.V., Olson, T.M., 2009. Mutations in ribonucleic acid binding protein gene cause familial dilated cardiomyopathy. J. Am. Coll. Cardiol. 54 (10), 930–941.
Brendel, V., Kleffe, J., 1998. Prediction of locally optimal splice sites in plant pre-mRNA with applications to gene identification in Arabidopsis thaliana genomic DNA. Nucleic Acids Res. 26 (20), 4748–4757.
Burge, C., Karlin, S., 1997. Prediction of complete gene structures in human genomic DNA. J. Mol. Biol. 268 (1), 78–94.
Byron, S.A., Van Keuren-Jensen, K.R., Engelthaler, D.M., Carpten, J.D., Craig, D.W., 2016. Translating RNA sequencing into clinical diagnostics: opportunities and challenges. Nat. Rev. Genet. 17 (5), 257–271.
Cao, L., Han, G., Lin, C., Gu, B., Gao, X., Moulton, H.M., Seow, Y., Yin, H., 2016. Fructose promotes uptake and activity of oligonucleotides with different chemistries in a context-dependent manner in mdx mice. Mol. Ther. Nucleic Acids 5, e329.
Cardani, R., Giagnacovo, M., Rossi, G., Renna, L.V., Bugiardini, E., Pizzamiglio, C., Botta, A., Meola, G., 2014. Progression of muscle histopathology but not of spliceopathy in myotonic dystrophy type 2. Neuromuscul. Disord. 24 (12), 1042–1053.
Carpenter, S., Ricci, E.P., Mercier, B.C., Moore, M.J., Fitzgerald, K.A., 2014. Post-transcriptional regulation of gene expression in innate immunity. Nat. Rev. Immunol. 14 (6), 361–376.
Carrigan, M., Duignan, E., Malone, C.P., Stephenson, K., Saad, T., McDermott, C., Green, A., Keegan, D., Humphries, P., Kenna, P.F., Farrar, G.J., 2016. Panel-based population next-generation sequencing for inherited retinal degenerations. Sci. Rep. 6, 33248.
Cartegni, L., Wang, J., Zhu, Z., Zhang, M.Q., Krainer, A.R., 2003. ESEfinder: a web resource to identify exonic splicing enhancers. Nucleic Acids Res. 31 (13), 3568–3571.
Cartegni, L., Hastings, M.L., Calarco, J.A., de Stanchina, E., Krainer, A.R., 2006. Determinants of exon 7 splicing in the spinal muscular atrophy genes, SMN1 and SMN2. Am. J. Hum. Genet. 78 (1), 63–77.
Chang, T.H., Huang, H.Y., Hsu, J.B., Weng, S.L., Horng, J.T., Huang, H.D., 2013. An enhanced computational platform for investigating the roles of regulatory RNA and for identifying functional RNA motifs. BMC Bioinf. 14 (Suppl. 2), S4.
Charton, K., Suel, L., Henriques, S.F., Moussu, J.P., Bovolenta, M., Taillepierre, M., Becker, C., Lipson, K., Richard, I., 2016. Exploiting the CRISPR/Cas9 system to study alternative splicing in vivo: application to titin. Hum. Mol. Genet. 25 (20), 4518–4532.
Chiriboga, A.S., Konno, T., van Gerpen, J.A., 2017. Personality changes, executive dysfunction, and motor and memory impairment. JAMA Neurol. 74 (2), 245–246.
Choi, Y.D., Grabowski, P.J., Sharp, P.A., Dreyfuss, G., 1986. Heterogeneous nuclear ribonucleoproteins: role in RNA splicing. Science 231 (4745), 1534–1539.
Chong, J.X., Buckingham, K.J., Jhangiani, S.N., Boehm, C., Sobreira, N., Smith, J.D., Harrell, T.M., McMillin, M.J., Wiszniewski, W., Gambin, T., Coban Akdemir, Z.H., Doheny, K., Scott, A.F., Avramopoulos, D., Chakravarti, A., Hoover-Fong, J., Mathews, D., Witmer, P.D., Ling, H., Hetrick, K., Watkins, L., Patterson, K.E.,

Reinier, F., Blue, E., Muzny, D., Kircher, M., Bilguvar, K., Lopez-Giraldez, F., Sutton, V.R., Tabor, H.K., Leal, S.M., Gunel, M., Mane, S., Gibbs, R.A., Boerwinkle, E., Hamosh, A., Shendure, J., Lupski, J.R., Lifton, R.P., Valle, D., Nickerson, D.A., Centers for Mendelian, G., Bamshad, M.J., 2015. The genetic basis of Mendelian phenotypes: discoveries, challenges, and opportunities. Am. J. Hum. Genet. 97 (2), 199–215.

Clayton, N.P., Nelson, C.A., Weeden, T., Taylor, K.M., Moreland, R.J., Scheule, R.K., Phillips, L., Leger, A.J., Cheng, S.H., Wentworth, B.M., 2014. Antisense oligonucleotide-mediated suppression of muscle glycogen synthase 1 synthesis as an approach for substrate reduction therapy of Pompe disease. Mol. Ther. Nucleic Acids 3, e206.

Consugar, M.B., Navarro-Gomez, D., Place, E.M., Bujakowska, K.M., Sousa, M.E., Fonseca-Kelly, Z.D., Taub, D.G., Janessian, M., Wang, D.Y., Au, E.D., Sims, K.B., Sweetser, D.A., Fulton, A.B., Liu, Q., Wiggs, J.L., Gai, X., Pierce, E.A., 2015. Panel-based genetic diagnostic testing for inherited eye diseases is highly accurate and reproducible, and more sensitive for variant detection, than exome sequencing. Genet. Med. 17 (4), 253–261.

Daxinger, L., Tapscott, S.J., van der Maarel, S.M., 2015. Genetic and epigenetic contributors to FSHD. Curr. Opin. Genet. Dev. 33, 56–61.

de Fougerolles, A., Vornlocher, H.P., Maraganore, J., Lieberman, J., 2007. Interfering with disease: a progress report on siRNA-based therapeutics. Nat. Rev. Drug Discov. 6 (6), 443–453.

Desmet, F.O., Hamroun, D., Lalande, M., Collod-Beroud, G., Claustres, M., Beroud, C., 2009. Human splicing finder: an online bioinformatics tool to predict splicing signals. Nucleic Acids Res. 37 (9), e67.

Di Giacomo, D., Gaildrat, P., Abuli, A., Abdat, J., Frebourg, T., Tosi, M., Martins, A., 2013. Functional analysis of a large set of BRCA2 exon 7 variants highlights the predictive value of hexamer scores in detecting alterations of exonic splicing regulatory elements. Hum. Mutat. 34 (11), 1547–1557.

Divina, P., Kvitkovicova, A., Buratti, E., Vorechovsky, I., 2009. Ab initio prediction of mutation-induced cryptic splice-site activation and exon skipping. Eur. J. Hum. Genet. 17 (6), 759–765.

Dogan, R.I., Getoor, L., Wilbur, W.J., Mount, S.M., 2007. SplicePort—an interactive splice-site analysis tool. Nucleic Acids Res. 35 (Web server issue), W285–91.

Doma, M.K., Parker, R., 2006. Endonucleolytic cleavage of eukaryotic mRNAs with stalls in translation elongation. Nature 440 (7083), 561–564.

Dominski, Z., Kole, R., 1993. Restoration of correct splicing in thalassemic pre-mRNA by antisense oligonucleotides. Proc. Natl. Acad. Sci. U.S.A. 90 (18), 8673–8677.

Dreyfuss, G., Matunis, M.J., Pinol-Roma, S., Burd, C.G., 1993. hnRNP proteins and the biogenesis of mRNA. Annu. Rev. Biochem. 62, 289–321.

Du, C., Shen, Z., Zang, R., Xie, H., Li, H., Chen, P., Hang, B., Xu, X., Tang, W., Xia, Y., 2016. Negative feedback circuitry between MIR143HG and RBM24 in Hirschsprung disease. Biochim. Biophys. Acta 1862 (11), 2127–2136.

Duell, P.B., Santos, R.D., Kirwan, B.A., Witztum, J.L., Tsimikas, S., Kastelein, J.J., 2016. Long-term mipomersen treatment is associated with a reduction in cardiovascular events in patients with familial hypercholesterolemia. J. Clin. Lipidol. 10 (4), 1011–1021.

Duff, M.O., Olson, S., Wei, X., Garrett, S.C., Osman, A., Bolisetty, M., Plocik, A., Celniker, S.E., Graveley, B.R., 2015. Genome-wide identification of zero nucleotide recursive splicing in Drosophila. Nature 521 (7552), 376–379.

d'Ydewalle, C., Ramos, D.M., Pyles, N.J., Ng, S.Y., Gorz, M., Pilato, C.M., Ling, K., Kong, L., Ward, A.J., Rubin, L.L., Rigo, F., Bennett, C.F., Sumner, C.J., 2017. The antisense transcript SMN-AS1 regulates SMN expression and is a novel therapeutic target for spinal muscular atrophy. Neuron 93 (1), 66–79.

Edery, P., Marcaillou, C., Sahbatou, M., Labalme, A., Chastang, J., Touraine, R., Tubacher, E., Senni, F., Bober, M.B., Nampoothiri, S., Jouk, P.S., Steichen, E., Berland, S., Toutain, A., Wise, C.A., Sanlaville, D., Rousseau, F., Clerget-Darpoux, F., Leutenegger, A.L., 2011. Association of TALS developmental disorder with defect in minor splicing component U4atac snRNA. Science 332 (6026), 240–243.

Elbashir, S.M., Lendeckel, W., Tuschl, T., 2001. RNA interference is mediated by 21- and 22-nucleotide RNAs. Genes Dev. 15 (2), 188–200.

Ellis, J.D., Barrios-Rodiles, M., Colak, R., Irimia, M., Kim, T., Calarco, J.A., Wang, X., Pan, Q., O'Hanlon, D., Kim, P.M., Wrana, J.L., Blencowe, B.J., 2012. Tissue-specific alternative splicing remodels protein–protein interaction networks. Mol. Cell 46 (6), 884–892.

Elsaid, M.F., Chalhoub, N., Ben-Omran, T., Kumar, P., Kamel, H., Ibrahim, K., Mohamoud, Y., Al-Dous, E., Al-Azwani, I., Malek, J.A., Suhre, K., Ross, M.E., Aleem, A.A., 2017. Mutation in noncoding RNA RNU12 causes early onset cerebellar ataxia. Ann. Neurol. 81 (1), 68–78.

Engstrom, P.G., Steijger, T., Sipos, B., Grant, G.R., Kahles, A., Ratsch, G., Goldman, N., Hubbard, T.J., Harrow, J., Guigo, R., Bertone, P., Consortium, R., 2013. Systematic evaluation of spliced alignment programs for RNA-seq data. Nat. Methods 10 (12), 1185–1191.

Faber, K., Glatting, K.H., Mueller, P.J., Risch, A., Hotz-Wagenblatt, A., 2011. Genome-wide prediction of splice-modifying SNPs in human genes using a new analysis pipeline called AASsites. BMC Bioinf. 12 (Suppl. 4), S2.

Fairbrother, W.G., Yeh, R.F., Sharp, P.A., Burge, C.B., 2002. Predictive identification of exonic splicing enhancers in human genes. Science 297 (5583), 1007–1013.

Farooqi, A.A., Rehman, Z.U., Muntane, J., 2014. Antisense therapeutics in oncology: current status. Onco Targets Ther. 7, 2035–2042.

Fu, X.D., Mayeda, A., Maniatis, T., Krainer, A.R., 1992. General splicing factors SF2 and SC35 have equivalent activities in vitro, and both affect alternative 5′ and 3′ splice site selection. Proc. Natl. Acad. Sci. U.S.A. 89 (23), 11224–11228.

Gabut, M., Mine, M., Marsac, C., Brivet, M., Tazi, J., Soret, J., 2005. The SR protein SC35 is responsible for aberrant splicing of the E1alpha pyruvate dehydrogenase mRNA in a case of mental retardation with lactic acidosis. Mol. Cell. Biol. 25 (8), 3286–3294.

Gao, Y., Tatavarty, V., Korza, G., Levin, M.K., Carson, J.H., 2008. Multiplexed dendritic targeting of alpha calcium calmodulin-dependent protein kinase II, neurogranin, and activity-regulated cytoskeleton-associated protein RNAs by the A2 pathway. Mol. Biol. Cell 19 (5), 2311–2327.

Gao, C., Ren, S., Lee, J.H., Qiu, J., Chapski, D.J., Rau, C.D., Zhou, Y., Abdellatif, M., Nakano, A., Vondriska, T.M., Xiao, X., Fu, X.D., Chen, J.N., Wang, Y., 2016. RBFox1-mediated RNA splicing regulates cardiac hypertrophy and heart failure. J. Clin. Invest. 126 (1), 195–206.

Garanto, A., Chung, D.C., Duijkers, L., Corral-Serrano, J.C., Messchaert, M., Xiao, R., Bennett, J., Vandenberghe, L.H., Collin, R.W., 2016. In vitro and in vivo rescue of aberrant splicing in CEP290-associated LCA by antisense oligonucleotide delivery. Hum. Mol. Genet. 25 (12), 2552–2563.

Gazzoli, I., Pulyakhina, I., Verwey, N.E., Ariyurek, Y., Laros, J.F., t Hoen, P.A., Aartsma-Rus, A., 2016. Non-sequential and multi-step splicing of the dystrophin transcript. RNA Biol. 13 (3), 290–305.

Geary, R.S., Norris, D., Yu, R., Bennett, C.F., 2015. Pharmacokinetics, biodistribution and cell uptake of antisense oligonucleotides. Adv. Drug Deliv. Rev. 87, 46–51.

Gehman, L.T., Meera, P., Stoilov, P., Shiue, L., O'Brien, J.E., Meisler, M.H., Ares Jr., M., Otis, T.S., Black, D.L., 2012. The splicing regulator Rbfox2 is required for both cerebellar development and mature motor function. Genes Dev. 26 (5), 445–460.

Gelfman, S., Ast, G., 2013. When epigenetics meets alternative splicing: the roles of DNA methylation and GC architecture. Epigenomics 5 (4), 351–353.

Gerstberger, S., Hafner, M., Tuschl, T., 2014. A census of human RNA-binding proteins. Nat. Rev. Genet. 15 (12), 829–845.

Geuens, T., Bouhy, D., Timmerman, V., 2016. The hnRNP family: insights into their role in health and disease. Hum. Genet. 135 (8), 851–867.

Giorgi, G., Casarin, A., Trevisson, E., Dona, M., Cassina, M., Graziano, C., Picci, L., Clementi, M., Salviati, L., 2015. Validation of CFTR intronic variants identified during cystic fibrosis population screening by a minigene splicing assay. Clin. Chem. Lab. Med. 53 (11), 1719–1723.

Giudice, J., Xia, Z., Wang, E.T., Scavuzzo, M.A., Ward, A.J., Kalsotra, A., Wang, W., Wehrens, X.H., Burge, C.B., Li, W., Cooper, T.A., 2014. Alternative splicing regulates vesicular trafficking genes in cardiomyocytes during postnatal heart development. Nat. Commun. 5, 3603.

Giudice, J., Loehr, J.A., Rodney, G.G., Cooper, T.A., 2016. Alternative splicing of four trafficking genes regulates myofiber structure and skeletal muscle physiology. Cell Rep. 17 (8), 1923–1933.

Gokben, S., Onay, H., Yilmaz, S., Atik, T., Serdaroglu, G., Tekin, H., Ozkinay, F., 2016. Targeted next generation sequencing: the diagnostic value in early-onset epileptic encephalopathy. Acta Neurol. Belg. 117, 131–138.

Gonzalez-Paredes, F.J., Ramos-Trujillo, E., Claverie-Martin, F., 2016. Three exonic mutations in polycystic kidney disease-2 gene (PKD2) alter splicing of its pre-mRNA in a minigene system. Gene 578 (1), 117–123.

Goodwin, S., McPherson, J.D., McCombie, W.R., 2016. Coming of age: ten years of next-generation sequencing technologies. Nat. Rev. Genet. 17 (6), 333–351.

Gorman, L., Suter, D., Emerick, V., Schumperli, D., Kole, R., 1998. Stable alteration of pre-mRNA splicing patterns by modified U7 small nuclear RNAs. Proc. Natl. Acad. Sci. U.S.A. 95 (9), 4929–4934.

Gorman, L., Mercatante, D.R., Kole, R., 2000. Restoration of correct splicing of thalassemic beta-globin pre-mRNA by modified U1 snRNAs. J. Biol. Chem. 275 (46), 35914–35919.

Goyenvalle, A., Griffith, G., Babbs, A., El Andaloussi, S., Ezzat, K., Avril, A., Dugovic, B., Chaussenot, R., Ferry, A., Voit, T., Amthor, H., Buhr, C., Schurch, S., Wood, M.J., Davies, K.E., Vaillend, C., Leumann, C., Garcia, L., 2015. Functional correction in mouse models of muscular dystrophy using exon-skipping tricyclo-DNA oligomers. Nat. Med. 21 (3), 270–275.

Guo, W., Schafer, S., Greaser, M.L., Radke, M.H., Liss, M., Govindarajan, T., Maatz, H., Schulz, H., Li, S., Parrish, A.M., Dauksaite, V., Vakeel, P., Klaassen, S., Gerull, B., Thierfelder, L., Regitz-Zagrosek, V., Hacker, T.A., Saupe, K.W., Dec, G.W., Ellinor, P.T., MacRae, C.A., Spallek, B., Fischer, R., Perrot, A., Ozcelik, C., Saar, K., Hubner, N., Gotthardt, M., 2012. RBM20, a gene for hereditary cardiomyopathy, regulates titin splicing. Nat. Med. 18 (5), 766–773.

Hafner, M., Landthaler, M., Burger, L., Khorshid, M., Hausser, J., Berninger, P., Rothballer, A., Ascano Jr., M., Jungkamp, A.C., Munschauer, M., Ulrich, A., Wardle, G.S., Dewell, S., Zavolan, M., Tuschl, T., 2010. Transcriptome-wide identification of RNA-binding protein and microRNA target sites by PAR-CLIP. Cell 141 (1), 129–141.

Hamada, N., Ito, H., Nishijo, T., Iwamoto, I., Morishita, R., Tabata, H., Momiyama, T., Nagata, K., 2016. Essential role of the nuclear isoform of RBFOX1, a candidate gene for autism spectrum disorders, in the brain development. Sci. Rep. 6, 30805.

Han, H., Irimia, M., Ross, P.J., Sung, H.K., Alipanahi, B., David, L., Golipour, A., Gabut, M., Michael, I.P., Nachman, E.N., Wang, E., Trcka, D., Thompson, T., O'Hanlon, D., Slobodeniuc, V., Barbosa-Morais, N.L., Burge, C.B., Moffat, J.,

Frey, B.J., Nagy, A., Ellis, J., Wrana, J.L., Blencowe, B.J., 2013. MBNL proteins repress ES-cell-specific alternative splicing and reprogramming. Nature 498 (7453), 241–245.
Han, G., Gu, B., Cao, L., Gao, X., Wang, Q., Seow, Y., Zhang, N., Wood, M.J., Yin, H., 2016. Hexose enhances oligonucleotide delivery and exon skipping in dystrophin-deficient mdx mice. Nat. Commun. 7, 10981.
Hatton, A.R., Subramaniam, V., Lopez, A.J., 1998. Generation of alternative Ultrabithorax isoforms and stepwise removal of a large intron by resplicing at exon–exon junctions. Mol. Cell 2 (6), 787–796.
He, X., Liu, F., Yan, J., Zhang, Y., Yan, J., Shang, H., Dou, Q., Zhao, Q., Song, Y., 2015. Trans-splicing repair of mutant p53 suppresses the growth of hepatocellular carcinoma cells in vitro and in vivo. Sci. Rep. 5, 8705.
Hebsgaard, S.M., Korning, P.G., Tolstrup, N., Engelbrecht, J., Rouze, P., Brunak, S., 1996. Splice site prediction in Arabidopsis thaliana pre-mRNA by combining local and global sequence information. Nucleic Acids Res. 24 (17), 3439–3452.
Hinze, F., Dieterich, C., Radke, M.H., Granzier, H., Gotthardt, M., 2016. Reducing RBM20 activity improves diastolic dysfunction and cardiac atrophy. J. Mol. Med. (Berl.) 94 (12), 1349–1358.
Ho, T.H., Kapur, P., Joseph, R.W., Serie, D.J., Eckel-Passow, J.E., Tong, P., Wang, J., Castle, E.P., Stanton, M.L., Cheville, J.C., Jonasch, E., Brugarolas, J., Parker, A.S., 2016. Loss of histone H3 lysine 36 trimethylation is associated with an increased risk of renal cell carcinoma-specific death. Mod. Pathol. 29 (1), 34–42.
Hollander, D., Naftelberg, S., Lev-Maor, G., Kornblihtt, A.R., Ast, G., 2016. How are short exons flanked by long introns defined and committed to splicing? Trends Genet. 32 (10), 596–606.
Hong, D.S., Kurzrock, R., Naing, A., Wheler, J.J., Falchook, G.S., Schiffman, J.S., Faulkner, N., Pilat, M.J., O'Brien, J., LoRusso, P., 2014. A phase I, open-label, single-arm, dose-escalation study of E7107, a precursor messenger ribonucleic acid (pre-mRNA) splicesome inhibitor administered intravenously on days 1 and 8 every 21 days to patients with solid tumors. Invest. New Drugs 32 (3), 436–444.
Hoskins, A.A., Moore, M.J., 2012. The spliceosome: a flexible, reversible macromolecular machine. Trends Biochem. Sci. 37 (5), 179–188.
Howard, J.M., Sanford, J.R., 2015. The RNAissance family: SR proteins as multifaceted regulators of gene expression. Wiley Interdiscip. Rev. RNA 6 (1), 93–110.
Hua, W.F., Zhong, Q., Xia, T.L., Chen, Q., Zhang, M.Y., Zhou, A.J., Tu, Z.W., Qu, C., Li, M.Z., Xia, Y.F., Wang, H.Y., Xie, D., Claret, F.X., Song, E.W., Zeng, M.S., 2016. RBM24 suppresses cancer progression by upregulating miR-25 to target MALAT1 in nasopharyngeal carcinoma. Cell Death Dis. 7 (9), e2352.
Igreja, S., Clarke, L.A., Botelho, H.M., Marques, L., Amaral, M.D., 2016. Correction of a cystic fibrosis splicing mutation by antisense oligonucleotides. Hum. Mutat. 37 (2), 209–215.
Imbert, M., Dias-Florencio, G., Goyenvalle, A., 2017. Viral vector-mediated antisense therapy for genetic diseases. Genes (Basel) 8 (2), 51.
Irimia, M., Weatheritt, R.J., Ellis, J.D., Parikshak, N.N., Gonatopoulos-Pournatzis, T., Babor, M., Quesnel-Vallieres, M., Tapial, J., Raj, B., O'Hanlon, D., Barrios-Rodiles, M., Sternberg, M.J., Cordes, S.P., Roth, F.P., Wrana, J.L., Geschwind, D.H., Blencowe, B.J., 2014. A highly conserved program of neuronal microexons is misregulated in autistic brains. Cell 159 (7), 1511–1523.
Iyombe-Engembe, J.P., Ouellet, D.L., Barbeau, X., Rousseau, J., Chapdelaine, P., Lague, P., Tremblay, J.P., 2016. Efficient restoration of the dystrophin gene reading frame and protein structure in DMD myoblasts using the CinDel method. Mol. Ther. Nucleic Acids 5, e283.
Izumikawa, K., Yoshikawa, H., Ishikawa, H., Nobe, Y., Yamauchi, Y., Philipsen, S., Simpson, R.J., Isobe, T., Takahashi, N., 2016. Chtop (Chromatin target of Prmt1)

auto-regulates its expression level via intron retention and nonsense-mediated decay of its own mRNA. Nucleic Acids Res. 44 (20), 9847–9859.
Jackson, A.L., Linsley, P.S., 2010. Recognizing and avoiding siRNA off-target effects for target identification and therapeutic application. Nat. Rev. Drug Discov. 9 (1), 57–67.
Jangi, M., Sharp, P.A., 2014. Building robust transcriptomes with master splicing factors. Cell 159 (3), 487–498.
Jangi, M., Boutz, P.L., Paul, P., Sharp, P.A., 2014. Rbfox2 controls autoregulation in RNA-binding protein networks. Genes Dev. 28 (6), 637–651.
Jian, X., Boerwinkle, E., Liu, X., 2014. In silico tools for splicing defect prediction: a survey from the viewpoint of end users. Genet. Med. 16 (7), 497–503.
Jiang, Y., Zhang, M., Qian, Y., Xu, E., Zhang, J., Chen, X., 2014. Rbm24, an RNA-binding protein and a target of p53, regulates p21 expression via mRNA stability. J. Biol. Chem. 289 (6), 3164–3175.
Jin, Y., Suzuki, H., Maegawa, S., Endo, H., Sugano, S., Hashimoto, K., Yasuda, K., Inoue, K., 2003. A vertebrate RNA-binding protein fox-1 regulates tissue-specific splicing via the pentanucleotide GCAUG. EMBO J. 22 (4), 905–912.
Jin, D., Hidaka, K., Shirai, M., Morisaki, T., 2010. RNA-binding motif protein 24 regulates myogenin expression and promotes myogenic differentiation. Genes Cells 15 (11), 1158–1167.
Jonkers, I., Kwak, H., Lis, J.T., 2014. Genome-wide dynamics of pol II elongation and its interplay with promoter proximal pausing, chromatin, and exons. eLife 3, e02407.
Kaida, D., Motoyoshi, H., Tashiro, E., Nojima, T., Hagiwara, M., Ishigami, K., Watanabe, H., Kitahara, T., Yoshida, T., Nakajima, H., Tani, T., Horinouchi, S., Yoshida, M., 2007. Spliceostatin A targets SF3b and inhibits both splicing and nuclear retention of pre-mRNA. Nat. Chem. Biol. 3 (9), 576–583.
Karlsson, K., Linnarsson, S., 2017. Single-cell mRNA isoform diversity in the mouse brain. BMC Genomics 18 (1), 126.
Kashima, T., Rao, N., David, C.J., Manley, J.L., 2007. hnRNP A1 functions with specificity in repression of SMN2 exon 7 splicing. Hum. Mol. Genet. 16 (24), 3149–3159.
Kastelein, J.J., Wedel, M.K., Baker, B.F., Su, J., Bradley, J.D., Yu, R.Z., Chuang, E., Graham, M.J., Crooke, R.M., 2006. Potent reduction of apolipoprotein B and low-density lipoprotein cholesterol by short-term administration of an antisense inhibitor of apolipoprotein B. Circulation 114 (16), 1729–1735.
Kelly, S., Georgomanolis, T., Zirkel, A., Diermeier, S., O'Reilly, D., Murphy, S., Langst, G., Cook, P.R., Papantonis, A., 2015. Splicing of many human genes involves sites embedded within introns. Nucleic Acids Res. 43 (9), 4721–4732.
Kircher, M., Witten, D.M., Jain, P., O'Roak, B.J., Cooper, G.M., Shendure, J., 2014. A general framework for estimating the relative pathogenicity of human genetic variants. Nat. Genet. 46 (3), 310–315.
Knecht, C., Mort, M., Junge, O., Cooper, D.N., Krawczak, M., Caliebe, A., 2017. IMHOTEP—a composite score integrating popular tools for predicting the functional consequences of non-synonymous sequence variants. Nucleic Acids Res. 45 (3), e13.
Kolasinska-Zwierz, P., Down, T., Latorre, I., Liu, T., Liu, X.S., Ahringer, J., 2009. Differential chromatin marking of introns and expressed exons by H3K36me3. Nat. Genet. 41 (3), 376–381.
Konig, J., Zarnack, K., Rot, G., Curk, T., Kayikci, M., Zupan, B., Turner, D.J., Luscombe, N.M., Ule, J., 2010. iCLIP reveals the function of hnRNP particles in splicing at individual nucleotide resolution. Nat. Struct. Mol. Biol. 17 (7), 909–915.
Kotterman, M.A., Schaffer, D.V., 2014. Engineering adeno-associated viruses for clinical gene therapy. Nat. Rev. Genet. 15 (7), 445–451.
Krainer, A.R., Conway, G.C., Kozak, D., 1990. The essential pre-mRNA splicing factor SF2 influences 5′ splice site selection by activating proximal sites. Cell 62 (1), 35–42.

Krawczak, M., Reiss, J., Cooper, D.N., 1992. The mutational spectrum of single base-pair substitutions in mRNA splice junctions of human genes: causes and consequences. Hum. Genet. 90 (1–2), 41–54.

Krecic, A.M., Swanson, M.S., 1999. hnRNP complexes: composition, structure, and function. Curr. Opin. Cell Biol. 11 (3), 363–371.

Krieg, A.M., 2011. Is RNAi dead? Mol. Ther. 19 (6), 1001–1002.

Kruger, S., Battke, F., Sprecher, A., Munz, M., Synofzik, M., Schols, L., Gasser, T., Grehl, T., Prudlo, J., Biskup, S., 2016. Rare variants in neurodegeneration associated genes revealed by targeted panel sequencing in a German ALS cohort. Front. Mol. Neurosci. 9, 92.

Kwak, H., Lis, J.T., 2013. Control of transcriptional elongation. Annu. Rev. Genet. 47, 483–508.

Laustriat, D., Gide, J., Barrault, L., Chautard, E., Benoit, C., Auboeuf, D., Boland, A., Battail, C., Artiguenave, F., Deleuze, J.F., Benit, P., Rustin, P., Franc, S., Charpentier, G., Furling, D., Bassez, G., Nissan, X., Martinat, C., Peschanski, M., Baghdoyan, S., 2015. In vitro and in vivo modulation of alternative splicing by the biguanide metformin. Mol. Ther. Nucleic Acids 4, e262.

Lee, S.C., Abdel-Wahab, O., 2016. Therapeutic targeting of splicing in cancer. Nat. Med. 22 (9), 976–986.

Lee, Y., Rio, D.C., 2015. Mechanisms and regulation of alternative pre-mRNA splicing. Annu. Rev. Biochem. 84, 291–323.

Lei, Q., Li, C., Zuo, Z., Huang, C., Cheng, H., Zhou, R., 2016. Evolutionary insights into RNA trans-splicing in vertebrates. Genome Biol. Evol. 8 (3), 562–577.

Lentz, J.J., Jodelka, F.M., Hinrich, A.J., McCaffrey, K.E., Farris, H.E., Spalitta, M.J., Bazan, N.G., Duelli, D.M., Rigo, F., Hastings, M.L., 2013. Rescue of hearing and vestibular function by antisense oligonucleotides in a mouse model of human deafness. Nat. Med. 19 (3), 345–350.

Li, S., Guo, W., Dewey, C.N., Greaser, M.L., 2013. Rbm20 regulates titin alternative splicing as a splicing repressor. Nucleic Acids Res. 41 (4), 2659–2672.

Li, Y.I., Sanchez-Pulido, L., Haerty, W., Ponting, C.P., 2015. RBFOX and PTBP1 proteins regulate the alternative splicing of micro-exons in human brain transcripts. Genome Res. 25 (1), 1–13.

Li, Y.I., van de Geijn, B., Raj, A., Knowles, D.A., Petti, A.A., Golan, D., Gilad, Y., Pritchard, J.K., 2016. RNA splicing is a primary link between genetic variation and disease. Science 352 (6285), 600–604.

Licatalosi, D.D., Mele, A., Fak, J.J., Ule, J., Kayikci, M., Chi, S.W., Clark, T.A., Schweitzer, A.C., Blume, J.E., Wang, X., Darnell, J.C., Darnell, R.B., 2008. HITS-CLIP yields genome-wide insights into brain alternative RNA processing. Nature 456 (7221), 464–469.

Lieber, M.R., 2010. The mechanism of double-strand DNA break repair by the nonhomologous DNA end-joining pathway. Annu. Rev. Biochem. 79, 181–211.

Lim, K.H., Fairbrother, W.G., 2012. Spliceman—a computational web server that predicts sequence variations in pre-mRNA splicing. Bioinformatics 28 (7), 1031–1032.

Linder, B., Dill, H., Hirmer, A., Brocher, J., Lee, G.P., Mathavan, S., Bolz, H.J., Winkler, C., Laggerbauer, B., Fischer, U., 2011. Systemic splicing factor deficiency causes tissue-specific defects: a zebrafish model for retinitis pigmentosa. Hum. Mol. Genet. 20 (2), 368–377.

Liu, H., Jin, T., Guan, J., Zhou, S., 2014. Histone modifications involved in cassette exon inclusions: a quantitative and interpretable analysis. BMC Genomics 15, 1148.

Llorian, M., Schwartz, S., Clark, T.A., Hollander, D., Tan, L.Y., Spellman, R., Gordon, A., Schweitzer, A.C., de la Grange, P., Ast, G., Smith, C.W., 2010. Position-dependent alternative splicing activity revealed by global profiling of alternative splicing events regulated by PTB. Nat. Struct. Mol. Biol. 17 (9), 1114–1123.

Long, J.C., Caceres, J.F., 2009. The SR protein family of splicing factors: master regulators of gene expression. Biochem. J. 417 (1), 15–27.

Long, C., Amoasii, L., Mireault, A.A., McAnally, J.R., Li, H., Sanchez-Ortiz, E., Bhattacharyya, S., Shelton, J.M., Bassel-Duby, R., Olson, E.N., 2016. Postnatal genome editing partially restores dystrophin expression in a mouse model of muscular dystrophy. Science 351 (6271), 400–403.

Lopez-Bigas, N., Audit, B., Ouzounis, C., Parra, G., Guigo, R., 2005. Are splicing mutations the most frequent cause of hereditary disease? FEBS Lett. 579 (9), 1900–1903.

Lou, C.H., Dumdie, J., Goetz, A., Shum, E.Y., Brafman, D., Liao, X., Mora-Castilla, S., Ramaiah, M., Cook-Andersen, H., Laurent, L., Wilkinson, M.F., 2016. Nonsense-mediated RNA decay influences human embryonic stem cell fate. Stem Cell Rep. 6 (6), 844–857.

Luco, R.F., Pan, Q., Tominaga, K., Blencowe, B.J., Pereira-Smith, O.M., Misteli, T., 2010. Regulation of alternative splicing by histone modifications. Science 327 (5968), 996–1000.

Maggio, I., Goncalves, M.A., 2015. Genome editing at the crossroads of delivery, specificity, and fidelity. Trends Biotechnol. 33 (5), 280–291.

Maggio, I., Stefanucci, L., Janssen, J.M., Liu, J., Chen, X., Mouly, V., Goncalves, M.A., 2016. Selection-free gene repair after adenoviral vector transduction of designer nucleases: rescue of dystrophin synthesis in DMD muscle cell populations. Nucleic Acids Res. 44 (3), 1449–1470.

Manley, J.L., Krainer, A.R., 2010. A rational nomenclature for serine/arginine-rich protein splicing factors (SR proteins). Genes Dev. 24 (11), 1073–1074.

Mansfield, S.G., Kole, J., Puttaraju, M., Yang, C.C., Garcia-Blanco, M.A., Cohn, J.A., Mitchell, L.G., 2000. Repair of CFTR mRNA by spliceosome-mediated RNA trans-splicing. Gene Ther. 7 (22), 1885–1895.

Martelli, P.L., D'Antonio, M., Bonizzoni, P., Castrignano, T., D'Erchia, A.M., D'Onorio De Meo, P., Fariselli, P., Finelli, M., Licciulli, F., Mangiulli, M., Mignone, F., Pavesi, G., Picardi, E., Rizzi, R., Rossi, I., Valletti, A., Zauli, A., Zambelli, F., Casadio, R., Pesole, G., 2011. ASPicDB: a database of annotated transcript and protein variants generated by alternative splicing. Nucleic Acids Res. 39 (Database issue), D80–5.

Martin, C.L., Duvall, J.A., Ilkin, Y., Simon, J.S., Arreaza, M.G., Wilkes, K., Alvarez-Retuerto, A., Whichello, A., Powell, C.M., Rao, K., Cook, E., Geschwind, D.H., 2007. Cytogenetic and molecular characterization of A2BP1/FOX1 as a candidate gene for autism. Am. J. Med. Genet. B Neuropsychiatr. Genet. 144B (7), 869–876.

Martinez-Contreras, R., Fisette, J.F., Nasim, F.U., Madden, R., Cordeau, M., Chabot, B., 2006. Intronic binding sites for hnRNP A/B and hnRNP F/H proteins stimulate pre-mRNA splicing. PLoS Biol. 4 (2), e21.

Methawasin, M., Hutchinson, K.R., Lee, E.J., Smith 3rd, J.E., Saripalli, C., Hidalgo, C.G., Ottenheijm, C.A., Granzier, H., 2014. Experimentally increasing titin compliance in a novel mouse model attenuates the Frank–Starling mechanism but has a beneficial effect on diastole. Circulation 129 (19), 1924–1936.

Miller, J.N., Pearce, D.A., 2014. Nonsense-mediated decay in genetic disease: friend or foe? Mutat. Res. Rev. Mutat. Res. 762, 52–64.

Miyamoto, S., Hidaka, K., Jin, D., Morisaki, T., 2009. RNA-binding proteins Rbm38 and Rbm24 regulate myogenic differentiation via p21-dependent and -independent regulatory pathways. Genes Cells 14 (11), 1241–1252.

Moore, M.J., Sharp, P.A., 1993. Evidence for two active sites in the spliceosome provided by stereochemistry of pre-mRNA splicing. Nature 365 (6444), 364–368.

Moraes, F., Goes, A., 2016. A decade of human genome project conclusion: scientific diffusion about our genome knowledge. Biochem. Mol. Biol. Educ. 44 (3), 215–223.

Motta-Mena, L.B., Heyd, F., Lynch, K.W., 2010. Context-dependent regulatory mechanism of the splicing factor hnRNP L. Mol. Cell 37 (2), 223–234.

Mucaki, E.J., Shirley, B.C., Rogan, P.K., 2013. Prediction of mutant mRNA splice isoforms by information theory-based exon definition. Hum. Mutat. 34 (4), 557–565.

Mudvari, P., Movassagh, M., Kowsari, K., Seyfi, A., Kokkinaki, M., Edwards, N.J., Golestaneh, N., Horvath, A., 2015. SNPlice: variants that modulate intron retention from RNA-sequencing data. Bioinformatics 31 (8), 1191–1198.

Mueller, W.F., Larsen, L.S., Garibaldi, A., Hatfield, G.W., Hertel, K.J., 2015. The silent sway of splicing by synonymous substitutions. J. Biol. Chem. 290 (46), 27700–27711.

Naldini, L., 2015. Gene therapy returns to centre stage. Nature 526 (7573), 351–360.

Naryshkin, N.A., Weetall, M., Dakka, A., Narasimhan, J., Zhao, X., Feng, Z., Ling, K.K., Karp, G.M., Qi, H., Woll, M.G., Chen, G., Zhang, N., Gabbeta, V., Vazirani, P., Bhattacharyya, A., Furia, B., Risher, N., Sheedy, J., Kong, R., Ma, J., Turpoff, A., Lee, C.S., Zhang, X., Moon, Y.C., Trifillis, P., Welch, E.M., Colacino, J.M., Babiak, J., Almstead, N.G., Peltz, S.W., Eng, L.A., Chen, K.S., Mull, J.L., Lynes, M.S., Rubin, L.L., Fontoura, P., Santarelli, L., Haehnke, D., McCarthy, K.D., Schmucki, R., Ebeling, M., Sivaramakrishnan, M., Ko, C.P., Paushkin, S.V., Ratni, H., Gerlach, I., Ghosh, A., Metzger, F., 2014. Motor neuron disease. SMN2 splicing modifiers improve motor function and longevity in mice with spinal muscular atrophy. Science 345 (6197), 688–693.

Nelson, C.E., Hakim, C.H., Ousterout, D.G., Thakore, P.I., Moreb, E.A., Castellanos Rivera, R.M., Madhavan, S., Pan, X., Ran, F.A., Yan, W.X., Asokan, A., Zhang, F., Duan, D., Gersbach, C.A., 2016. In vivo genome editing improves muscle function in a mouse model of Duchenne muscular dystrophy. Science 351 (6271), 403–407.

Neveling, K., Mensenkamp, A.R., Derks, R., Kwint, M., Ouchene, H., Steehouwer, M., van Lier, B., Bosgoed, E., Rikken, A., Tychon, M., Zafeiropoulou, D., Castelein, S., Hehir-Kwa, J., Tjwan Thung, D., Hofste, T., Lelieveld, S.H., Bertens, S.M., Adan, I.B., Eijkelenboom, A., Tops, B.B., Yntema, H., Stokowy, T., Knappskog, P.M., Hoberg-Vetti, H., Steen, V.M., Boyle, E., Martin, B., Ligtenberg, M.J., Shendure, J., Nelen, M.R., Hoischen, A., 2017. BRCA testing by single-molecule molecular inversion probes. Clin. Chem. 63 (2), 503–512.

Niks, E.H., Aartsma-Rus, A., 2017. Exon skipping: a first in class strategy for Duchenne muscular dystrophy. Expert Opin. Biol. Ther. 17 (2), 225–236.

Nilsen, T.W., Graveley, B.R., 2010. Expansion of the eukaryotic proteome by alternative splicing. Nature 463 (7280), 457–463.

Nishiguchi, K.M., Tearle, R.G., Liu, Y.P., Oh, E.C., Miyake, N., Benaglio, P., Harper, S., Koskiniemi-Kuendig, H., Venturini, G., Sharon, D., Koenekoop, R.K., Nakamura, M., Kondo, M., Ueno, S., Yasuma, T.R., Beckmann, J.S., Ikegawa, S., Matsumoto, N., Terasaki, H., Berson, E.L., Katsanis, N., Rivolta, C., 2013. Whole genome sequencing in patients with retinitis pigmentosa reveals pathogenic DNA structural changes and NEK2 as a new disease gene. Proc. Natl. Acad. Sci. U.S.A. 110 (40), 16139–16144.

Nutter, C.A., Jaworski, E.A., Verma, S.K., Deshmukh, V., Wang, Q., Botvinnik, O.B., Lozano, M.J., Abass, I.J., Ijaz, T., Brasier, A.R., Garg, N.J., Wehrens, X.H., Yeo, G.W., Kuyumcu-Martinez, M.N., 2016. Dysregulation of RBFOX2 is an early event in cardiac pathogenesis of diabetes. Cell Rep. 15 (10), 2200–2213.

Oh, H., Lee, E., Jang, H.N., Lee, J., Moon, H., Sheng, Z., Jun, Y., Loh, T.J., Cho, S., Zhou, J., Green, M.R., Zheng, X., Shen, H., 2013. hnRNP A1 contacts exon 5 to promote exon 6 inclusion of apoptotic Fas gene. Apoptosis 18 (7), 825–835.

Ohe, K., Hagiwara, M., 2015. Modulation of alternative splicing with chemical compounds in new therapeutics for human diseases. ACS Chem. Biol. 10 (4), 914–924.

Okunola, H.L., Krainer, A.R., 2009. Cooperative-binding and splicing-repressive properties of hnRNP A1. Mol. Cell. Biol. 29 (20), 5620–5631.

Osman, E.Y., Washington 3rd, C.W., Kaifer, K.A., Mazzasette, C., Patitucci, T.N., Florea, K.M., Simon, M.E., Ko, C.P., Ebert, A.D., Lorson, C.L., 2016. Optimization of morpholino antisense oligonucleotides targeting the Intronic repressor element1 in spinal muscular atrophy. Mol. Ther. 24 (9), 1592–1601.

Osorio, F.G., Navarro, C.L., Cadinanos, J., Lopez-Mejia, I.C., Quiros, P.M., Bartoli, C., Rivera, J., Tazi, J., Guzman, G., Varela, I., Depetris, D., de Carlos, F., Cobo, J., Andres, V., De Sandre-Giovannoli, A., Freije, J.M., Levy, N., Lopez-Otin, C., 2011. Splicing-directed therapy in a new mouse model of human accelerated aging. Sci. Transl. Med. 3 (106), 106ra107.

Palacino, J., Swalley, S.E., Song, C., Cheung, A.K., Shu, L., Zhang, X., Van Hoosear, M., Shin, Y., Chin, D.N., Keller, C.G., Beibel, M., Renaud, N.A., Smith, T.M., Salcius, M., Shi, X., Hild, M., Servais, R., Jain, M., Deng, L., Bullock, C., McLellan, M., Schuierer, S., Murphy, L., Blommers, M.J., Blaustein, C., Berenshteyn, F., Lacoste, A., Thomas, J.R., Roma, G., Michaud, G.A., Tseng, B.S., Porter, J.A., Myer, V.E., Tallarico, J.A., Hamann, L.G., Curtis, D., Fishman, M.C., Dietrich, W.F., Dales, N.A., Sivasankaran, R., 2015. SMN2 splice modulators enhance U1-pre-mRNA association and rescue SMA mice. Nat. Chem. Biol. 11 (7), 511–517.

Pan, Q., Shai, O., Lee, L.J., Frey, B.J., Blencowe, B.J., 2008. Deep surveying of alternative splicing complexity in the human transcriptome by high-throughput sequencing. Nat. Genet. 40 (12), 1413–1415.

Pandit, S., Zhou, Y., Shiue, L., Coutinho-Mansfield, G., Li, H., Qiu, J., Huang, J., Yeo, G.W., Ares Jr., M., Fu, X.D., 2013. Genome-wide analysis reveals SR protein cooperation and competition in regulated splicing. Mol. Cell 50 (2), 223–235.

Park, J.W., Graveley, B.R., 2007. Complex alternative splicing. Adv. Exp. Med. Biol. 623, 50–63.

Park, S.J., Lee, H., Jo, D.S., Jo, Y.K., Shin, J.H., Kim, H.B., Seo, H.M., Rubinsztein, D.C., Koh, J.Y., Lee, E.K., Cho, D.H., 2015. Heterogeneous nuclear ribonucleoprotein A1 post-transcriptionally regulates Drp1 expression in neuroblastoma cells. Biochim. Biophys. Acta 1849 (12), 1423–1431.

Park, S.A., Ahn, S.I., Gallo, J.M., 2016. Tau mis-splicing in the pathogenesis of neurodegenerative disorders. BMB Rep. 49 (8), 405–413.

Pertea, M., Lin, X., Salzberg, S.L., 2001. GeneSplicer: a new computational method for splice site prediction. Nucleic Acids Res. 29 (5), 1185–1190.

Petrilli, R., Eloy, J.O., Marchetti, J.M., Lopez, R.F., Lee, R.J., 2014. Targeted lipid nanoparticles for antisense oligonucleotide delivery. Curr. Pharm. Biotechnol. 15 (9), 847–855.

Pont, A.R., Sadri, N., Hsiao, S.J., Smith, S., Schneider, R.J., 2012. mRNA decay factor AUF1 maintains normal aging, telomere maintenance, and suppression of senescence by activation of telomerase transcription. Mol. Cell 47 (1), 5–15.

Pradeepa, M.M., Sutherland, H.G., Ule, J., Grimes, G.R., Bickmore, W.A., 2012. Psip1/Ledgf p52 binds methylated histone H3K36 and splicing factors and contributes to the regulation of alternative splicing. PLoS Genet. 8 (5), e1002717.

Proost, D., Saenen, J., Vandeweyer, G., Rotthier, A., Alaerts, M., Van Craenenbroeck, E.M., Van Crombruggen, J., Mortier, G., Wuyts, W., Vrints, C., Del Favero, J., Loeys, B., Van Laer, L., 2017. Targeted next-generation sequencing of 51 genes involved in primary electrical disease. J. Mol. Diagn. 19 (3), 445–459.

Raj, B., Blencowe, B.J., 2015. Alternative splicing in the mammalian nervous system: recent insights into mechanisms and functional roles. Neuron 87 (1), 14–27.

Ran, F.A., Cong, L., Yan, W.X., Scott, D.A., Gootenberg, J.S., Kriz, A.J., Zetsche, B., Shalem, O., Wu, X., Makarova, K.S., Koonin, E.V., Sharp, P.A., Zhang, F., 2015. In vivo genome editing using Staphylococcus aureus Cas9. Nature 520 (7546), 186–191.
Raponi, M., Kralovicova, J., Copson, E., Divina, P., Eccles, D., Johnson, P., Baralle, D., Vorechovsky, I., 2011. Prediction of single-nucleotide substitutions that result in exon skipping: identification of a splicing silencer in BRCA1 exon 6. Hum. Mutat. 32 (4), 436–444.
Ratni, H., Karp, G.M., Weetall, M., Naryshkin, N.A., Paushkin, S.V., Chen, K.S., McCarthy, K.D., Qi, H., Turpoff, A., Woll, M.G., Zhang, X., Zhang, N., Yang, T., Dakka, A., Vazirani, P., Zhao, X., Pinard, E., Green, L., David-Pierson, P., Tuerck, D., Poirier, A., Muster, W., Kirchner, S., Mueller, L., Gerlach, I., Metzger, F., 2016. Specific correction of alternative survival motor neuron 2 splicing by small molecules: discovery of a potential novel medicine to treat spinal muscular atrophy. J. Med. Chem. 59 (13), 6086–6100.
Reese, M.G., Eeckman, F.H., Kulp, D., Haussler, D., 1997. Improved splice site detection in genie. J. Comput. Biol. 4 (3), 311–323.
Rhoads, A., Au, K.F., 2015. PacBio sequencing and its applications. Genomics Proteomics Bioinformatics 13 (5), 278–289.
Rindt, H., Yen, P.F., Thebeau, C.N., Peterson, T.S., Weisman, G.A., Lorson, C.L., 2012. Replacement of huntingtin exon 1 by trans-splicing. Cell. Mol. Life Sci. 69 (24), 4191–4204.
Rodor, J., FitzPatrick, D.R., Eyras, E., Caceres, J.F., 2017. The RNA-binding landscape of RBM10 and its role in alternative splicing regulation in models of mouse early development. RNA Biol. 14 (1), 45–57.
Rogalska, M.E., Tajnik, M., Licastro, D., Bussani, E., Camparini, L., Mattioli, C., Pagani, F., 2016. Therapeutic activity of modified U1 core spliceosomal particles. Nat. Commun. 7, 11168.
Rogozin, I.B., Milanesi, L., 1997. Analysis of donor splice sites in different eukaryotic organisms. J. Mol. Evol. 45 (1), 50–59.
Ruzickova, S., Stanek, D., 2016. Mutations in spliceosomal proteins and retina degeneration. RNA Biol. 1–9.
Salton, M., Misteli, T., 2016. Small molecule modulators of pre-mRNA splicing in cancer therapy. Trends Mol. Med. 22 (1), 28–37.
Saul, M.J., Stein, S., Grez, M., Jakobsson, P.J., Steinhilber, D., Suess, B., 2016. UPF1 regulates myeloid cell functions and S100A9 expression by the hnRNP E2/miRNA-328 balance. Sci. Rep. 6, 31995.
Sawyer, S.L., Hartley, T., Dyment, D.A., Beaulieu, C.L., Schwartzentruber, J., Smith, A., Bedford, H.M., Bernard, G., Bernier, F.P., Brais, B., Bulman, D.E., Warman Chardon, J., Chitayat, D., Deladoey, J., Fernandez, B.A., Frosk, P., Geraghty, M.T., Gerull, B., Gibson, W., Gow, R.M., Graham, G.E., Green, J.S., Heon, E., Horvath, G., Innes, A.M., Jabado, N., Kim, R.H., Koenekoop, R.K., Khan, A., Lehmann, O.J., Mendoza-Londono, R., Michaud, J.L., Nikkel, S.M., Penney, L.S., Polychronakos, C., Richer, J., Rouleau, G.A., Samuels, M.E., Siu, V.M., Suchowersky, O., Tarnopolsky, M.A., Yoon, G., Zahir, F.R., Consortium, F.C., Care4Rare Canada, C., Majewski, J., Boycott, K.M., 2016. Utility of whole-exome sequencing for those near the end of the diagnostic odyssey: time to address gaps in care. Clin. Genet. 89 (3), 275–284.
Schleit, J., Bailey, S.S., Tran, T., Chen, D., Stowers, S., Schwarze, U., Byers, P.H., 2015. Molecular outcome, prediction, and clinical consequences of splice variants in COL1A1, which encodes the proalpha1(I) chains of type I procollagen. Hum. Mutat. 36 (7), 728–739.

Schultz, A.S., Preussner, M., Bunse, M., Karni, R., Heyd, F., 2016. Activation-dependent TRAF3 exon 8 alternative splicing is controlled by CELF2 and hnRNP C binding to an upstream intronic element. Mol. Cell. Biol. 37, e00488–16.

Schumperli, D., Pillai, R.S., 2004. The special Sm core structure of the U7 snRNP: far-reaching significance of a small nuclear ribonucleoprotein. Cell. Mol. Life Sci. 61 (19–20), 2560–2570.

Schwartz, S., Hall, E., Ast, G., 2009a. SROOGLE: webserver for integrative, user-friendly visualization of splicing signals. Nucleic Acids Res. 37 (Web server issue), W189–92.

Schwartz, S., Meshorer, E., Ast, G., 2009b. Chromatin organization marks exon-intron structure. Nat. Struct. Mol. Biol. 16 (9), 990–995.

Schwarz, J.M., Rodelsperger, C., Schuelke, M., Seelow, D., 2010. MutationTaster evaluates disease-causing potential of sequence alterations. Nat. Methods 7 (8), 575–576.

Schwarz, J.M., Cooper, D.N., Schuelke, M., Seelow, D., 2014. MutationTaster2: mutation prediction for the deep-sequencing age. Nat. Methods 11 (4), 361–362.

Sebat, J., Lakshmi, B., Malhotra, D., Troge, J., Lese-Martin, C., Walsh, T., Yamrom, B., Yoon, S., Krasnitz, A., Kendall, J., Leotta, A., Pai, D., Zhang, R., Lee, Y.H., Hicks, J., Spence, S.J., Lee, A.T., Puura, K., Lehtimaki, T., Ledbetter, D., Gregersen, P.K., Bregman, J., Sutcliffe, J.S., Jobanputra, V., Chung, W., Warburton, D., King, M.C., Skuse, D., Geschwind, D.H., Gilliam, T.C., Ye, K., Wigler, M., 2007. Strong association of de novo copy number mutations with autism. Science 316 (5823), 445–449.

Shabanpoor, F., Hammond, S.M., Abendroth, F., Hazell, G., Wood, M.J., Gait, M.J., 2017. Identification of a peptide for systemic brain delivery of a morpholino oligonucleotide in mouse models of spinal muscular atrophy. Nucleic Acid Ther. 27, 130–143.

Shan, J., Munro, T.P., Barbarese, E., Carson, J.H., Smith, R., 2003. A molecular mechanism for mRNA trafficking in neuronal dendrites. J. Neurosci. 23 (26), 8859–8866.

Shapiro, M.B., Senapathy, P., 1987. RNA splice junctions of different classes of eukaryotes: sequence statistics and functional implications in gene expression. Nucleic Acids Res. 15 (17), 7155–7174.

Sharon, D., Tilgner, H., Grubert, F., Snyder, M., 2013. A single-molecule long-read survey of the human transcriptome. Nat. Biotechnol. 31 (11), 1009–1014.

Shibata, A., Okuno, T., Rahman, M.A., Azuma, Y., Takeda, J., Masuda, A., Selcen, D., Engel, A.G., Ohno, K., 2016. IntSplice: prediction of the splicing consequences of intronic single-nucleotide variations in the human genome. J. Hum. Genet. 61 (7), 633–640.

Shindo, Y., Nozaki, T., Saito, R., Tomita, M., 2013. Computational analysis of associations between alternative splicing and histone modifications. FEBS Lett. 587 (5), 516–521.

Sibley, C.R., Emmett, W., Blazquez, L., Faro, A., Haberman, N., Briese, M., Trabzuni, D., Ryten, M., Weale, M.E., Hardy, J., Modic, M., Curk, T., Wilson, S.W., Plagnol, V., Ule, J., 2015. Recursive splicing in long vertebrate genes. Nature 521 (7552), 371–375.

Sierakowska, H., Sambade, M.J., Agrawal, S., Kole, R., 1996. Repair of thalassemic human beta-globin mRNA in mammalian cells by antisense oligonucleotides. Proc. Natl. Acad. Sci. U.S.A. 93 (23), 12840–12844.

Simms, C.L., Thomas, E.N., Zaher, H.S., 2017. Ribosome-based quality control of mRNA and nascent peptides. Wiley Interdiscip. Rev. RNA 8 (1), e1366.

Simon, J.M., Hacker, K.E., Singh, D., Brannon, A.R., Parker, J.S., Weiser, M., Ho, T.H., Kuan, P.F., Jonasch, E., Furey, T.S., Prins, J.F., Lieb, J.D., Rathmell, W.K., Davis, I.J., 2014. Variation in chromatin accessibility in human kidney cancer links H3K36 methyltransferase loss with widespread RNA processing defects. Genome Res. 24 (2), 241–250.

Sims 3rd, R.J., Millhouse, S., Chen, C.F., Lewis, B.A., Erdjument-Bromage, H., Tempst, P., Manley, J.L., Reinberg, D., 2007. Recognition of trimethylated histone H3 lysine 4 facilitates the recruitment of transcription postinitiation factors and pre-mRNA splicing. Mol. Cell 28 (4), 665–676.

Singh, R.K., Cooper, T.A., 2012. Pre-mRNA splicing in disease and therapeutics. Trends Mol. Med. 18 (8), 472–482.

Singh, N.K., Singh, N.N., Androphy, E.J., Singh, R.N., 2006. Splicing of a critical exon of human survival motor neuron is regulated by a unique silencer element located in the last intron. Mol. Cell. Biol. 26 (4), 1333–1346.

Singh, N.N., Lee, B.M., DiDonato, C.J., Singh, R.N., 2015. Mechanistic principles of antisense targets for the treatment of spinal muscular atrophy. Future Med. Chem. 7 (13), 1793–1808.

Sironi, M., Menozzi, G., Riva, L., Cagliani, R., Comi, G.P., Bresolin, N., Giorda, R., Pozzoli, U., 2004. Silencer elements as possible inhibitors of pseudoexon splicing. Nucleic Acids Res. 32 (5), 1783–1791.

Soldati, D., Schumperli, D., 1988. Structural and functional characterization of mouse U7 small nuclear RNA active in 3′ processing of histone pre-mRNA. Mol. Cell. Biol. 8 (4), 1518–1524.

Soukarieh, O., Gaildrat, P., Hamieh, M., Drouet, A., Baert-Desurmont, S., Frebourg, T., Tosi, M., Martins, A., 2016. Exonic splicing mutations are more prevalent than currently estimated and can be predicted by using in silico tools. PLoS Genet. 12 (1), e1005756.

Stanislaw, C., Xue, Y., Wilcox, W.R., 2016. Genetic evaluation and testing for hereditary forms of cancer in the era of next-generation sequencing. Cancer Biol. Med. 13 (1), 55–67.

Stark, H., Dube, P., Luhrmann, R., Kastner, B., 2001. Arrangement of RNA and proteins in the spliceosomal U1 small nuclear ribonucleoprotein particle. Nature 409 (6819), 539–542.

Sterne-Weiler, T., Howard, J., Mort, M., Cooper, D.N., Sanford, J.R., 2011. Loss of exon identity is a common mechanism of human inherited disease. Genome Res. 21 (10), 1563–1571.

Stoilov, P., Lin, C.H., Damoiseaux, R., Nikolic, J., Black, D.L., 2008. A high-throughput screening strategy identifies cardiotonic steroids as alternative splicing modulators. Proc. Natl. Acad. Sci. U.S.A. 105 (32), 11218–11223.

Supek, F., Minana, B., Valcarcel, J., Gabaldon, T., Lehner, B., 2014. Synonymous mutations frequently act as driver mutations in human cancers. Cell 156 (6), 1324–1335.

Suzuki, K., Tsunekawa, Y., Hernandez-Benitez, R., Wu, J., Zhu, J., Kim, E.J., Hatanaka, F., Yamamoto, M., Araoka, T., Li, Z., Kurita, M., Hishida, T., Li, M., Aizawa, E., Guo, S., Chen, S., Goebl, A., Soligalla, R.D., Qu, J., Jiang, T., Fu, X., Jafari, M., Esteban, C.R., Berggren, W.T., Lajara, J., Nunez-Delicado, E., Guillen, P., Campistol, J.M., Matsuzaki, F., Liu, G.H., Magistretti, P., Zhang, K., Callaway, E.M., Zhang, K., Belmonte, J.C., 2016. In vivo genome editing via CRISPR/Cas9 mediated homology-independent targeted integration. Nature 540 (7631), 144–149.

Tabebordbar, M., Zhu, K., Cheng, J.K., Chew, W.L., Widrick, J.J., Yan, W.X., Maesner, C., Wu, E.Y., Xiao, R., Ran, F.A., Cong, L., Zhang, F., Vandenberghe, L.H., Church, G.M., Wagers, A.J., 2016. In vivo gene editing in dystrophic mouse muscle and muscle stem cells. Science 351 (6271), 407–411.

Takata, A., Matsumoto, N., Kato, T., 2017. Genome-wide identification of splicing QTLs in the human brain and their enrichment among schizophrenia-associated loci. Nat. Commun. 8, 14519.

Tebas, P., Stein, D., Tang, W.W., Frank, I., Wang, S.Q., Lee, G., Spratt, S.K., Surosky, R.T., Giedlin, M.A., Nichol, G., Holmes, M.C., Gregory, P.D., Ando, D.G., Kalos, M., Collman, R.G., Binder-Scholl, G., Plesa, G., Hwang, W.T.,

Levine, B.L., June, C.H., 2014. Gene editing of CCR5 in autologous CD4 T cells of persons infected with HIV. N. Engl. J. Med. 370 (10), 901–910.
Tekin, D., Yan, D., Bademci, G., Feng, Y., Guo, S., Foster 2nd, J., Blanton, S., Tekin, M., Liu, X., 2016. A next-generation sequencing gene panel (MiamiOtoGenes) for comprehensive analysis of deafness genes. Hear. Res. 333, 179–184.
Tilgner, H., Grubert, F., Sharon, D., Snyder, M.P., 2014. Defining a personal, allele-specific, and single-molecule long-read transcriptome. Proc. Natl. Acad. Sci. U.S.A. 111 (27), 9869–9874.
Tress, M.L., Abascal, F., Valencia, A., 2017. Alternative splicing may not be the key to proteome complexity. Trends Biochem. Sci. 42 (2), 98–110.
Tsuboi, T., Kuroha, K., Kudo, K., Makino, S., Inoue, E., Kashima, I., Inada, T., 2012. Dom34:hbs1 plays a general role in quality-control systems by dissociation of a stalled ribosome at the 3′ end of aberrant mRNA. Mol. Cell 46 (4), 518–529.
Turunen, J.J., Niemela, E.H., Verma, B., Frilander, M.J., 2013. The significant other: splicing by the minor spliceosome. Wiley Interdiscip. Rev. RNA 4 (1), 61–76.
Ule, J., Jensen, K.B., Ruggiu, M., Mele, A., Ule, A., Darnell, R.B., 2003. CLIP identifies Nova-regulated RNA networks in the brain. Science 302 (5648), 1212–1215.
Van der Wal, E., Bergsma, A.J., Pijnenburg, J.M., Van der Ploeg, A.T., Pijnappel, W.W.M., 2017a. Antisense oligonucleotides promote exon inclusion and correct the common c.-32-13T > G *GAA* splicing variant in Pompe disease. Mol. Ther. Nucleic Acids 7, 90–100.
Van der Wal, E., Bergsma, A.J., Van Gestel, T.J.M., In 't Groen, S.L., Zaehres, H., Araúzo-Bravo, M.J., Schöler, H.R., Van der Ploeg, A.T., Pijnappel, W.W.M., 2017b. GAA deficiency in Pompe disease is alleviated by exon inclusion in iPSC-derived skeletal muscle cells. Mol. Ther. Nucleic Acids 7, 101–115.
Van Nostrand, E.L., Pratt, G.A., Shishkin, A.A., Gelboin-Burkhart, C., Fang, M.Y., Sundararaman, B., Blue, S.M., Nguyen, T.B., Surka, C., Elkins, K., Stanton, R., Rigo, F., Guttman, M., Yeo, G.W., 2016. Robust transcriptome-wide discovery of RNA-binding protein binding sites with enhanced CLIP (eCLIP). Nat. Methods 13 (6), 508–514.
Venables, J.P., Lapasset, L., Gadea, G., Fort, P., Klinck, R., Irimia, M., Vignal, E., Thibault, P., Prinos, P., Chabot, B., Abou Elela, S., Roux, P., Lemaitre, J.M., Tazi, J., 2013. MBNL1 and RBFOX2 cooperate to establish a splicing programme involved in pluripotent stem cell differentiation. Nat. Commun. 4, 2480.
Verma, S.K., Deshmukh, V., Nutter, C.A., Jaworski, E., Jin, W., Wadhwa, L., Abata, J., Ricci, M., Lincoln, J., Martin, J.F., Yeo, G.W., Kuyumcu-Martinez, M.N., 2016. Rbfox2 function in RNA metabolism is impaired in hypoplastic left heart syndrome patient hearts. Sci. Rep. 6, 30896.
Villarroya-Beltri, C., Gutierrez-Vazquez, C., Sanchez-Cabo, F., Perez-Hernandez, D., Vazquez, J., Martin-Cofreces, N., Martinez-Herrera, D.J., Pascual-Montano, A., Mittelbrunn, M., Sanchez-Madrid, F., 2013. Sumoylated hnRNPA2B1 controls the sorting of miRNAs into exosomes through binding to specific motifs. Nat. Commun. 4, 2980.
Voineagu, I., Wang, X., Johnston, P., Lowe, J.K., Tian, Y., Horvath, S., Mill, J., Cantor, R.M., Blencowe, B.J., Geschwind, D.H., 2011. Transcriptomic analysis of autistic brain reveals convergent molecular pathology. Nature 474 (7351), 380–384.
Vulin, A., Barthelemy, I., Goyenvalle, A., Thibaud, J.L., Beley, C., Griffith, G., Benchaouir, R., le Hir, M., Unterfinger, Y., Lorain, S., Dreyfus, P., Voit, T., Carlier, P., Blot, S., Garcia, L., 2012. Muscle function recovery in golden retriever muscular dystrophy after AAV1-U7 exon skipping. Mol. Ther. 20 (11), 2120–2133.

Wagner, S.D., Struck, A.J., Gupta, R., Farnsworth, D.R., Mahady, A.E., Eichinger, K., Thornton, C.A., Wang, E.T., Berglund, J.A., 2016. Dose-dependent regulation of alternative splicing by MBNL proteins reveals biomarkers for myotonic dystrophy. PLoS Genet. 12 (9), e1006316.

Wampfler, J., Federzoni, E.A., Torbett, B.E., Fey, M.F., Tschan, M.P., 2016. The RNA binding proteins RBM38 and DND1 are repressed in AML and have a novel function in APL differentiation. Leuk. Res. 41, 96–102.

Wang, E.T., Sandberg, R., Luo, S., Khrebtukova, I., Zhang, L., Mayr, C., Kingsmore, S.F., Schroth, G.P., Burge, C.B., 2008. Alternative isoform regulation in human tissue transcriptomes. Nature 456 (7221), 470–476.

Wang, Y., Gogol-Doring, A., Hu, H., Frohler, S., Ma, Y., Jens, M., Maaskola, J., Murakawa, Y., Quedenau, C., Landthaler, M., Kalscheuer, V., Wieczorek, D., Wang, Y., Hu, Y., Chen, W., 2013. Integrative analysis revealed the molecular mechanism underlying RBM10-mediated splicing regulation. EMBO Mol. Med. 5 (9), 1431–1442.

Wang, Y., Yang, Y., Liu, J., Chen, X.C., Liu, X., Wang, C.Z., He, X.Y., 2014. Whole dystrophin gene analysis by next-generation sequencing: a comprehensive genetic diagnosis of Duchenne and Becker muscular dystrophy. Mol. Genet. Genomics 289 (5), 1013–1021.

Wang, T., Xiao, G., Chu, Y., Zhang, M.Q., Corey, D.R., Xie, Y., 2015. Design and bioinformatics analysis of genome-wide CLIP experiments. Nucleic Acids Res. 43 (11), 5263–5274.

Wang, Z., Lin, Y., Qiu, L., Zheng, D., Yan, A., Zeng, J., Lan, F., 2016. Hybrid minigene splicing assay verified the pathogenicity of a novel splice site variant in the dystrophin gene of a Chinese patient with typical Duchenne muscular dystrophy phenotype. Clin. Chem. Lab. Med. 54 (9), 1435–1440.

Weatheritt, R.J., Sterne-Weiler, T., Blencowe, B.J., 2016. The ribosome-engaged landscape of alternative splicing. Nat. Struct. Mol. Biol. 23 (12), 1117–1123.

Weyn-Vanhentenryck, S.M., Mele, A., Yan, Q., Sun, S., Farny, N., Zhang, Z., Xue, C., Herre, M., Silver, P.A., Zhang, M.Q., Krainer, A.R., Darnell, R.B., Zhang, C., 2014. HITS-CLIP and integrative modeling define the Rbfox splicing-regulatory network linked to brain development and autism. Cell Rep. 6 (6), 1139–1152.

Woll, M.G., Qi, H., Turpoff, A., Zhang, N., Zhang, X., Chen, G., Li, C., Huang, S., Yang, T., Moon, Y.C., Lee, C.S., Choi, S., Almstead, N.G., Naryshkin, N.A., Dakka, A., Narasimhan, J., Gabbeta, V., Welch, E., Zhao, X., Risher, N., Sheedy, J., Weetall, M., Karp, G.M., 2016. Discovery and optimization of small molecule splicing modifiers of survival motor neuron 2 as a treatment for spinal muscular atrophy. J. Med. Chem. 59 (13), 6070–6085.

Wong, J.J., Au, A.Y., Ritchie, W., Rasko, J.E., 2016. Intron retention in mRNA: no longer nonsense: known and putative roles of intron retention in normal and disease biology. Bioessays 38 (1), 41–49.

Wright, C.F., Fitzgerald, T.W., Jones, W.D., Clayton, S., McRae, J.F., van Kogelenberg, M., King, D.A., Ambridge, K., Barrett, D.M., Bayzetinova, T., Bevan, A.P., Bragin, E., Chatzimichali, E.A., Gribble, S., Jones, P., Krishnappa, N., Mason, L.E., Miller, R., Morley, K.I., Parthiban, V., Prigmore, E., Rajan, D., Sifrim, A., Swaminathan, G.J., Tivey, A.R., Middleton, A., Parker, M., Carter, N.P., Barrett, J.C., Hurles, M.E., FitzPatrick, D.R., Firth, H.V., study, D.D.D., 2015. Genetic diagnosis of developmental disorders in the DDD study: a scalable analysis of genome-wide research data. Lancet 385 (9975), 1305–1314.

Wu, H., Sun, L., Wen, Y., Liu, Y., Yu, J., Mao, F., Wang, Y., Tong, C., Guo, X., Hu, Z., Sha, J., Liu, M., Xia, L., 2016. Major spliceosome defects cause male infertility and are associated with nonobstructive azoospermia in humans. Proc. Natl. Acad. Sci. U.S.A. 113 (15), 4134–4139.

Xiong, H.Y., Alipanahi, B., Lee, L.J., Bretschneider, H., Merico, D., Yuen, R.K., Hua, Y., Gueroussov, S., Najafabadi, H.S., Hughes, T.R., Morris, Q., Barash, Y., Krainer, A.R., Jojic, N., Scherer, S.W., Blencowe, B.J., Frey, B.J., 2015. RNA splicing. The human splicing code reveals new insights into the genetic determinants of disease. Science 347 (6218), 1254806.

Xu, L., Park, K.H., Zhao, L., Xu, J., El Refaey, M., Gao, Y., Zhu, H., Ma, J., Han, R., 2016. CRISPR-mediated genome editing restores dystrophin expression and function in mdx mice. Mol. Ther. 24 (3), 564–569.

Yang, Y., Muzny, D.M., Reid, J.G., Bainbridge, M.N., Willis, A., Ward, P.A., Braxton, A., Beuten, J., Xia, F., Niu, Z., Hardison, M., Person, R., Bekheirnia, M.R., Leduc, M.S., Kirby, A., Pham, P., Scull, J., Wang, M., Ding, Y., Plon, S.E., Lupski, J.R., Beaudet, A.L., Gibbs, R.A., Eng, C.M., 2013. Clinical whole-exome sequencing for the diagnosis of mendelian disorders. N. Engl. J. Med. 369 (16), 1502–1511.

Yang, J., Hung, L.H., Licht, T., Kostin, S., Looso, M., Khrameeva, E., Bindereif, A., Schneider, A., Braun, T., 2014. RBM24 is a major regulator of muscle-specific alternative splicing. Dev. Cell 31 (1), 87–99.

Ye, Z., Chen, Z., Lan, X., Hara, S., Sunkel, B., Huang, T.H., Elnitski, L., Wang, Q., Jin, V.X., 2014. Computational analysis reveals a correlation of exon-skipping events with splicing, transcription and epigenetic factors. Nucleic Acids Res. 42 (5), 2856–2869.

Yeo, G., Burge, C.B., 2004. Maximum entropy modeling of short sequence motifs with applications to RNA splicing signals. J. Comput. Biol. 11 (2–3), 377–394.

Yeo, G., Hoon, S., Venkatesh, B., Burge, C.B., 2004. Variation in sequence and organization of splicing regulatory elements in vertebrate genes. Proc. Natl. Acad. Sci. U.S.A. 101 (44), 15700–15705.

Young, C.S., Hicks, M.R., Ermolova, N.V., Nakano, H., Jan, M., Younesi, S., Karumbayaram, S., Kumagai-Cresse, C., Wang, D., Zack, J.A., Kohn, D.B., Nakano, A., Nelson, S.F., Miceli, M.C., Spencer, M.J., Pyle, A.D., 2016. A single CRISPR-Cas9 deletion strategy that targets the majority of DMD patients restores dystrophin function in hiPSC-derived muscle cells. Cell Stem Cell 18 (4), 533–540.

Yu, R.Z., Grundy, J.S., Geary, R.S., 2013. Clinical pharmacokinetics of second generation antisense oligonucleotides. Expert Opin. Drug Metab. Toxicol. 9 (2), 169–182.

Zarnack, K., Konig, J., Tajnik, M., Martincorena, I., Eustermann, S., Stevant, I., Reyes, A., Anders, S., Luscombe, N.M., Ule, J., 2013. Direct competition between hnRNP C and U2AF65 protects the transcriptome from the exonization of Alu elements. Cell 152 (3), 453–466.

Zarnegar, B.J., Flynn, R.A., Shen, Y., Do, B.T., Chang, H.Y., Khavari, P.A., 2016. irCLIP platform for efficient characterization of protein–RNA interactions. Nat. Methods 13 (6), 489–492.

Zhang, C., Zhang, Z., Castle, J., Sun, S., Johnson, J., Krainer, A.R., Zhang, M.Q., 2008. Defining the regulatory network of the tissue-specific splicing factors fox-1 and fox-2. Genes Dev. 22 (18), 2550–2563.

Zhang, Y., Schmid, B., Nielsen, T.T., Nielsen, J.E., Clausen, C., Hyttel, P., Holst, B., Freude, K.K., 2016. Generation of a human induced pluripotent stem cell line via CRISPR-Cas9 mediated integration of a site-specific homozygous mutation in CHMP2B. Stem Cell Res. 17 (1), 151–153.

Zhou, Z., Fu, X.D., 2013. Regulation of splicing by SR proteins and SR protein-specific kinases. Chromosoma 122 (3), 191–207.

CHAPTER FOUR

Transcriptional Regulation of Metabolism by SIRT1 and SIRT7

Kazuya Yamagata[1], Tatsuya Yoshizawa
Faculty of Life Sciences, Kumamoto University, Kumamoto, Japan
[1]Corresponding author: e-mail address: k-yamaga@kumamoto-u.ac.jp

Contents

Abstract

Sirtuins are a family of evolutionally conserved nicotinamide adenine dinucleotide (NAD^+)-dependent protein deacetylases/deacylases that regulate metabolism. The mammalian sirtuin family consists of seven sirtuins (SIRT1–7). Recent findings have identified critical roles for SIRT1 and SIRT7 in glucose/lipid metabolism in multiple tissues. This review focuses on the metabolic roles of these two sirtuins and the benefits of modulating the activity of sirtuins for the treatment of metabolic diseases such as type 2 diabetes.

1. INTRODUCTION

Sirtuins are a family of evolutionally conserved nicotinamide adenine dinucleotide (NAD^+)-dependent protein deacetylases/deacylases that regulate a wide variety of biological processes, such as metabolism, inflammation, genomic stability, tumorigenesis, and aging (Guarente, 2000). To date, seven

International Review of Cell and Molecular Biology, Volume 335
ISSN 1937-6448
https://doi.org/10.1016/bs.ircmb.2017.07.009

sirtuins (SIRT1–7) have been identified in mammals, which share a highly conserved NAD^+-binding and catalytic core domain. SIRT1, SIRT6, and SIRT7 are predominantly located in the nucleus, while SIRT2 is cytoplasmically localized. SIRT3, SIRT4, and SIRT5 are located in the mitochondria. The yeast sirtuin silent information regulator 2 (Sir2), which functions as a transcriptional silencing regulator, was identified as an NAD^+-dependent deacetylase (Houtkooper et al., 2012; Imai et al., 2000). Consistently, SIRT1–3 have strong NAD^+-dependent deacetylase activity. In contrast, SIRT4–6 have either weak or undetectable deacetylase activity. SIRT4 has ADP-ribosyltransferase activity (Haigis et al., 2006), and SIRT5 has both demalonylase and desuccinylase activity (Du et al., 2011). SIRT6 preferentially hydrolyzes long-chain fatty acyl groups (e.g., myristoyl) on target proteins (Jiang et al., 2013). SIRT7 has selective deacetylase activity for histone protein (Barber et al., 2012). Sirtuins regulate gene expression in response to the energetic state of the cell by deacetylating/deacylating transcription factors, coregulators, and histones (Feige and Auwerx, 2008; Feige et al., 2008).

SIRT1, the most extensively studied mammalian sirtuin, is expressed in multiple metabolic tissues including the liver, white adipose tissue (WAT), and skeletal muscle. SIRT1 activation by small molecule activators (Baur et al., 2006; Feige and Auwerx, 2008; Feige et al., 2008; Milne et al., 2007) or SIRT1 overexpression (Banks et al., 2008; Pfluger et al., 2008) improves metabolic disorders such as diet-induced diabetes. Furthermore, activation of SIRT1 by increasing NAD^+ availability also improves glucose tolerance and/or lipid profiles (Bai et al., 2011a,b; Cantó et al., 2012; Yoshino et al., 2011). These findings support the hypothesis that SIRT1 plays a pivotal role in metabolism and is an appropriate molecular target for the treatment of obesity and type 2 diabetes (Imai and Yoshino, 2013; Li and Kazgan, 2011). In this review, we focus on the metabolic function of SIRT1 in various tissues.

SIRT7 is another nuclear sirtuin. The enzymatic activity and functional roles, especially in metabolism, of SIRT7 were poorly defined until recently. However, recent studies have clarified that SIRT7 also controls glucose and lipid metabolism. We review the roles of SIRT7 in metabolism. A comparison of the metabolic roles of SIRT1 and SIRT7 is also discussed.

2. METABOLIC FUNCTIONS OF SIRT1 IN THE LIVER

The liver plays important roles in the regulation of lipid metabolism. During fasting/starvation, free fatty acids are released from adipose tissue and oxidized (β-oxidation) to acetyl-CoA to produce ATP. In the fed condition, fatty acids are synthesized from acetyl-CoA in the liver (lipogenesis) and transferred to adipocytes for storage. Peroxisome proliferator-activated receptor α (PPARα) is a key transcription factor for fatty acid oxidation (β-oxidation). Proliferator-activated receptor-γ coactivator 1α (PGC-1α), a coactivator of PPARα, is a direct substrate of SIRT1, and deacetylation by SIRT1 leads to its activation (Fig. 1A). Thus, SIRT1 regulates fatty acid oxidation-related genes, including medium chain acyl-CoA dehydrogenase (*Mcadh*), acyl CoA oxidase (*Aox*), and enoyl-CoA hydratase/3-hydroxyacyl CoA dehydrogenase (*Ehhadh*), through the activation of PGC-1α (Purushotham et al., 2009). Hepatic deletion of SIRT1 in mice results in hepatic lipid accumulation and development of liver steatosis after high-fat diet (HFD) feeding or prolonged fasting (Li et al., 2014; Purushotham et al., 2009). Conversely, transgenic mice overexpressing SIRT1 are protected from HFD-induced hepatic steatosis (Li et al., 2011; Pfluger et al., 2008). Sterol regulatory element-binding protein (SREBP)-1c is a key lipogenic transcription factor that regulates the expression

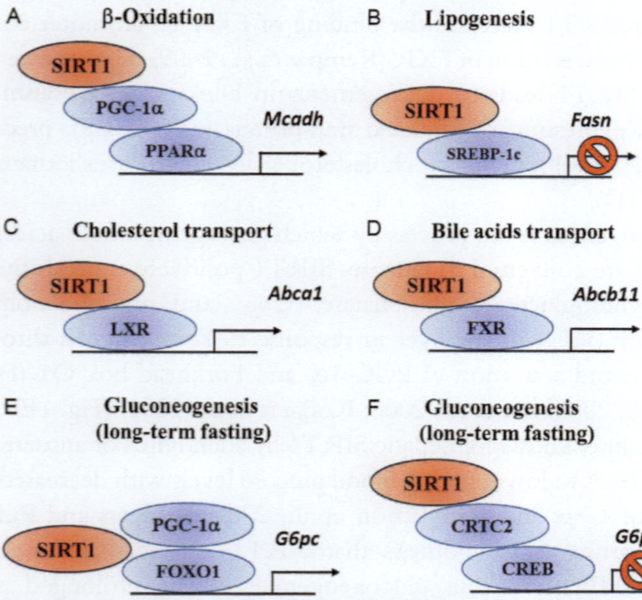

Fig. 1 Transcriptional regulation of metabolism-related genes by SIRT1 in the liver. (A) and (B) SIRT1 enhances β-oxidation via deacetylation of PGC-1α/PPARα, while SIRT1 attenuates lipogenesis by inhibiting SREBP-1c. (C) and (D) SIRT1 activates reverse cholesterol transport by deacetylating LXR and increases the export of bile acids by increasing the expression of transporters via FXR. (E) and (F) In addition, SIRT1 regulates gluconeogenesis through the deacetylation of PGC-1α, FOXO1, and CRTC2.

of multiple target genes, including fatty acid synthase (*Fasn*), and elongation of very long-chain fatty acids-like 6 (*Elovl6*). SIRT1 deacetylates and inhibits SREBP-1c activity in the liver (Ponugoti et al., 2010; Walker et al., 2010) (Fig. 1B). Therefore, hepatic SIRT1 seems to stimulate fat oxidation and inhibit fat synthesis, but inconsistent findings have also been reported (Chen et al., 2008).

The liver also plays an important role in cholesterol and bile acid homeostasis. Liver X receptor (LXR) regulates the expression levels of ATP-binding cassette transporter A1 (*Abca1*) and *Abcg1*, which mediate cholesterol efflux. Thus, LXR plays an important role in cholesterol transport from peripheral tissues to the liver (reverse cholesterol transport) via high-density lipoprotein-cholesterol biogenesis. SIRT1 deacetylates and increases LXR activity, leading to the enhancement of reverse cholesterol transport (Li et al., 2007) (Fig. 1C), suggesting that activation of LXR by SIRT1 may protect against atherosclerosis. Farnesoid X receptor (FXR) is a nuclear bile acid receptor that regulates the expression of bile acid transporters, such as *Abcb11*/bile salt export pump (*Bsep*). Acetylation of FXR inhibits its activity, and SIRT1 increases the binding of FXR to promoter chromatin through the deacetylation of FXR (Kemper et al., 2009). Hepatocyte-specific deletion of SIRT1 leads to derangements in bile acid metabolism due to the reduced expression of bile acid transporters (e.g., BSEP), predisposing mice to the development of cholesterol gallstones (Purushotham et al., 2012) (Fig. 1D).

Gluconeogenesis is the process by which glucogenic amino acids, lactate, and glycerol are converted to glucose. SIRT1 positively controls the gluconeogenic genes glucose-6-phosphatase (*G6pc*) and phosphoenolpyruvate carboxykinase (*Pck1*) in the liver in response to fasting signals through the deacetylation and activation of PGC-1α, and Forkhead box O1 (FOXO1) (Frescas et al., 2005; Liu et al., 2008; Rodgers et al., 2005) (Fig. 1E). Consistently, acute knockdown of hepatic SIRT1 by adenovirus or antisense oligonucleotides leads to lower fasting blood glucose levels with decreased hepatic expression of *G6pc* and *Pck1* (Erion et al., 2009; Rodgers and Puigserver, 2007), supporting the hypothesis that SIRT1 enhances gluconeogenesis. Conversely, SIRT1 inhibits gluconeogenesis in the prolonged phase of fasting by promoting the ubiquitin-mediated degradation of CREB-regulated transcription coactivator 2 (CRTC2), a coactivator of gluconeogenesis (Liu et al., 2008) (Fig. 1F). However, *G6pc* and *Pck1* gene expression and fasting glucose levels are reportedly normal in hepatocyte-specific SIRT1 knockout mice (Purushotham et al., 2009). Another line of liver-specific

SIRT1 knockout mice shows increased hepatic glucose production (Wang et al., 2011). Thus, the regulation of hepatic gluconeogenesis by SIRT1 is complex.

3. METABOLIC FUNCTION OF SIRT1 IN ADIPOCYTES

WAT is the main storage compartment of triacylglycerol in the body. In the fasting condition, the rate of lipolysis of triacylglycerol and release of free fatty acids into the blood is increased. In contrast, insulin inhibits the lipolysis and release of free fatty acids in the fed condition. PPARγ is a master regulator of adipocyte differentiation and plays an essential role in the regulation of lipid storage. PPARγ increases fatty acid uptake via the transcriptional regulation of fatty acid-binding protein 4 (*Fabp4*)/adipocyte protein 2 (*Ap2*) and lipoprotein lipase (*Lpl*). SIRT1 represses PPARγ activity by docking with nuclear receptor corepressor (NCoR), leading to inhibition of adipocyte differentiation (Picard et al., 2004) (Fig. 2A). SIRT1 enhances lipolysis in differentiated adipocytes. Indeed, the release of free fatty acids from WAT is reduced in *Sirt1*$^{+/-}$ mice (Picard et al., 2004).

Excess lipid storage in obesity causes chronic inflammation in adipose tissue, which induces insulin resistance in the liver and skeletal muscle via the overproduction of free fatty acids from WAT. In addition, WAT inflammation causes dysregulation of adipocytokine secretion such as an increase of

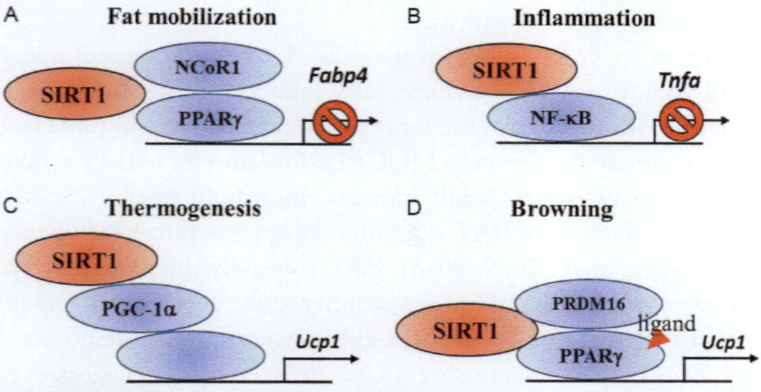

Fig. 2 Transcriptional regulation of metabolism-related genes by SIRT1 in adipocytes. (A) SIRT1 inhibits adipogenesis and favors lipolysis by the suppression of PPARγ. (B) SIRT1 deacetylates and inhibits NF-κB and reduces inflammation in WAT. (C) SIRT1 stimulates thermogenesis via deacetylation of PGC-1α in BAT. (D) PPARγ ligand binding induces a conformational change and recruits SIRT1. SIRT1 promotes browning of adipocytes by PPARγ deacetylation and PRDM16 recruitment.

tumor necrosis factor α (TNF-α) and decrease of adiponectin (Samuel and Shulman, 2012; Suganami et al., 2012). Nuclear factor (NF)-κB is a central transcription factor that controls the expression of genes involved in inflammation, and SIRT1 represses the activity of NF-κB by deacetylating the p65 subunit of NF-κB (Yeung et al., 2004) (Fig. 2B). SIRT1 exerts antiinflammatory effects in adipocytes and improves glucose tolerance and insulin resistance in animal models (Chalkiadaki and Guarente, 2012; Gillum et al., 2011; Yoshizaki et al., 2009, 2010). During diet-induced obesity, macrophages play important roles in the inflammation of adipose tissue (Weisberg et al., 2003). SIRT1, in macrophages, also inhibits NF-κB-mediated transcription, resulting in reduced inflammation and improved insulin sensitivity in adipose tissue (Schug et al., 2010; Yoshizaki et al., 2010). Conversely, SIRT1 downregulation also reportedly attenuates long-term HFD-induced adipose tissue inflammation by enhancing PPARγ activity (Mayoral et al., 2015). Thus, not only the activation but also the suppression of SIRT1 activity in adipocytes may be beneficial for the treatment of obesity and diabetes mellitus.

Nicotinamide phosphoribosyltransferase (NAMPT) is a key enzyme for NAD^+ biosynthesis. SIRT1 deacetylates NAMPT and promotes its secretion from adipocytes into the blood (Yoon et al., 2015). Extracellular NAMPT secreted from adipose tissue promotes NAD^+ synthesis, SIRT1 activity, and neural activity in the hypothalamus (Yoon et al., 2015). Thus, adipocyte SIRT1 regulates the function of remote tissues through the NAMPT/NAD^+/SIRT1 pathway.

In contrast to WAT, which is specialized for the storage of excess fat, brown adipose tissue (BAT) oxidizes fatty acids to produce heat by catalyzing the leak of protons across the inner membrane of mitochondria through the action of uncoupling protein 1 (UCP1). The amount of BAT is inversely correlated with obesity in adult humans. Increasing energy expenditure through the activation of BAT is a potential approach to treat obesity and diabetes (Cypess et al., 2009, 2015). BAT is heavily innervated by sympathetic nerves and is responsible for adaptive thermogenesis in response to exposure to cold. SIRT1 enhances BAT transcriptional responses (e.g., increase of *Ucp1*) by decreasing PGC-1α acetylation (Lagouge et al., 2006) (Fig. 2C). Mice with moderate SIRT1 overexpression display higher energy expenditure due to increased heat production and improved glucose tolerance (Boutant et al., 2014). SIRT1 activation by SRT1720 also stimulates energy expenditure from BAT (Feige and Auwerx, 2008; Feige et al., 2008). Furthermore, SIRT1 induces browning (i.e., remodeling of white adipocytes to brown adipocytes with increased *Ucp1* expression) in subcutaneous WAT through deacetylation of PPARγ (Qiang et al., 2012) (Fig. 2D). Deacetylated PPARγ interacts with PR domain-containing 16 (PRDM16), an important cofactor for browning, to promote thermogenesis (energy expenditure) and improve insulin sensitivity (Qiang et al., 2012). These data indicate that SIRT1 activation has therapeutic potential against obesity and diabetes by promoting energy expenditure over energy storage via the activation of brown adipocytes.

4. METABOLIC FUNCTION OF SIRT1 IN SKELETAL MUSCLE AND THE HEART

Under the fed condition, insulin promotes glucose uptake in skeletal muscle, and excess glucose is stored as glycogen. A defect of insulin action (insulin resistance) in muscle is a key component of the etiology of type 2 diabetes. Conversely, skeletal muscle switches from carbohydrates to lipids as its main energy source during fasting and exercise. Nutrient deprivation (fasting) or increased energy demand (exercise) activates AMP-activated protein kinase (AMPK), a metabolic sensor, in muscle. AMPK activation increases the transcription of *Nampt*, an NAD^+ biosynthetic enzyme, and raises the intracellular levels of NAD^+ (Fulco et al., 2008), leading to the enhancement of SIRT1 activity. SIRT1, in turn, deacetylates PGC-1α and FOXO1 and increases the expression of downstream target genes, such as carnitine palmitoyltransferase 1b (*Cpt1b*), *Mcadh*, pyruvate dehydrogenase kinase 4 (*Pdk4*), cytochrome *C* (*Cytc*), and mitofusin-2 (Cantó et al., 2009, 2010; Gerhart-Hines et al., 2007) (Fig. 3A and B). Thus, AMPK-mediated SIRT1 activation enhances mitochondrial function and lipid oxidation in skeletal muscle. Adiponectin, an antidiabetic adipocytokine secreted from adipocytes, improves glucose tolerance through the activation of the AMPK/SIRT1 signaling pathway in skeletal muscle (Iwabu et al., 2010).

Protein-tyrosine phosphatase 1B (PTP1B) is a negative regulator of the insulin signaling pathway, and SIRT1 reportedly improves skeletal muscle insulin sensitivity by repressing *Ptp1b* gene transcription at the chromatin level (Sun et al., 2007) (Fig. 3C). SIRT1 also enhances insulin sensitivity in skeletal muscle by decreasing the expression of p55a/p50a, which are regulatory subunits of phosphoinositide 3-kinase (PI3K) (Schenk et al., 2011). These findings suggest that muscle SIRT1 is a possible target for the treatment of diabetes. However, it has been reported that muscle-specific overexpression of SIRT1 in mice cannot prevent the HFD-induced impairment

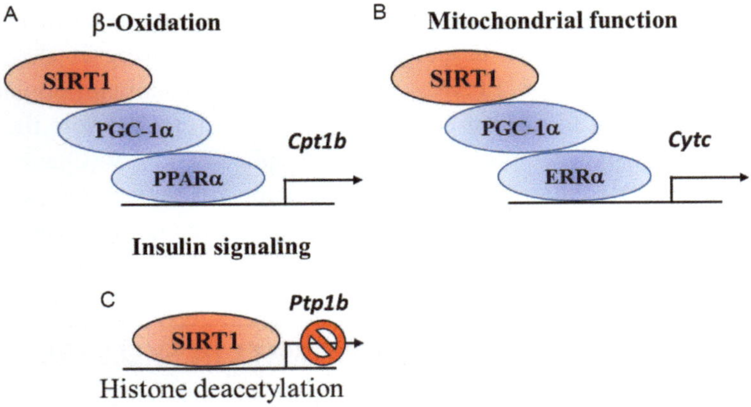

Fig. 3 Transcriptional regulation of metabolism-related genes by SIRT1 in skeletal muscle. (A) and (B) SIRT1 enhances β-oxidation and mitochondrial function through the deacetylation of PGC-1α. (C) SIRT1 promotes insulin sensitivity by decreasing the gene expression of insulin signaling inhibitors (PTP1B).

of insulin sensitivity and glucose intolerance (White et al., 2014). Furthermore, skeletal muscle endurance and wheel running-induced mitochondrial biogenesis are not impaired in muscle-specific SIRT1 knockout mice (Philp et al., 2011). Thus, further studies are necessary to prove the beneficial metabolic roles of SIRT1 in skeletal muscle.

In addition, SIRT1 has a role in cardioprotection. Myocardial ischemia/reperfusion is an important cause of myocardial injury. SIRT1 increases the expression of superoxide dismutase 2 (SOD2) by activating FOXO1 and suppresses ischemia/reperfusion-induced oxidative stress in cardiac myocytes (Hsu et al., 2010). SIRT1 also displays a cardioprotective effect by suppressing cardiac complement component 3 expression (Yamamoto et al., 2016).

5. METABOLIC FUNCTION OF SIRT1 IN PANCREATIC β-CELLS

Pancreatic β-cells play a central role in glucose homeostasis by secreting insulin. A defect in insulin secretion from β-cells is essential for the etiology of both type 1 diabetes (autoimmune disease) and type 2 diabetes (lifestyle-related disease). SIRT1 increases insulin secretion by repressing the transcription of uncoupling protein 2 (*Ucp2*), which uncouples ATP synthesis in mitochondria and reduces insulin secretion (Bordone et al., 2006; Moynihan et al., 2005) (Fig. 4A). SIRT1 also protects β-cells from

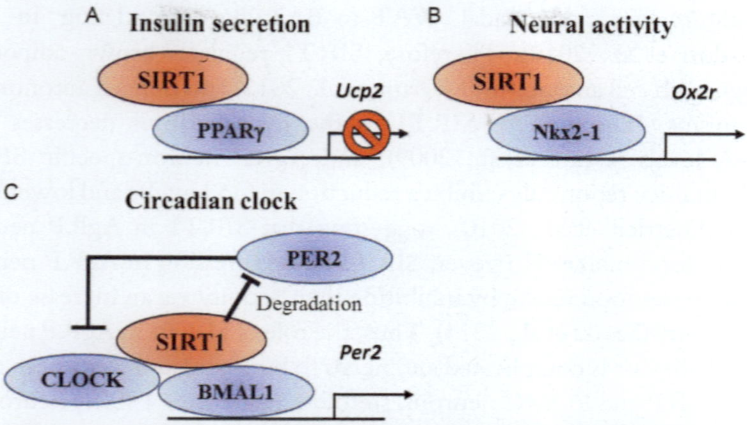

Fig. 4 Transcriptional regulation of metabolism-related genes by SIRT1 in pancreatic β-cells and brain. (A) SIRT1 increases insulin secretion by inhibiting the expression of *Ucp2*. (B) SIRT1 deacetylates Nkx2-1 and increases the expression of Ox2r, leading to neuronal activation. (C) SIRT1 also regulates the circadian clock via deacetylation and degradation of PER2.

cytokine-mediated damage by suppressing the NF-κB signaling pathway (Lee et al., 2009). These findings indicate that SIRT1 plays an important role in the maintenance of glucose-stimulated insulin secretion.

Interestingly, a missense (L107P) mutation in the *SIRT1* gene reportedly causes a familial form of autoimmune diabetes (Biason-Lauber et al., 2013). Regulatory T (Treg) cells are important for the maintenance of immune homeostasis and protection against autoimmunity. Since SIRT1 influences the development of Treg cells (Beier et al., 2012), this mutation in *SIRT1* may underlie the development of the autoimmune form of diabetes in humans.

6. METABOLIC FUNCTION OF SIRT1 IN THE HYPOTHALAMUS

Hypothalamic neurons play critical roles in controlling energy balance and glucose homeostasis. Proopiomelanocortin (POMC) neurons in the arcuate nucleus (ARC) suppress feeding. In contrast, agouti-related peptide (AgRP) neurons in the ARC promote food intake. Leptin, an adipocytokine secreted from adipocytes, inhibits feeding by activating POMC neurons and inhibiting AgRP neurons (Coppari, 2012). SIRT1 in POMC neurons does not regulate food intake, but controls energy expenditure via perigonadal WAT-to-BAT-like remodeling in vivo (Ramadori et al., 2010). Therefore, SIRT1 regulates brown adipocytes through both cell autonomous (Qiang et al., 2012) and noncell autonomous mechanisms. Inhibition of SIRT1 in the hypothalamus decreases *Agrp* mRNA levels (Cakir et al., 2009), and AgRP neuron-specific SIRT1 knockout mice reportedly exhibit a reduction in food intake and lower body weight (Dietrich et al., 2010), suggesting that SIRT1 in AgRP neurons enhances food intake. However, SIRT1 overexpression in AgRP neurons also suppresses food intake by inhibiting their activity via an increase of leptin sensitivity (Sasaki et al., 2014). Thus, the role of SIRT1 in AgRP neurons in food behavior is complicated and needs to be elucidated further. In addition to AgRP and POMC neurons, steroidogenic factor 1 (SF1) neurons in the ventromedial hypothalamic nucleus play critical roles in metabolic regulation (Dhillon et al., 2006). SIRT1 in SF1 neurons regulates skeletal muscle insulin sensitivity, and loss of SIRT1 in SF1 neurons results in glucose intolerance (Ramadori et al., 2011).

Orexin is a hypothalamic neuropeptide that has an important role in the regulation of arousal, motivational components of physical activity, and metabolism. Diet restriction increases SIRT1 protein levels in the dorsomedial and lateral hypothalamic nuclei (DMH and LH), and SIRT1 in the DMH and LH enhances orexin type 2 receptor (*Ox2r*) expression by deacetylating the homeodomain transcription factor Nkx2-1 (Satoh et al., 2010, 2013) (Fig. 4B). The activation of these OX2R-positive DMH and LH neurons leads to the stimulation of thermogenesis and the elevation of physical activity under chronic energy limitation (Satoh et al., 2010). Increased neural activity in the DMH and LH is also critical for the preservation of sleep quality. These findings imply that hypothalamic SIRT1 contributes to the neurobehavioral adaptive response to search for food and to function in diet-restricted environments.

The suprachiasmatic nucleus is responsible for controlling circadian rhythm. The circadian clock system consists of a feedback loop in which the transcription factors CLOCK and BMAL activate the transcription of Cryptochrome (*Cry*) and Period (*Per*), leading to the subsequent repression of CLOCK/BMAL complexes by CRY and PER proteins. The circadian clock plays a role in the regulation of glucose/lipid metabolism, and disruption of circadian rhythm increases the risk of metabolic diseases (Panda, 2016). SIRT1 regulates circadian clock gene expression through PER2 deacetylation and degradation (Asher et al., 2008) (Fig. 4C). SIRT1 also deacetylates BMAL and regulates the circadian cycle of clock genes, such as *Per2* (Nakahata et al., 2008). In addition, CLOCK/BMAL regulates the circadian expression of NAMPT, a rate-limiting enzyme in the NAD^+ salvage pathway (Nakahata et al., 2009; Ramsey et al., 2009). Thus, NAD^+ levels display circadian oscillation. These oscillations in NAD^+ levels modulate the activity of SIRT1 and the circadian clock feedback cycle.

7. METABOLIC FUNCTION OF SIRT1 IN BONE

Osteoporosis is a representative age-related disease characterized by low bone mass, bone fragility, and increased risk of fracture. SIRT1 also controls bone metabolism. SIRT1 activators or overexpression of SIRT1 attenuate the loss of bone mass with age (Herranz et al., 2010; Mercken et al., 2014). Conversely, female $Sirt1^{+/-}$ mice exhibit a reduction in bone mass (Cohen-Kfir et al., 2011). Osteoblasts are the major cellular component of bone and are responsible for the formation and mineralization of the bone matrix. Mice with a *Sirt1* deletion in osteoprogenitor cells exhibit decreased bone formation resulting from increased β-catenin sequestration by FOXOs (Iyer et al., 2014). These results suggest that the beneficial effects of SIRT1 on bone are mediated, at least in part, by the promotion and survival of osteoprogenitor cells. Thus, SIRT1 in osteoprogenitor cells could be a therapeutic target for osteoporosis.

In summary, these recent findings indicate that SIRT1 promotes lipolysis from adipose tissue and increases the utilization of free fatty acids in the liver and skeletal muscle in low-energy conditions. In addition, hypothalamic SIRT1 regulates food behavior and maintains physical activity and body temperature under diet-restricted conditions. Therefore, SIRT1 appears to be important for executing adaptive responses to a low-energy state. Interestingly, SIRT1 also exerts protective roles against metabolic diseases including obesity and type 2 diabetes through the improvement of both insulin sensitivity in peripheral tissues and insulin secretion by pancreatic β-cells. Furthermore, SIRT1 also has a protective role against osteoporosis.

8. BIOLOGICAL FUNCTION OF SIRT7

SIRT7 is ubiquitously expressed in all tissues (Ford et al., 2006). Although the enzymatic activity and physiological functions of SIRT7 are poorly defined, recent studies have revealed important biological roles for SIRT7. First, Barber et al. identified that SIRT7 is an NAD^+-dependent deacetylase with high selectivity for acetylated lysine 18 of histone H3

(H3K18) and represses the transcription of tumor suppressor genes (e.g., *NME1* and *COPS2*) in a transcription factor ELK4-dependent manner (Fig. 5A). Thus, SIRT7 plays a critical role in maintaining the transformed state of cancer cells (Barber et al., 2012). Indeed, SIRT7 expression is upregulated in several human cancers (Kim et al., 2013; Yu et al., 2014; Zhang et al., 2015). Myb-binding protein 1a (Mybbp1a), which has tumor suppressor activity, binds to SIRT7 and inhibits the deacetylation of H3K18 (Karim et al., 2013).

SIRT7 is present in both the nucleolus and nucleoplasm (Yamagata et al., 2015). SIRT7 functions as a positive regulator of rRNA transcription in the

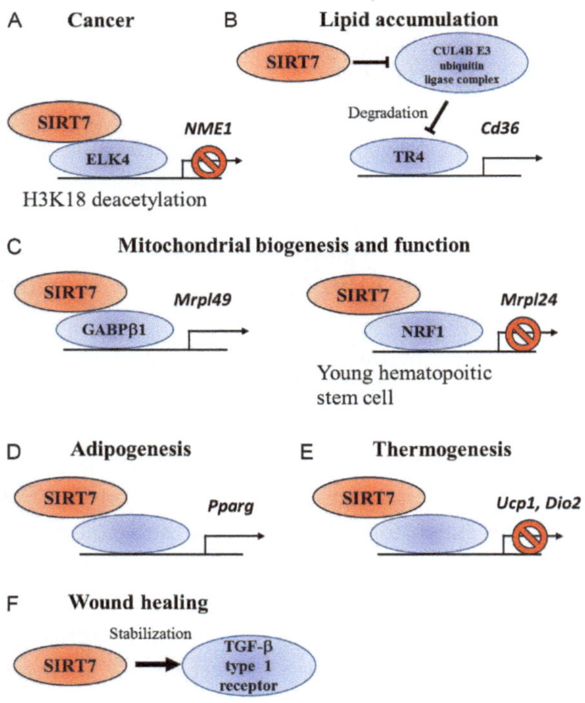

Fig. 5 Transcriptional regulation of genes by SIRT7. (A) SIRT7 represses the transcription of tumor suppressor genes by deacetylating histone H3K18. (B) SIRT7 inhibits the degradation of TR4 via the DCAF1/DDB1/CUL4B E3 ubiquitin ligase complex. Thus, SIRT7 increases hepatic lipid accumulation by increasing the expression of *Cd36* and other genes. (C) SIRT7 also controls mitochondrial biogenesis and function by interaction with GABPβ1 and NRF1. (D) and (E) SIRT7 promotes adipogenesis of white adipocytes and suppresses thermogenesis in brown adipocytes. (F) Furthermore, SIRT7 is involved in susceptibility to cardiac rupture after acute myocardial infarction by regulating the expression of TGF-β type 1 receptor protein.

nucleolus by deacetylating PAF53, a subunit of RNA polymerase I (Pol I) (Chen et al., 2013; Ford et al., 2006). SIRT7 also controls pre-rRNA processing through the deacetylation of U3-55k protein, a core component of the U3 snoRNP complex (Chen et al., 2015). Furthermore, SIRT7 is recruited to DNA double-strand break (DSB) sites and promotes DSB repair through deacetylation of H3K18 or desuccinylation of H3K122 (Li et al., 2016; Vazquez et al., 2016).

Moreover, SIRT7 plays important roles in the functional regulation of mitochondria. Nuclear respiratory factor 1 (NRF1) is known as a master regulator of mitochondria. SIRT7 binds to the proximal promoters of the genes encoding mitochondrial ribosomal proteins (e.g., MRPL24) and mitochondrial translation factors (e.g., GFM2) in hematopoietic stem cells (Mohrin et al., 2015). SIRT7 suppresses mitochondrial biogenesis and energy metabolism by inhibiting the activity of NRF1. The suppressive activity by SIRT7 is required for maintaining the regenerative capacity of hematopoietic stem cells (Mohrin et al., 2015).

9. METABOLIC FUNCTION OF SIRT7 IN THE LIVER

As described earlier, hepatic deletion of SIRT1 in mice results in hepatic lipid accumulation after HFD feeding (Purushotham et al., 2009). In striking contrast, we have shown that SIRT7 knockout mice are resistant to HFD-induced fatty liver (Yoshizawa et al., 2014). The nuclear receptor TR4 plays a critical role in lipid homeostasis in the liver by regulating the expression of *Cd36*, cell death-inducing DFFA-like effector a (*Cidea*), *Cidec*, monoacylglycerol *O*-acyltransferase 1 (*Mogat1*), and *Pparg*. SIRT7 increases TR4 protein expression by inhibiting the degradation of TR4 via the DCAF1/DDB1/CUL4B E3 ubiquitin ligase complex, leading to hepatic lipid accumulation (Karim et al., 2017; Yoshizawa et al., 2014) (Fig. 5B). Thus, SIRT7 and SIRT1 apparently play opposite roles in hepatic lipid homeostasis. Like SIRT7, SIRT4 increases hepatic lipid accumulation by repressing PPARα (Laurent et al., 2013). Thus, sirtuins have various roles in lipid homeostasis. Shin et al. reported that SIRT7 prevents the development of fatty liver by repressing the activity of the transcription factor Myc (Shin et al., 2013). In their study, SIRT7 knockout mice fed a chow diet exhibit fatty liver due to the induction of endoplasmic reticulum (ER) stress (Shin et al., 2013). However, the expression of ER stress-related genes such as *Grp78*, *Atf4*, and *Chop* is not increased in the liver of our SIRT7 knockout mice (unpublished data). The reasons for the different phenotypes of the two lines of SIRT7 knockout mice are unclear.

Aged SIRT7 knockout mice reportedly display hepatic microvesicular steatosis due to mitochondrial dysfunction (Ryu et al., 2014). GA-binding protein (GABP) is a nuclear transcription factor that is composed of GABPα and GABPβ. GABP controls mitochondrial biogenesis and function. SIRT7 enhances the transcriptional activity of GABP via GABPβ1 deacetylation (Ryu et al., 2014) (Fig. 5C). The expression of MRPL49, a component of the mitochondrial large ribosomal subunit, is reduced in the liver of SIRT7 knockout mice. SIRT7 induces mitochondrial biogenesis in hepatocytes, while it suppresses mitochondrial function in young hematopoietic stem cells (Mohrin et al., 2015) (Fig. 5C). Thus, SIRT7 exerts heterogeneous effects on mitochondria in different cells.

10. METABOLIC FUNCTION OF SIRT7 IN OTHER TISSUES

SIRT7 knockdown results in impaired adipogenic differentiation in vitro (Cioffi et al., 2015). Furthermore, expression of *Pparg*, a master gene of adipogenesis, is decreased in the WAT of SIRT7 knockout mice (Cioffi et al., 2015), indicating that SIRT7 promotes adipogenesis (Fig. 5D). Although the exact mechanism is unknown, these findings imply that SIRT1 and SIRT7 control adipogenesis in opposite directions (Picard et al., 2004).

BAT plays an essential role in thermogenesis and affects body weight and insulin sensitivity by regulating energy expenditure. SIRT7 also contributes to the regulation of BAT activity. The body temperature of SIRT7 knockout mice fed an HFD is significantly higher than that of control mice (Yoshizawa et al., 2014). In addition, the expression of the genes encoding *Ucp1* and deiodinase, iodothyronine, type 2 (*Dio2*), which are involved in thermogenesis, is increased in SIRT7 knockout mice. These findings indicate that SIRT7 regulates whole-body energy expenditure by suppressing thermogenesis in BAT (Fig. 5E). SIRT7 knockout mice are resistant to HFD-induced obesity and glucose intolerance (Yoshizawa et al., 2014). The increased energy expenditure of SIRT7 knockout mice may be involved in their improved metabolic phenotype.

Recent studies also clarified an important role for SIRT7 in the heart. SIRT7 knockout mice show susceptibility to cardiac rupture after acute myocardial infarction (Araki et al., 2015). Transforming growth factor beta

(TGF-β) plays a central role in wound healing, and SIRT7 promotes wound healing by stabilizing TGF-β type 1 receptor protein, enabling efficient TGF-β signaling (Fig. 5F). Inhibition of fibroblast differentiation and reduction of fibrosis after myocardial infarction are reported in SIRT7 knockout mice (Araki et al., 2015). Conversely, aged SIRT7 knockout mice exhibit extensive myocardial fibrosis with excessive collagen accumulation and increased inflammation under physiological conditions (Vakhrusheva et al., 2008). Thus, the roles of cardiac SIRT7 in response to acute injury (myocardial infarction) and aging seem to be different.

In summary, SIRT7 plays a critical role in the regulation of metabolism. Interestingly, SIRT7 and SIRT1 seem to control hepatic lipid accumulation and adipogenesis in opposite manners. However, it is unclear whether this assumption is applicable to other metabolic aspects. Further studies are required to establish the functional roles of SIRT7 in other metabolic tissues.

11. CONCLUSIONS

Both SIRT1 and SIRT7 are nuclear sirtuins and function as NAD^+-dependent deacetylases/deacylases. As described in this review, recent studies have improved our understanding of the metabolic role of SIRT1. SIRT1 regulates metabolism in multiple tissues through the deacetylation of several target substrates, including PGC-1α, FOXO1, PPARγ, and NF-κB. In contrast, the metabolic roles of SIRT7 are still limited, and only a small number of SIRT7 substrates have been identified to date. Interestingly, recent studies have shown that DNA and RNA can increase the catalytic activity of SIRT7 and that SIRT7 can remove long-chain fatty acyl groups more efficiently than it can remove acetyl groups (Tong et al., 2016, 2017). Further research is needed to clarify the functional roles of SIRT7. Both SIRT1 and SIRT7 sense NAD^+ levels and exert their catalytic activity. However, the metabolic roles of SIRT1 and SIRT7 are opposite in some cases (Fig. 6). How do SIRT1 and SIRT7 communicate and coordinate their actions? The identification of the molecular mechanisms underlying the integration of different sirtuins is necessary. Small molecule activators of SIRT1, such as resveratrol, improve metabolic abnormalities in mice, but the beneficial effect of resveratrol in humans is under debate (Timmers et al., 2011, Yoshino et al., 2012). More studies are needed to assess the potential benefits of the supplementation of sirtuin activators

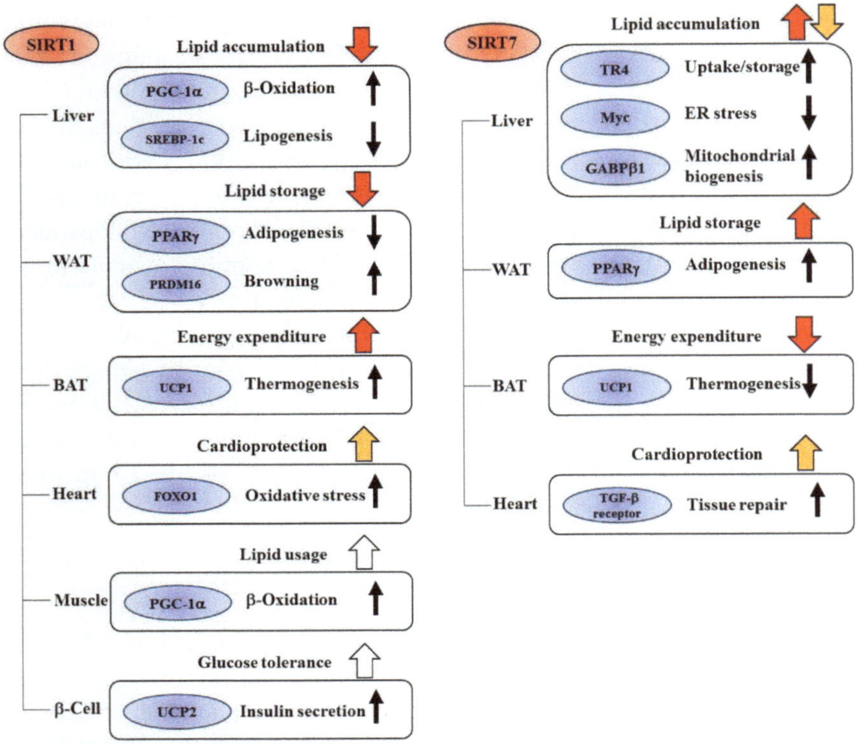

Fig. 6 Comparison of the metabolic roles of SIRT1 and SIRT7. SIRT1 lowers hepatic lipid accumulation by promoting lipid oxidation and inhibiting lipogenesis (Ponugoti et al., 2010; Purushotham et al., 2009). SIRT7 increases hepatic lipid accumulation by regulating lipid uptake and triglyceride storage (Yoshizawa et al., 2014), although the hepatic roles of SIRT7 are under debate (Ryu et al., 2014; Shin et al., 2013). SIRT1 reduces lipid storage in WAT and increases thermogenesis in BAT (Lagouge et al., 2006; Picard et al., 2004). In contrast, SIRT7 increases lipid storage in WAT and reduces thermogenesis in BAT (Cioffi et al., 2015; Yoshizawa et al., 2014). Thus, SIRT1 and SIRT7 seem to have opposite metabolic roles in adipocytes. However, both SIRT1 and SIRT7 are cardioprotective (Araki et al., 2015; Hsu et al., 2010). The *red* and *yellow arrows* indicate the opposite and similar effects, respectively, of SIRT1 and SIRT7. The metabolic functions of SIRT7 in skeletal muscle and pancreatic β-cells are unknown (*white arrow*).

and NAD^+ intermediates (e.g., nicotinamide mononucleotide) against metabolic diseases in humans. Finally, SIRT7-specific inhibitors may offer additional benefits in the treatment of metabolic diseases. A better understanding of the enzymatic characteristics of SIRT7 and the development of SIRT7 inhibitors are exciting future challenges.

ACKNOWLEDGMENTS

The authors apologize to all the researchers whose work could not be included due to space limitations. This work was supported by JSPS KAKENHI (16H05328).

REFERENCES

Araki, S., Izumiya, Y., Rokutanda, T., Ianni, A., Hanatani, S., Kimura, Y., Onoue, Y., Senokuchi, T., Yoshizawa, T., Yasuda, O., Koitabashi, N., Kurabayashi, M., Braun, T., Bober, E., Yamagata, K., Ogawa, H., 2015. Sirt7 contributes to myocardial tissue repair by maintaining transforming growth factor-β signaling pathway. Circulation 132, 1081–1093.

Asher, G., Gatfield, D., Stratmann, M., Reinke, H., Dibner, C., Kreppel, F., Mostoslavsky, R., Alt, F.W., Schibler, U., 2008. SIRT1 regulates circadian clock gene expression through PER2 deacetylation. Cell 134, 317–328.

Bai, P., Canto, C., Brunyánszki, A., Huber, A., Szántó, M., Cen, Y., Yamamoto, H., Houten, S.M., Kiss, B., Oudart, H., Gergely, P., Menissier-de Murcia, J., Schreiber, V., Sauve, A.A., Auwerx, J., 2011a. PARP-2 regulates SIRT1 expression and whole-body energy expenditure. Cell Metab. 13, 450–460.

Bai, P., Cantó, C., Oudart, H., Brunyánszki, A., Cen, Y., Thomas, C., Yamamoto, H., Huber, A., Kiss, B., Houtkooper, R.H., Schoonjans, K., Schreiber, V., Sauve, A.A., Menissier-de Murcia, J., Auwerx, J., 2011b. PARP-1 inhibition increases mitochondrial metabolism through SIRT1 activation. Cell Metab. 13, 461–468.

Banks, A.S., Kon, N., Knight, C., Matsumoto, M., Gutiérrez-Juárez, R., Rossetti, L., Gu, W., Accili, D., 2008. SirT1 gain of function increases energy efficiency and prevents diabetes in mice. Cell Metab. 8, 333–841.

Barber, M.F., Michishita-Kioi, E., Xi, Y., Tasselli, L., Kioi, M., Moqtaderi, Z., Tennen, R.I., Paredes, S., Young, N.L., Chen, K., Struhl, K., Garcia, B.A., Gozani, O., Li, W., Chua, K.F., 2012. SIRT7 links H3K18 deacetylation to maintenance of oncogenic transformation. Nature 487, 114–118.

Baur, J.A., Pearson, K.J., Price, N.L., Jamieson, H.A., Lerin, C., Kalra, A., Prabhu, V.V., Allard, J.S., Lopez-Lluch, G., Lewis, K., Pistell, P.J., Poosala, S., Becker, K.G., Boss, O., Gwinn, D., Wang, M., Ramaswamy, S., Fishbein, K.W., Spencer, R.G., Lakatta, E.G., Le Couteur, D., Shaw, R.J., Navas, P., Puigserver, P., Ingram, D.K., de Cabo, R., Sinclair, D.A., 2006. Resveratrol improves health and survival of mice on a high-calorie diet. Nature 444, 337–342.

Beier, U.H., Wang, L., Han, R., Akimova, T., Liu, Y., Hancock, W.W., 2012. Histone deacetylases 6 and 9 and sirtuin-1 control Foxp3+ regulatory T cell function through shared and isoform-specific mechanisms. Sci. Signal. 5, ra45.

Biason-Lauber, A., Böni-Schnetzler, M., Hubbard, B.P., Bouzakri, K., Brunner, A., Cavelti-Weder, C., Keller, C., Meyer-Böni, M., Meier, D.T., Brorsson, C., Timper, K., Leibowitz, G., Patrignani, A., Bruggmann, R., Boily, G., Zulewski, H., Geier, A., Cermak, J.M., Elliott, P., Ellis, J.L., Westphal, C., Knobel, U., Eloranta, J.J., Kerr-Conte, J., Pattou, F., Konrad, D., Matter, C.M., Fontana, A., Rogler, G., Schlapbach, R., Regairaz, C., Carballido, J.M., Glaser, B., McBurney, M.W., Pociot, F., Sinclair, D.A., Donath, M.Y., 2013. Identification of a SIRT1 mutation in a family with type 1 diabetes. Cell Metab. 17, 448–455.

Bordone, L., Motta, M.C., Picard, F., Robinson, A., Jhala, U.S., Apfeld, J., McDonagh, T., Lemieux, M., McBurney, M., Szilvasi, A., Easlon, E.J., Lin, S.J., Guarente, L., 2006. Sirt1 regulates insulin secretion by repressing UCP2 in pancreatic beta cells. PLoS Biol. 4, e31.

Boutant, M., Joffraud, M., Kulkarni, S.S., García-Casarrubios, E., García-Roves, P.M., Ratajczak, J., Fernández-Marcos, P.J., Valverde, A.M., Serrano, M., Cantó, C., 2014. SIRT1 enhances glucose tolerance by potentiating brown adipose tissue function. Mol. Metab. 4, 118–131.

Cakir, I., Perello, M., Lansari, O., Messier, N.J., Vaslet, C.A., Nillni, E.A., 2009. Hypothalamic Sirt1 regulates food intake in a rodent model system. PLoS One 4, e8322.

Cantó, C., Gerhart-Hines, Z., Feige, J.N., Lagouge, M., Noriega, L., Milne, J.C., Elliott, P.J., Puigserver, P., Auwerx, J., 2009. AMPK regulates energy expenditure by modulating NAD+ metabolism and SIRT1 activity. Nature 458, 1056–1060.

Cantó, C., Jiang, L.Q., Deshmukh, A.S., Mataki, C., Coste, A., Lagouge, M., Zierath, J.R., Auwerx, J., 2010. Interdependence of AMPK and SIRT1 for metabolic adaptation to fasting and exercise in skeletal muscle. Cell Metab. 11, 213–219.

Cantó, C., Houtkooper, R.H., Pirinen, E., Youn, D.Y., Oosterveer, M.H., Cen, Y., Fernandez-Marcos, P.J., Yamamoto, H., Andreux, P.A., Cettour-Rose, P., Gademann, K., Rinsch, C., Schoonjans, K., Sauve, A.A., Auwerx, J., 2012. The NAD(+) precursor nicotinamide riboside enhances oxidative metabolism and protects against high-fat diet-induced obesity. Cell Metab. 15, 838–847.

Chalkiadaki, A., Guarente, L., 2012. High-fat diet triggers inflammation-induced cleavage of SIRT1 in adipose tissue to promote metabolic dysfunction. Cell Metab. 16, 180–188.

Chen, D., Bruno, J., Easlon, E., Lin, S.J., Cheng, H.L., Alt, F.W., Guarente, L., 2008. Tissue-specific regulation of SIRT1 by calorie restriction. Genes Dev. 22, 1753–1757.

Chen, S., Seiler, J., Santiago-Reichelt, M., Felbel, K., Grummt, I., Voit, R., 2013. Repression of RNA polymerase I upon stress is caused by inhibition of RNA-dependent deacetylation of PAF53 by SIRT7. Mol. Cell 52, 303–313.

Chen, S., Blank, M.F., Iyer, A., Huang, B., Wang, L., Grummt, I., Voit, R., 2015. SIRT7-dependent deacetylation of the U3-55k protein controls pre-rRNA processing. Nat. Commun. 7, 10734.

Cioffi, M., Vallespinos-Serrano, M., Trabulo, S.M., Fernandez-Marcos, P.J., Firment, A.N., Vazquez, B.N., Vieira, C.R., Mulero, F., Camara, J.A., Cronin, U.P., Perez, M., Soriano, J., G Galvez, B., Castells-Garcia, A., Haage, V., Raj, D., Megias, D., Hahn, S., Serrano, L., Moon, A., Aicher, A., Heeschen, C., 2015. MiR-93 controls adiposity via inhibition of Sirt7 and Tbx3. Cell Rep. 12, 1594–1605.

Cohen-Kfir, E., Artsi, H., Levin, A., Abramowitz, E., Bajayo, A., Gurt, I., Zhong, L., D'Urso, A., Toiber, D., Mostoslavsky, R., Dresner-Pollak, R., 2011. Sirt1 is a regulator of bone mass and a repressor of Sost encoding for sclerostin, a bone formation inhibitor. Endocrinology 152, 4514–4524.

Coppari, R., 2012. Metabolic actions of hypothalamic SIRT1. Trends Endocrinol. Metab. 23, 179–185.

Cypess, A.M., Lehman, S., Williams, G., Tal, I., Rodman, D., Goldfine, A.B., Kuo, F.C., Palmer, E.L., Tseng, Y.H., Doria, A., Kolodny, G.M., Kahn, C.R., 2009. Identification and importance of brown adipose tissue in adult humans. N. Engl. J. Med. 360, 1509–15017.

Cypess, A.M., Weiner, L.S., Roberts-Toler, C., Franquet Elía, E., Kessler, S.H., Kahn, P.A., English, J., Chatman, K., Trauger, S.A., Doria, A., Kolodny, G.M., 2015. Activation of human brown adipose tissue by a β3-adrenergic receptor agonist. Cell Metab. 21, 33–38.

Dhillon, H., Zigman, J.M., Ye, C., Lee, C.E., McGovern, R.A., Tang, V., Kenny, C.D., Christiansen, L.M., White, R.D., Edelstein, E.A., Coppari, R., Balthasar, N., Cowley, M.A., Chua Jr., S., Elmquist, J.K., Lowell, B.B., 2006. Leptin directly activates SF1 neurons in the VMH, and this action by leptin is required for normal body-weight homeostasis. Neuron 49, 191–203.

Dietrich, M.O., Antunes, C., Geliang, G., Liu, Z.W., Borok, E., Nie, Y., Xu, A.W., Souza, D.O., Gao, Q., Diano, S., Gao, X.B., Horvath, T.L., 2010. Agrp neurons

mediate Sirt1's action on the melanocortin system and energy balance: roles for Sirt1 in neuronal firing and synaptic plasticity. J. Neurosci. 30, 11815–11825.
Du, J., Zhou, Y., Su, X., Yu, J.J., Khan, S., Jiang, H., Kim, J., Woo, J., Kim, J.H., Choi, B.H., He, B., Chen, W., Zhang, S., Cerione, R.A., Auwerx, J., Hao, Q., Lin, H., 2011. Sirt5 is a NAD-dependent protein lysine demalonylase and desuccinylase. Science 334, 806–809.
Erion, D.M., Yonemitsu, S., Nie, Y., Nagai, Y., Gillum, M.P., Hsiao, J.J., Iwasaki, T., Stark, R., Weismann, D., Yu, X.X., Murray, S.F., Bhanot, S., Monia, B.P., Horvath, T.L., Gao, Q., Samuel, V.T., Shulman, G.I., 2009. SirT1 knockdown in liver decreases basal hepatic glucose production and increases hepatic insulin responsiveness in diabetic rats. Proc. Natl. Acad. Sci. U. S. A. 106, 11288–11293.
Feige, J.N., Auwerx, J., 2008. Transcriptional targets of sirtuins in the coordination of mammalian physiology. Curr. Opin. Cell Biol. 20, 303–309.
Feige, J.N., Lagouge, M., Canto, C., Strehle, A., Houten, S.M., Milne, J.C., Lambert, P.D., Mataki, C., Elliott, P.J., Auwerx, J., 2008. Specific SIRT1 activation mimics low energy levels and protects against diet-induced metabolic disorders by enhancing fat oxidation. Cell Metab. 8, 347–358.
Ford, E., Voit, R., Liszt, G., Magin, C., Grummt, I., Guarente, L., 2006. Mammalian Sir2 homolog SIRT7 is an activator of RNA polymerase I transcription. Genes Dev. 20, 1075–1080.
Frescas, D., Valenti, L., Accili, D., 2005. Nuclear trapping of the forkhead transcription factor FoxO1 via Sirt-dependent deacetylation promotes expression of glucogenetic genes. J. Biol. Chem. 280, 20589–20595.
Fulco, M., Cen, Y., Zhao, P., Hoffman, E.P., McBurney, M.W., Sauve, A.A., Sartorelli, V., 2008. Glucose restriction inhibits skeletal myoblast differentiation by activating SIRT1 through AMPK-mediated regulation of Nampt. Dev. Cell 14, 661–673.
Gerhart-Hines, Z., Rodgers, J.T., Bare, O., Lerin, C., Kim, S.H., Mostoslavsky, R., Alt, F.W., Wu, Z., Puigserver, P., 2007. Metabolic control of muscle mitochondrial function and fatty acid oxidation through SIRT1/PGC-1alpha. EMBO J. 26, 1913–1923.
Gillum, M.P., Kotas, M.E., Erion, D.M., Kursawe, R., Chatterjee, P., Nead, K.T., Muise, E.S., Hsiao, J.J., Frederick, D.W., Yonemitsu, S., Banks, A.S., Qiang, L., Bhanot, S., Olefsky, J.M., Sears, D.D., Caprio, S., Shulman, G.I., 2011. SirT1 regulates adipose tissue inflammation. Diabetes 60, 3235–3245.
Guarente, L., 2000. Sir2 links chromatin silencing, metabolism, and aging. Genes Dev. 14, 1021–1026.
Haigis, M.C., Mostoslavsky, R., Haigis, K.M., Fahie, K., Christodoulou, D.C., Murphy, A.J., Valenzuela, D.M., Yancopoulos, G.D., Karow, M., Blander, G., Wolberger, C., Prolla, T.A., Weindruch, R., Alt, F.W., Guarente, L., 2006. SIRT4 inhibits glutamate dehydrogenase and opposes the effects of calorie restriction in pancreatic beta cells. Cell 126, 941–954.
Herranz, D., Muñoz-Martin, M., Cañamero, M., Mulero, F., Martinez-Pastor, B., Fernandez-Capetillo, O., Serrano, M., 2010. Sirt1 improves healthy ageing and protects from metabolic syndrome-associated cancer. Nat. Commun. 1, 3.
Houtkooper, R.H., Pirinen, E., Auwerx, J., 2012. Sirtuins as regulators of metabolism and healthspan. Nat. Rev. Mol. Cell Biol. 13, 225–238.
Hsu, C.P., Zhai, P., Yamamoto, T., Maejima, Y., Matsushima, S., Hariharan, N., Shao, D., Takagi, H., Oka, S., Sadoshima, J., 2010. Silent information regulator 1 protects the heart from ischemia/reperfusion. Circulation 122, 2170–2182.
Imai, S., Yoshino, J., 2013. The importance of NAMPT/NAD/SIRT1 in the systemic regulation of metabolism and ageing. Diabetes Obes. Metab. 15, 26–33.
Imai, S., Armstrong, C.M., Kaeberlein, M., Guarente, L., 2000. Transcriptional silencing and longevity protein Sir2 is an NAD-dependent histone deacetylase. Nature 403, 795–800.

Iwabu, M., Yamauchi, T., Okada-Iwabu, M., Sato, K., Nakagawa, T., Funata, M., Yamaguchi, M., Namiki, S., Nakayama, R., Tabata, M., Ogata, H., Kubota, N., Takamoto, I., Hayashi, Y.K., Yamauchi, N., Waki, H., Fukayama, M., Nishino, I., Tokuyama, K., Ueki, K., Oike, Y., Ishii, S., Hirose, K., Shimizu, T., Touhara, K., Kadowaki, T., 2010. Adiponectin and AdipoR1 regulate PGC-1alpha and mitochondria by Ca(2+) and AMPK/SIRT1. Nature 464, 1313–1319.

Iyer, S., Han, L., Bartell, S.M., Kim, H.N., Gubrij, I., de Cabo, R., O'Brien, C.A., Manolagas, S.C., Almeida, M., 2014. Sirtuin1 (Sirt1) promotes cortical bone formation by preventing β-catenin sequestration by FoxO transcription factors in osteoblast progenitors. J. Biol. Chem. 289, 24069–24078.

Jiang, H., Khan, S., Wang, Y., Charron, G., He, B., Sebastian, C., Du, J., Kim, R., Ge, E., Mostoslavsky, R., Hang, H.C., Hao, Q., Lin, H., 2013. SIRT6 regulates TNF-α secretion through hydrolysis of long-chain fatty acyl lysine. Nature 496, 110–113.

Karim, M.F., Yoshizawa, T., Sato, Y., Sawa, T., Tomizawa, K., Akaike, T., Yamagata, K., 2013. Inhibition of H3K18 deacetylation of Sirt7 by Myb-binding protein 1a (Mybbp1a). Biochem. Biophys. Res. Commun. 441, 157–163.

Karim, M.F., Yoshizawa, T., Sobuz, S.U., Sato, Y., Yamagata, K., 2017. Sirtuin-7 dependent deacetylation ofDDB1 regulates the expression of nuclear receptor TR4. Biochem. Biophys. Res. Commun. 490, 423–428.

Kemper, J.K., Xiao, Z., Ponugoti, B., Miao, J., Fang, S., Kanamaluru, D., Tsang, S., Wu, S.Y., Chiang, C.M., Veenstra, T.D., 2009. FXR acetylation is normally dynamically regulated by p300 and SIRT1 but constitutively elevated in metabolic disease states. Cell Metab. 10, 392–404.

Kim, J.K., Noh, J.H., Jung, K.H., Eun, J.W., Bae, H.J., Kim, M.G., Chang, Y.G., Shen, Q., Park, W.S., Lee, J.Y., Borlak, J., Nam, S.W., 2013. Sirtuin7 oncogenic potential in human hepatocellular carcinoma and its regulation by the tumor suppressors MiR-125a-5p and MiR-125b. Hepatology 57, 1055–1067.

Lagouge, M., Argmann, C., Gerhart-Hines, Z., Meziane, H., Lerin, C., Daussin, F., Messadeq, N., Milne, J., Lambert, P., Elliott, P., Geny, B., Laakso, M., Puigserver, P., Auwerx, J., 2006. Resveratrol improves mitochondrial function and protects against metabolic disease by activating SIRT1 and PGC-1alpha. Cell 127, 1109–1122.

Laurent, G., de Boer, V.C., Finley, L.W., Sweeney, M., Lu, H., Schug, T.T., Cen, Y., Jeong, S.M., Li, X., Sauve, A.A., Haigis, M.C., 2013. SIRT4 represses peroxisome proliferator-activated receptor α activity to suppress hepatic fat oxidation. Mol. Cell. Biol. 33, 4552–4561.

Lee, J.H., Song, M.Y., Song, E.K., Kim, E.K., Moon, W.S., Han, M.K., Park, J.W., Kwon, K.B., Park, B.H., 2009. Overexpression of SIRT1 protects pancreatic beta-cells against cytokine toxicity by suppressing the nuclear factor-kappaB signaling pathway. Diabetes 58, 344–351.

Li, X., Kazgan, N., 2011. Mammalian sirtuins and energy metabolism. Int. J. Biol. Sci. 7, 575–587.

Li, X., Zhang, S., Blander, G., Tse, J.G., Krieger, M., Guarente, L., 2007. SIRT1 deacetylates and positively regulates the nuclear receptor LXR. Mol. Cell 28, 91–106.

Li, Y., Xu, S., Giles, A., Nakamura, K., Lee, J.W., Hou, X., Donmez, G., Li, J., Luo, Z., Walsh, K., Guarentem, L., Zang, M., 2011. Hepatic overexpression of SIRT1 in mice attenuates endoplasmic reticulum stress and insulin resistance in the liver. FASEB J. 25, 1664–1679.

Li, Y., Wong, K., Giles, A., Jiang, J., Lee, J.W., Adams, A.C., Kharitonenkov, A., Yang, Q., Gao, B., Guarente, L., Zang, M., 2014. Hepatic SIRT1 attenuates hepatic steatosis and controls energy balance in mice by inducing fibroblast growth factor 21. Gastroenterology 146, 539–549.

Li, L., Shi, L., Yang, S., Yan, R., Zhang, D., Yang, J., He, L., Li, W., Yi, X., Sun, L., Liang, J., Cheng, Z., Shi, L., Shang, Y., Yu, W., 2016. SIRT7 is a histone desuccinylase that functionally links to chromatin compaction and genome stability. Nat. Commun. 7, 12235.

Liu, Y., Dentin, R., Chen, D., Hedrick, S., Ravnskjaer, K., Schenk, S., Milne, J., Meyers, D.J., Cole, P., Yates 3rd, J., Olefsky, J., Guarente, L., Montminy, M., 2008. A fasting inducible switch modulates gluconeogenesis via activator/coactivator exchange. Nature 456, 269–273.

Mayoral, R., Osborn, O., McNelis, J., Johnson, A.M., Oh, D.Y., Izquierdo, C.L., Chung, H., Li, P., Traves, P.G., Bandyopadhyay, G., Pessentheiner, A.R., Ofrecio, J.M., Cook, J.R., Qiang, L., Accili, D., Olefsky, J.M., 2015. Adipocyte SIRT1 knockout promotes PPARγ activity, adipogenesis and insulin sensitivity in chronic-HFD and obesity. Mol. Metab. 4, 378–391.

Mercken, E.M., Mitchell, S.J., Martin-Montalvo, A., Minor, R.K., Almeida, M., Gomes, A.P., Scheibye-Knudsen, M., Palacios, H.H., Licata, J.J., Zhang, Y., Becker, K.G., Khraiwesh, H., González-Reyes, J.A., Villalba, J.M., Baur, J.A., Elliott, P., Westphal, C., Vlasuk, G.P., Ellis, J.L., Sinclair, D.A., Bernier, M., de Cabo, R., 2014. SRT2104 extends survival of male mice on a standard diet and preserves bone and muscle mass. Aging Cell 13, 787–796.

Milne, J.C., Lambert, P.D., Schenk, S., Carney, D.P., Smith, J.J., Gagne, D.J., Jin, L., Boss, O., Perni, R.B., Vu, C.B., Bemis, J.E., Xie, R., Disch, J.S., Ng, P.Y., Nunes, J.J., Lynch, A.V., Yang, H., Galonek, H., Israelian, K., Choy, W., Iffland, A., Lavu, S., Medvedik, O., Sinclair, D.A., Olefsky, J.M., Jirousek, M.R., Elliott, P.J., Westphal, C.H., 2007. Small molecule activators of SIRT1 as therapeutics for the treatment of type 2 diabetes. Nature 450, 712–716.

Mohrin, M., Shin, J., Liu, Y., Brown, K., Luo, H., Xi, Y., Haynes, C.M., Chen, D., 2015. Stem cell aging. A mitochondrial UPR-mediated metabolic checkpoint regulates hematopoietic stem cell aging. Science 347, 1374–1377.

Moynihan, K.A., Grimm, A.A., Plueger, M.M., Bernal-Mizrachi, E., Ford, E., Cras-Méneur, C., Permutt, M.A., Imai, S., 2005. Increased dosage of mammalian Sir2 in pancreatic beta cells enhances glucose-stimulated insulin secretion in mice. Cell Metab. 2, 105–117.

Nakahata, Y., Kaluzova, M., Grimaldi, B., Sahar, S., Hirayama, J., Chen, D., Guarente, L.P., Sassone-Corsi, P., 2008. The NAD+-dependent deacetylase SIRT1 modulates CLOCK-mediated chromatin remodeling and circadian control. Cell 134, 329–340.

Nakahata, Y., Sahar, S., Astarita, G., Kaluzova, M., Sassone-Corsi, P., 2009. Circadian control of the NAD+ salvage pathway by CLOCK-SIRT1. Science 324, 654–657.

Panda, S., 2016. Circadian physiology of metabolism. Science 354, 1008–1015.

Pfluger, P.T., Herranz, D., Velasco-Miguel, S., Serrano, M., Tschöp, M.H., 2008. Sirt1 protects against high-fat diet-induced metabolic damage. Proc. Natl. Acad. Sci. U. S. A. 105, 9793–9798.

Philp, A., Chen, A., Lan, D., Meyer, G.A., Murphy, A.N., Knapp, A.E., Olfert, I.M., McCurdy, C.E., Marcotte, G.R., Hogan, M.C., Baar, K., Schenk, S., 2011. Sirtuin 1 (SIRT1) deacetylase activity is not required for mitochondrial biogenesis or peroxisome proliferator-activated receptor-gamma coactivator-1alpha (PGC-1alpha) deacetylation following endurance exercise. J. Biol. Chem. 286, 30561–30570.

Picard, F., Kurtev, M., Chung, N., Topark-Ngarm, A., Senawong, T., Machado De Oliveira, R., Leid, M., McBurney, M.W., Guarente, L., 2004. Sirt1 promotes fat mobilization in white adipocytes by repressing PPAR-gamma. Nature 429, 771–776.

Ponugoti, B., Kim, D.H., Xiao, Z., Smith, Z., Miao, J., Zang, M., Wu, S.Y., Chiang, C.M., Veenstra, T.D., Kemper, J.K., 2010. SIRT1 deacetylates and inhibits SREBP-1C activity in regulation of hepatic lipid metabolism. J. Biol. Chem. 285, 33959–33970.

Purushotham, A., Schug, T.T., Xu, Q., Surapureddi, S., Guo, X., Li, X., 2009. Hepatocyte-specific deletion of SIRT1 alters fatty acid metabolism and results in hepatic steatosis and inflammation. Cell Metab. 9, 327–338.

Purushotham, A., Xu, Q., Lu, J., Foley, J.F., Yan, X., Kim, D.H., Kemper, J.K., Li, X., 2012. Hepatic deletion of SIRT1 decreases hepatocyte nuclear factor 1α/farnesoid X receptor signaling and induces formation of cholesterol gallstones in mice. Mol. Cell. Biol. 32, 1226–1236.

Qiang, L., Wang, L., Kon, N., Zhao, W., Lee, S., Zhang, Y., Rosenbaum, M., Zhao, Y., Gu, W., Farmer, S.R., Accili, D., 2012. Brown remodeling of white adipose tissue by SirT1-dependent deacetylation of Pparγ. Cell 150, 620–632.

Ramadori, G., Fujikawa, T., Fukuda, M., Anderson, J., Morgan, D.A., Mostoslavsky, R., Stuart, R.C., Perello, M., Vianna, C.R., Nillni, E.A., Rahmouni, K., Coppari, R., 2010. SIRT1 deacetylase in POMC neurons is required for homeostatic defenses against diet-induced obesity. Cell Metab. 12, 78–87.

Ramadori, G., Fujikawa, T., Anderson, J., Berglund, E.D., Frazao, R., Michán, S., Vianna, C.R., Sinclair, D.A., Elias, C.F., Coppari, R., 2011. SIRT1 deacetylase in SF1 neurons protects against metabolic imbalance. Cell Metab. 14, 301–312.

Ramsey, K.M., Yoshino, J., Brace, C.S., Abrassart, D., Kobayashi, Y., Marcheva, B., Hong, H.K., Chong, J.L., Buhr, E.D., Lee, C., Takahashi, J.S., Imai, S., Bass, J., 2009. Circadian clock feedback cycle through NAMPT-mediated NAD+ biosynthesis. Science 324, 651–654.

Rodgers, J.T., Puigserver, P., 2007. Fasting-dependent glucose and lipid metabolic response through hepatic sirtuin 1. Proc. Natl. Acad. Sci. U. S. A. 104, 12861–12866.

Rodgers, J.T., Lerin, C., Haas, W., Gygi, S.P., Spiegelman, B.M., Puigserver, P., 2005. Nutrient control of glucose homeostasis through a complex of PGC-1alpha and SIRT1. Nature 434, 113–118.

Ryu, D., Jo, Y.S., Lo Sasso, G., Stein, S., Zhang, H., Perino, A., Lee, J.U., Zeviani, M., Romand, R., Hottiger, M.O., Schoonjans, K., Auwerx, J., 2014. A SIRT7-dependent acetylation switch of GABPβ1 controls mitochondrial function. Cell Metab. 20, 856–869.

Samuel, V.T., Shulman, G.I., 2012. Mechanisms for insulin resistance: common threads and missing links. Cell 148, 852–871.

Sasaki, T., Kikuchi, O., Shimpuku, M., Susanti, V.Y., Yokota-Hashimoto, H., Taguchi, R., Shibusawa, N., Sato, T., Tang, L., Amano, K., Kitazumi, T., Kuroko, M., Fujita, Y., Maruyama, J., Lee, Y.S., Kobayashi, M., Nakagawa, T., Minokoshi, Y., Harada, A., Yamada, M., Kitamura, T., 2014. Hypothalamic SIRT1 prevents age-associated weight gain by improving leptin sensitivity in mice. Diabetologia 57, 819–831.

Satoh, A., Brace, C.S., Ben-Josef, G., West, T., Wozniak, D.F., Holtzman, D.M., Herzog, E.D., Imai, S., 2010. SIRT1 promotes the central adaptive response to diet restriction through activation of the dorsomedial and lateral nuclei of the hypothalamus. J. Neurosci. 30, 10220–10232.

Satoh, A., Brace, C.S., Rensing, N., Cliften, P., Wozniak, D.F., Herzog, E.D., Yamada, K.A., Imai, S., 2013. Sirt1 extends life span and delays aging in mice through the regulation of Nk2 homeobox 1 in the DMH and LH. Cell Metab. 18, 416–430.

Schenk, S., McCurdy, C.E., Philp, A., Chen, M.Z., Holliday, M.J., Bandyopadhyay, G.K., Osborn, O., Baar, K., Olefsky, J.M., 2011. Sirt1 enhances skeletal muscle insulin sensitivity in mice during caloric restriction. J. Clin. Invest. 121, 4281–4288.

Schug, T.T., Xu, Q., Gao, H., Peres-da-Silva, A., Draper, D.W., Fessler, M.B., Purushotham, A., Li, X., 2010. Myeloid deletion of SIRT1 induces inflammatory signaling in response to environmental stress. Mol. Cell. Biol. 30, 4712–4721.

Shin, J., He, M., Liu, Y., Paredes, S., Villanova, L., Brown, K., Qiu, X., Nabavi, N., Mohrin, M., Wojnoonski, K., Li, P., Cheng, H.L., Murphy, A.J., Valenzuela, D.M.,

Luo, H., Kapahi, P., Krauss, R., Mostoslavsky, R., Yancopoulos, G.D., Alt, F.W., Chua, K.F., Chen, D., 2013. SIRT7 represses Myc activity to suppress ER stress and prevent fatty liver disease. Cell Rep. 5, 654–665.

Suganami, T., Tanaka, M., Ogawa, Y., 2012. Adipose tissue inflammation and ectopic lipid accumulation. Endocr. J. 59, 849–857.

Sun, C., Zhang, F., Ge, X., Yan, T., Chen, X., Shi, X., Zhai, Q., 2007. SIRT1 improves insulin sensitivity under insulin-resistant conditions by repressing PTP1B. Cell Metab. 6, 307–319.

Timmers, S., Konings, E., Bilet, L., Houtkooper, R.H., van de Weijer, T., Goossens, G.H., Hoeks, J., van der Krieken, S., Ryu, D., Kersten, S., Moonen-Kornips, E., Hesselink, M.K., Kunz, I., Schrauwen-Hinderling, V.B., Blaak, E.E., Auwerx, J., Schrauwen, P., 2011. Calorie restriction-like effects of 30 days of resveratrol supplementation on energy metabolism and metabolic profile in obese humans. Cell Metab. 14, 612–622.

Tong, Z., Wang, Y., Zhang, X., Kim, D.D., Sadhukhan, S., Hao, Q., Lin, H., 2016. SIRT7 is activated by DNA and deacetylates histone H3 in the chromatin context. ACS Chem. Biol. 11, 742–747.

Tong, Z., Wang, M., Wang, Y., Kim, D.D., Grenier, J.K., Cao, J., Sadhukhan, S., Hao, Q., Lin, H., 2017. SIRT7 is an RNA-activated protein lysine deacylase. ACS Chem. Biol. 12, 300–310.

Vakhrusheva, O., Smolka, C., Gajawada, P., Kostin, S., Boettger, T., Kubin, T., Braun, T., Bober, E., 2008. Sirt7 increases stress resistance of cardiomyocytes and prevents apoptosis and inflammatory cardiomyopathy in mice. Circ. Res. 102, 703–710.

Vazquez, B.N., Thackray, J.K., Simonet, N.G., Kane-Goldsmith, N., Martinez-Redondo, P., Nguyen, T., Bunting, S., Vaquero, A., Tischfield, J.A., Serrano, L., 2016. SIRT7 promotes genome integrity and modulates non-homologous end joining DNA repair. EMBO J. 35, 1488–1503.

Walker, A.K., Yang, F., Jiang, K., Ji, J.Y., Watts, J.L., Purushotham, A., Boss, O., Hirsch, M.L., Ribich, S., Smith, J.J., Israelian, K., Westphal, C.H., Rodgers, J.T., Shioda, T., Elson, S.L., Mulligan, P., Najafi-Shoushtari, H., Black, J.C., Thakur, J.K., Kadyk, L.C., Whetstine, J.R., Mostoslavsky, R., Puigserver, P., Li, X., Dyson, N.J., Hart, A.C., Näär, A.M., 2010. Conserved role of SIRT1 orthologs in fasting-dependent inhibition of the lipid/cholesterol regulator SREBP. Genes Dev. 24, 1403–1417.

Wang, R.H., Kim, H.S., Xiao, C., Xu, X., Gavrilova, O., Deng, C.X., 2011. Hepatic Sirt1 deficiency in mice impairs mTorc2/Akt signaling and results in hyperglycemia, oxidative damage, and insulin resistance. J. Clin. Invest. 121, 4477–4490.

Weisberg, S.P., McCann, D., Desai, M., Rosenbaum, M., Leibel, R.L., Ferrante Jr., A.W., 2003. Obesity is associated with macrophage accumulation in adipose tissue. J. Clin. Invest. 112, 4712–4721.

White, A.T., Philp, A., Fridolfsson, H.N., Schilling, J.M., Murphy, A.N., Hamilton, D.L., McCurdy, C.E., Patel, H.H., Schenk, S., 2014. High-fat diet-induced impairment of skeletal muscle insulin sensitivity is not prevented by SIRT1 overexpression. Am. J. Physiol. Endocrinol. Metab. 307, E764–772.

Yamagata, K., Karim, M.F., Sato, Y., Yoshizawa, T., 2015. Role of SIRT7 in hepatic lipid metabolism. Diabetol. Int. 6, 193–196.

Yamamoto, T., Tamaki, K., Shirakawa, K., Ito, K., Yan, X., Katsumata, Y., Anzai, A., Matsuhashi, T., Endo, J., Inaba, T., Tsubota, K., Sano, M., Fukuda, K., Shinmura, K., 2016. Cardiac Sirt1 mediates the cardioprotective effect of caloric restriction by suppressing local complement system activation after ischemia-reperfusion. Am. J. Physiol. Heart Circ. Physiol. 310, H1003–1014.

Yeung, F., Hoberg, J.E., Ramsey, C.S., Keller, M.D., Jones, D.R., Frye, R.A., Mayo, M.W., 2004. Modulation of NF-kappaB-dependent transcription and cell survival by the SIRT1 deacetylase. EMBO J. 23, 2369–2380.

Yoon, M.J., Yoshida, M., Johnson, S., Takikawa, A., Usui, I., Tobe, K., Nakagawa, T., Yoshino, J., Imai, S., 2015. SIRT1-mediated eNAMPT secretion from adipose tissue regulates hypothalamic NAD+ and function in mice. Cell Metab. 21, 706–717.

Yoshino, J., Mills, K.F., Yoon, M.J., Imai, S., 2011. Nicotinamide mononucleotide, a key NAD(+) intermediate, treats the pathophysiology of diet- and age-induced diabetes in mice. Cell Metab. 14, 528–536.

Yoshino, J., Conte, C., Fontana, L., Mittendorfer, B., Imai, S., Schechtman, K.B., Gu, C., Kunz, I., Rossi Fanelli, F., Patterson, B.W., Klein, S., 2012. Resveratrol supplementation does not improve metabolic function in nonobese women with normal glucose tolerance. Cell Metab. 16, 658–664.

Yoshizaki, T., Milne, J.C., Imamura, T., Schenk, S., Sonoda, N., Babendure, J.L., Lu, J.C., Smith, J.J., Jirousek, M.R., Olefsky, J.M., 2009. SIRT1 exerts anti-inflammatory effects and improves insulin sensitivity in adipocytes. Mol. Cell. Biol. 29, 1363–1374.

Yoshizaki, T., Schenk, S., Imamura, T., Babendure, J.L., Sonoda, N., Bae, E.J., Oh, D.Y., Lu, M., Milne, J.C., Westphal, C., Bandyopadhyay, G., Olefsky, J.M., 2010. SIRT1 inhibits inflammatory pathways in macrophages and modulates insulin sensitivity. Am. J. Physiol. Endocrinol. Metab. 298, E419–428.

Yoshizawa, T., Karim, M.F., Sato, Y., Senokuchi, T., Miyata, K., Fukuda, T., Go, C., Tasaki, M., Uchimura, K., Kadomatsu, T., Tian, Z., Smolka, C., Sawa, T., Takeya, M., Tomizawa, K., Ando, Y., Araki, E., Akaike, T., Braun, T., Oike, Y., Bober, E., Yamagata, K., 2014. SIRT7 controls hepatic lipid metabolism by regulating the ubiquitin-proteasome pathway. Cell Metab. 19, 712–721.

Yu, H., Ye, W., Wu, J., Meng, X., Liu, R.Y., Ying, X., Zhou, Y., Wang, H., Pan, C., Huang, W., 2014. Overexpression of sirt7 exhibits oncogenic property and serves as a prognostic factor in colorectal cancer. Clin. Cancer Res. 20, 3434–3445.

Zhang, S., Chen, P., Huang, Z., Hu, X., Chen, M., Hu, S., Hu, Y., Cai, T., 2015. Sirt7 promotes gastric cancer growth and inhibits apoptosis by epigenetically inhibiting miR-34a. Sci. Rep. 5, 9787.

CHAPTER FIVE

Regulation of Gene Transcription Following Stimulation of Transient Receptor Potential (TRP) Channels

Gerald Thiel[1], Andrea Lesch, Sandra Rubil, Tobias M. Backes, Oliver G. Rössler

Saarland University Medical Faculty, Homburg, Germany

[1]Corresponding author: e-mail address: gerald.thiel@uks.eu

Contents

Abstract

Transient receptor potential (TRP) channels belong to a heterogeneous superfamily of cation channels that are involved in the regulation of numerous biological functions, including regulation of Ca^{2+} and glucose homeostasis, tumorigenesis, temperature, and pain sensation. To understand the functions of TRP channels, their associated intracellular signaling pathways and molecular targets have to be identified on the cellular level. Stimulation of TRP channels frequently induces an influx of Ca^{2+} ions into the cells and the subsequent activation of protein kinases. These intracellular signal transduction pathways ultimately induce changes in the gene expression pattern of the cells. Here, we review the effects of TRPC6, TRPM3, and TRPV1 channel stimulation on the activation of the stimulus-responsive transcription factors AP-1, CREB, Egr-1, Elk-1, and NFAT. Following activation, these transcription factors induce the transcription of delayed response genes. We propose that many biological functions of TRP channels can be explained by the activation of stimulus-responsive transcription factors and their delayed response genes. The proteins encoded by those delayed response genes may be responsible for the biochemical and physiological changes following TRP channel activation.

International Review of Cell and Molecular Biology, Volume 335
ISSN 1937-6448
https://doi.org/10.1016/bs.ircmb.2017.07.010

1. INTRODUCTION

Transient receptor potential (TRP) channels belong to a heterogeneous superfamily of cation channels that are separated into six subfamilies: TRPC (canonical), TRPV (vanilloid), TRPM (melastatin), TRPML (mucolipins), TRPP (polycystin), and TRPA (ankyrin). All TRP channels share a basic architecture with six transmembrane regions and a pore region between transmembrane regions five and six. Ankyrin repeats are found in the N-terminal portion of TRPA, TRPC, and TRPV channels. A typical feature of most TRP channels is their permeability to Ca^{2+}, with TRPM3 and TRPA1 acting as relatively efficient Ca^{2+} channels (Nilius and Flockerzi, 2014). TRP channels regulate sensory transduction pathways, including pain sensation, thermosensation, mechanosensation, and perception of pungent chemicals. Other biological functions include cardiovascular regulation, Ca^{2+} homeostasis, tumorigenesis, neurotransmission, cell proliferation, and cell death (Nilius, 2007; Nilius and Szallasi, 2014; Nilius et al., 2007). Research on TRP channels has been rapidly growing within the last years. The modular structures of the channel proteins have been elucidated together with biophysical analyses of channel conductances. Modulators of TRP channels such as calmodulin and phosphoinositides have been discovered and characterized. In addition, numerous seminal papers have been published, describing the results of gain-of-function and loss-of-function mutations of particular TRP channels. The intracellular signaling pathways triggered by TRP channel activation have not been analyzed in detail. Here, we review the data describing how TRP channel activation leads to changes in the gene expression pattern of the cells by activating stimulus-responsive transcription factors and inducing transcription of their delayed response genes. Stimulus-responsive transcription factors are rapidly but transiently activated as response to cellular stimulation. The activation may be executed via stimulation of their biosynthesis, phosphorylation/dephosphorylation, or nuclear translocation. The activated transcription factors bind to genetic elements of delayed response genes and stimulate transcription of these genes. The gene products of the delayed response genes trigger the biochemical and physiological changes observed as a result of cellular stimulation.

TRP channels can be activated by a wide range of stimuli, including temperature, mechanical or osmotic stress, redox metabolites, naturally derived products such as dietary compounds or plant toxins, endogenous lipids, or synthetic ligands. The physiological ligands are often not yet known, or may be not entirely TRP channel specific. TRPC channels are, for instance, activated by lipids that are generated as a result of Gαq-coupled receptor stimulation. Thus, two signaling pathways are simultaneously activated, one as a result of G protein-coupled receptor stimulation and the other as a result of TRP channel activation. Likewise, *N*-arachidonoyldopamine functions as a ligand for both TRPV1 channels and CB_1 cannabinoid receptors. To analyze TRP channel-mediated signaling and to avoid interference with other signaling pathways, plant-derived metabolites or synthetic compounds are frequently employed to stimulate particular TRP channels. Here, we will focus on the TRP channels TRPC6, TRPM3, and TRPV1 that can specifically be activated by plant-derived or pharmacological ligands.

2. TRP CHANNEL STIMULATION ACTIVATES STIMULUS-RESPONSIVE TRANSCRIPTION FACTORS

2.1 TRPC6

TRPC6 channels belong to the subfamily of transient receptor potential canonical (TRPC) channels. The modular structure of TRPC6 is depicted in Fig. 1A. TRPC6 channels can be activated by stimulation of G protein-coupled receptors or directly by diacylglycerol that is synthesized in the cells via phospholipase C-catalyzed hydrolysis of phosphatidylinositol 4,5-bisphosphate. The membrane-permeable diacylglycerol analogue 1-oleoyl-2-acetyl-*sn*-glycerol (OAG) is often used as an activator of TRPC6 channels (Hofmann et al., 1999). In addition, hyperforin, a main ingredient of *Hypericum perforatum* (St. John's wort), has been identified as a pharmacological activator of TRPC6 (Leuner et al., 2007). Recent results confirmed that hyperforin functions as a ligand for TRPC6 channels (Soni and Abebiyi, 2016; Thiel and Rössler, 2017).

TRPC6 channels are found in the kidney, in smooth muscle containing tissues, in the brain, the immune system, and a variety of biological functions have been connected with the activation of TRPC6 channels (Bouron et al., 2016; Dietrich and Gudermann, 2014). TRPC6 channels are expressed in podocytes and are a component of the glomerular slit diaphragm (Reiser et al., 2005). Mutations in the TRPC6 gene have been associated with focal segmental glomerulosclerosis (Winn et al., 2005). Activation of TRPC6 channels regulates acute hypoxic pulmonary vasoconstriction and alveolar gas exchange (Weissmann et al., 2006) and smooth muscle contractility in smooth muscle tissues (Dietrich et al., 2005). In the nervous system, TRPC6 channel activation promotes dendritic growth in hippocampal cultures and

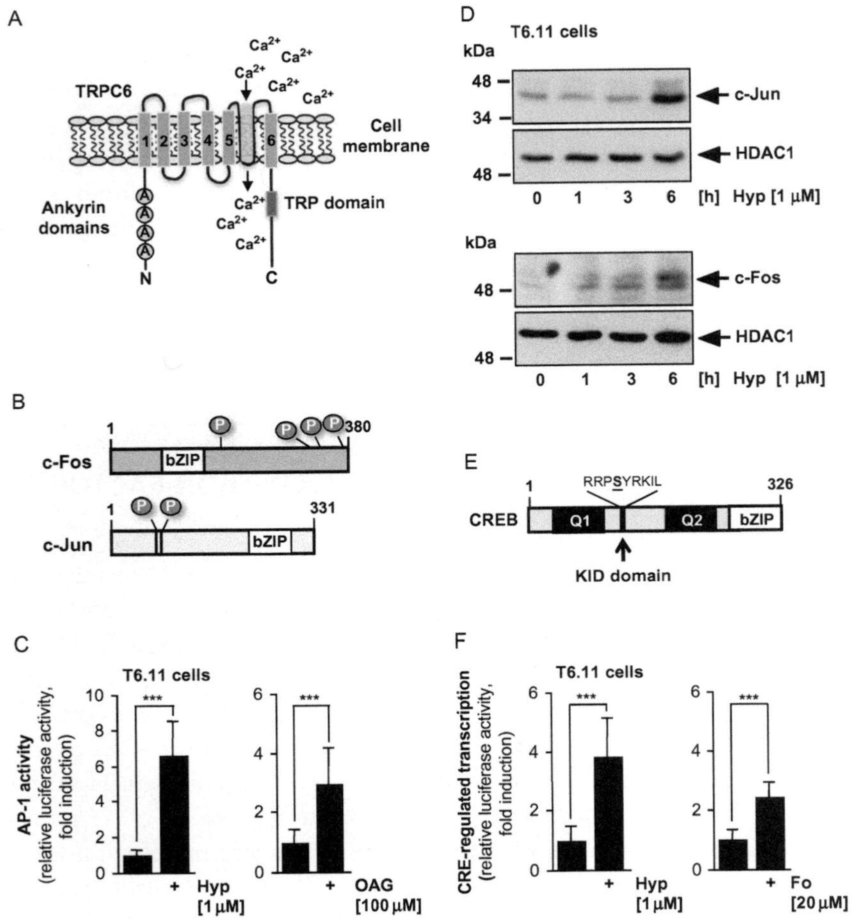

Fig. 1 Activation of AP-1 and CRE-regulated gene transcription following stimulation of TRPC6 channels. (A) Schematic representation of the modular structure of the TRPC6 channel. TRPC6 has six transmembrane domains with a pore-forming domain between transmembrane regions five and six. Both the N- and C-termini are located in the cytosol. The N-terminus contains four ankyrin domains which may function as a protein–protein interaction platform. TRPC6 channels have a TRP box, which is found in several TRP channel proteins. In addition, binding sites for calmodulin, phosphoinositides, and the IP_3 receptor have been identified. (B) Schematic representation of the modular structure of the basic region leucine zipper (bZIP) proteins c-Fos and c-Jun. The bZIP domains are indicated as well as phosphorylation-regulated activation domains. (C) Stimulation of TRPC6-expressing T6.11 cells with the TRPC6 ligands hyperforin (Hyp) or 1-oleoyl-2-acetyl-*sn*-glycerol (OAG) activates the transcription factor AP-1 (***, $P < 0.001$). (D) Induction of c-Jun and c-Fos biosynthesis in hyperforin-stimulated T6.11 cells. Nuclear extracts of stimulated cells were prepared and subjected to Western blot analysis using antibodies directed against either c-Jun or c-Fos. An antibody

the formation of excitatory synapses. TRPC6 may also be involved in the development of human glioma (Ding et al., 2010; Tai et al., 2008; Zhou et al., 2008).

TRPC6 is a nonselective cation channel that allows the influx of Ca^{2+} and Na^{+} ions into the cells via a pore between the fifth and sixth transmembrane domains. The channel is six times more permeable to Ca^{2+} than to Na^{+} (Dietrich and Gudermann, 2014). Thus, stimulation of TRPC6 channels results in elevated intracellular Ca^{2+} ion concentrations (Dietrich and Gudermann, 2014; Gibon et al., 2011; Tu et al., 2008). The channel properties of TRPC6 are influenced by calmodulin, phosphoinositides, and steroids (Dietrich and Gudermann, 2014; Miehe et al., 2012; Thiel and Rössler, 2017). The rise in intracellular Ca^{2+}, triggered by the stimulation of TRPC6 channels, results in the activation of Ca^{2+}/calmodulin-dependent protein kinase CaMKIV. In addition, MAP kinases are activated, in particular the MAP kinase extracellular signal-regulated protein kinase (ERK) (Chiluiza et al., 2013; Heiser et al., 2013).

Stimulation of TRPC6 channels leads to the activation of the stimulus-responsive transcription factor activator protein-1 (AP-1), initially identified as a heterodimer of the basic region leucine zipper (bZIP) proteins c-Jun and c-Fos (Chiu et al., 1988). The bZIP proteins dimerize with their leucine zipper domain and bind to the acidic DNA with their basic DNA-binding region. AP-1 is a cellular convergence point for intracellular signaling cascades that are initiated by the stimulation of cells with cytokines, mitogens, hormones, neurotransmitters, nutrients, and stressors such as UV light or arsenite. The activation of AP-1 is involved in the tissue-specific regulation of many biological functions, e.g., cellular proliferation, transformation, differentiation, and apoptosis (Shaulian and Karin, 2002). Fig. 1B shows the

directed against HDAC1 was used as a loading control. (E) Schematic representation of the modular structure of the CRE-binding protein (CREB). CREB has a C-terminal bZIP domain responsible for DNA binding and dimerization. Transcriptional activation is mediated via two constitutively active glutamine-rich activation domains (Q1 and Q2) and the kinase-inducible domain (KID) which is a phosphorylation-regulated transcriptional activation domain. (F) Stimulation of TRPC6 channels with hyperforin (1 μM) activates transcription via the cyclic AMP-response element (CRE). As a control, cells were treated with forskolin (Fo, 20 μM) (***, $P < 0.001$). *Reproduced with modifications from Thiel, G., Rössler, O.G., 2017. Hyperforin activates gene transcription involving transient receptor potential C6 channels. Biochem. Pharmacol. 129, 96–107 with kind permission from Elsevier Inc.*

modular structures of the bZIP proteins c-Jun and c-Fos, including the bZIP and the transactivation domains. The N-terminal transcriptional activation domain of c-Jun contains the serine residues 63 and 73 that are phosphorylated by c-Jun N-terminal kinase (JNK). Phosphorylation of JNK is necessary for activating c-Jun (Morton et al., 2003).

Stimulation of TRPC6-expressing T6.11 cells with the TRPC6 ligands hyperforin or OAG significantly increased cellular AP-1 activity (Fig. 1C). Mutational analysis identified the AP-1 motif encompassing the sequence 5′-TGAGTCA-3′ as the responsible genetic element. These results were corroborated by the finding that stimulation of TRPC6 channels with hyperforin results in an upregulation of c-Jun and c-Fos biosynthesis (Fig. 1D) that together form the AP-1 transcription factor complex (Thiel and Rössler, 2017).

AP-1 binds to the regulatory regions of delayed response genes and activates transcription of these genes. Earlier observations showed that hyperforin activates expression of the proinflammatory cytokine interleukin-8 (IL-8), also termed CXCL8, in human intestinal epithelial cells (Zhou et al., 2004). IL-8 is a potent chemoattractant and neutrophil activator and IL-8 secretion has been connected with numerous inflammatory diseases. We recently confirmed that TRPC6-initiated AP-1 activation stimulates IL-8 gene transcription, indicating that the IL-8 gene is an AP-1-regulated delayed response gene (S. Rubil, A. Lesch, N. Mukaida, G. Thiel, manuscript in preparation).

TRPC6 channels have been shown to be necessary for brain-derived neurotrophic factor-induced CREB phosphorylation and cAMP-response element (CRE)-mediated reporter gene transcription in cerebellar granule neurons (Jia et al., 2007). The CRE encompasses the palindromic sequence 5′-TGACGTCA-3′ and functions as a binding site for the CRE-binding protein (CREB), a member of the family of bZIP proteins (Fig. 1E). CREB translates increased intracellular cAMP and Ca^{2+} levels into an altered gene expression program and integrates signaling pathways activated by the stimulation of the cells with hormones, neurotransmitters, neurotrophins, cytokines, or mitogens (Barco and Kandel, 2006; Johannessen et al., 2004; Mayr and Montminy, 2001). TRPC6 regulates the formation of excitatory synapses. This activity relies on the activation of Ca^{2+}/calmodulin-dependent protein kinase CaMKIV and phosphorylation of CREB (Zhou et al., 2008). Moreover, in a cell culture system, stimulation of TRPC6 channels has been shown to activate CRE-mediated transcription (Fig. 1F) (Thiel and Rössler, 2017). Hyperforin was even a stronger activator of CRE-mediated gene transcription than forskolin, an adenylate cyclase activator (Fig. 1F).

The rise in intracellular Ca^{2+}, triggered by the stimulation of TRPC6 channels, is accompanied by the activation of the Ca^{2+}-dependent protein phosphatase calcineurin that has been correlated with neonatal glomerular mesangial cell apoptosis, myoblast transformation, and pathologic cardiac modeling (Kuwahara et al., 2006; Soni and Abebiyi, 2016). Calcineurin dephosphorylates the cytoplasmic subunits of NFATc transcription factors, leading to their nuclear translocation. Therefore, NFATc proteins bind to DNA together with nuclear partner proteins and stimulate transcription of NFAT-regulated genes. Accordingly, stimulation of TRPC6 channels causes nuclear translocation of NFATc1 in neonatal glomerular mesangial cells (Soni and Abebiyi, 2016). Likewise, expression of gain-of-function mutations of TRPC6 channels triggered a constitutive activation of NFAT-regulated gene transcription and has been associated with focal segmental glomerulosclerosis (Schlöndorff et al., 2009). In support of this, it has been shown that activation of NFAT in podocytes causes glomerulosclerosis (Wang et al., 2010), indicating that NFAT activation represents an intermediate step connecting mutant TRPC6 channels with the development of focal segmental glomerulosclerosis. Fig. 2 summarizes the signaling pathway of TRPC6, leading to the activation of the transcription factors AP-1, CREB, and NFATc.

2.2 TRPM3

TRPM3 channels belong to the subfamily of transient receptor potential melastatin (TRPM) channels that have exceptionally long N-terminal and C-terminal domains. The modular structure, depicted in Fig. 3A, shows that TRPM3 channels have the typical features of TRP channels including six transmembrane domains and a pore domain between the fifth and sixth transmembrane domains. Both the amino and carboxy termini are located in the cytosol. TRPM3 channels have a TRP domain located on the C-terminal side of the sixth transmembrane domain. Two calmodulin-binding sites have been detected within the N-terminal cytosolic portion of TRPM3 (Holakovska et al., 2012). In addition, phosphatidylinositol 4,5-bisphosphate, which is required for TRPM3 activity, interacts with the N-terminal calmodulin-binding sites (Badheka et al., 2015; Holendova et al., 2012).

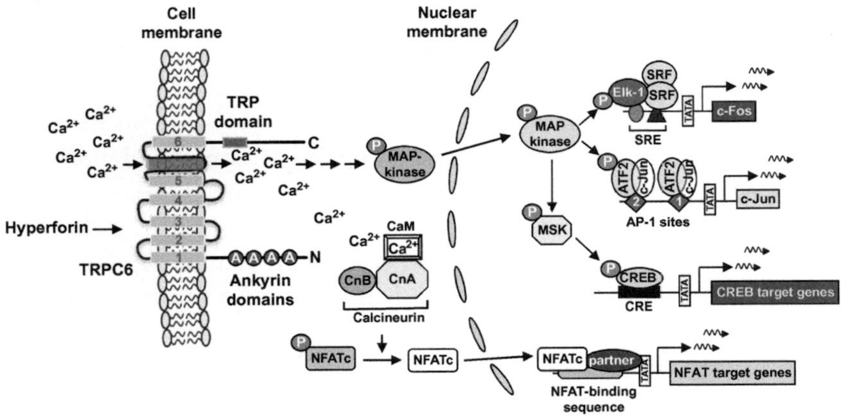

Fig. 2 Stimulation of TRPC6 channels triggers an intracellular signaling cascade, leading to enhanced gene transcription. Stimulation of TRPC6 channels results in an influx of Ca^{2+} ions into the cells. Elevation of the intracellular Ca^{2+} concentration leads to the activation of MAP kinases that translocate into the nucleus following activation. In the nucleus, TRPC6 signaling activates transcription of the c-Jun and c-Fos genes that encode bZIP transcription factors constituting the AP-1 transcription factor complex. Elevated Ca^{2+} levels, together with the activation of MAP kinases ERK and stress-activated kinase MSK, stimulate gene transcription via the CRE which functions as a Ca^{2+}-response element. The influx of Ca^{2+} ions into the cells also activates the Ca^{2+}-dependent protein phosphatase calcineurin that dephosphorylates the catalytic subunit of the NFATc transcription factor in the cytosol. Dephosphorylated NFATc translocates into the nucleus and activates gene transcription of NFAT-controlled genes together with nuclear partners. In podocytes, NFAT-controlled genes are important for podocyte function and maintenance. *Reproduced with modifications from Thiel, G., Rössler, O.G., 2017. Hyperforin activates gene transcription involving transient receptor potential C6 channels. Biochem. Pharmacol. 129, 96–107 with kind permission from Elsevier Inc.*

Many TRPM3 isoforms are generated via alternative splicing and show different activity in terms of their Ca^{2+} permeabilities. The endogeneous ligands for TRPM3 are not yet identified. However, the neurosteroid pregnenolone sulfate has been shown to function as a pharmacological ligand for TRPM3 channels (Lesch et al., 2014; Naylor et al., 2010; Wagner et al., 2008). A synthetic TRPM3 ligand, CIM0216, has recently been described, exhibiting higher potency and affinity in comparison to pregnenolone sulfate (Held et al., 2015). Studies performed with neurons derived from TRPM3-deficient mice confirmed that TRPM3 is the major receptor for pregnenolone sulfate or CIM0216 in sensory neurons (Held et al., 2015; Vriens et al., 2011).

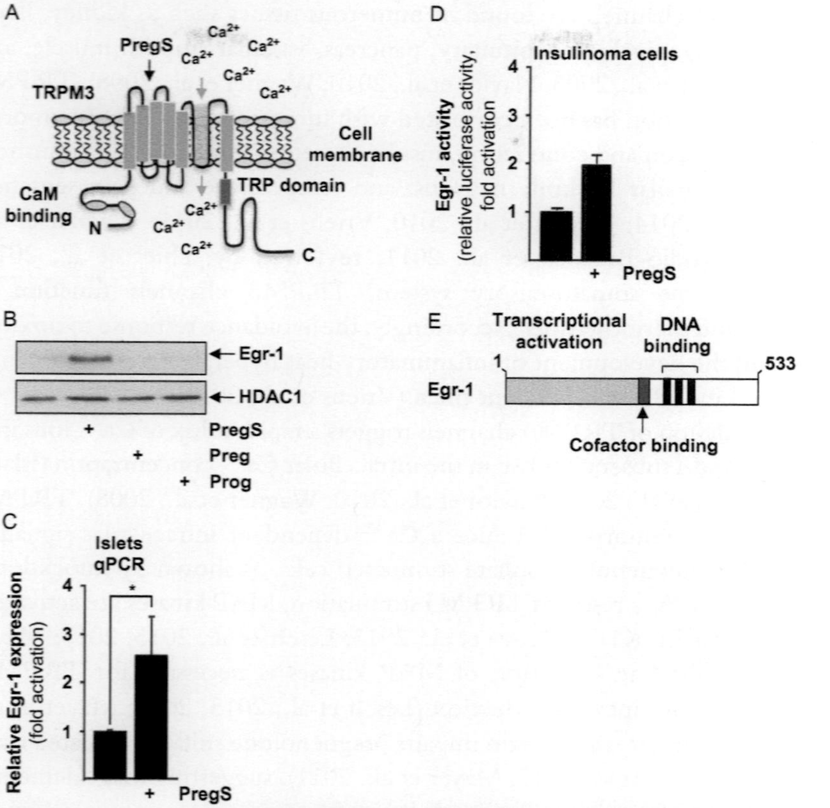

Fig. 3 Stimulation of Egr-1 biosynthesis and upregulation of Egr-1 activity following stimulation of TRPM3 channels. (A) Schematic representation of the modular structure of TRPM3 channels. TRPM3 has six transmembrane domains with a pore-forming domain between transmembrane regions five and six. The N-terminus contains two calmodulin-binding sites encompassing amino acids 35–124 and 291–382. The C-terminus contains the TRP domain on the C-terminal side of the sixth transmembrane domain. (B) Induction of Egr-1 biosynthesis in pregnenolone sulfate-stimulated INS-1 insulinoma cells that express endogenous TRPM3 channels. Cells were stimulated with either pregnenolone sulfate (PregS, 50 μM), pregnenolone (Preg, 50 μM), or progesterone (Prog, 50 μM). Egr-1 expression was assessed by Western blotting. (C) Induction of Egr-1 biosynthesis in pregnenolone sulfate-stimulated pancreatic islets. Quantitative real-time PCR of RNA isolated from stimulated (PregS, 50 μM) or control pancreatic islet cells in primary short-term culture. The Egr-1 signal was normalized to the β-actin signal and the fold-stimulation was calculated using the $2^{-\Delta\Delta Ct}$ equation (*, $P < 0.05$). (D) The newly synthesized Egr-1 protein is biologically active. INS-1 insulinoma cells were infected with a recombinant lentivirus encoding an Egr-1-responsive reporter gene. The infected cells were stimulated with pregnenolone sulfate for 24 h. (E) Modular structure of the zinc finger protein Egr-1. *Reproduced with modifications from Mayer, S.I.,*

(Continued)

TRPM3 channels are found in numerous tissues such as kidney, liver, ovary, brain, spinal cord, pituitary, pancreas, vascular smooth muscle, and testis (Grimm et al., 2003, Naylor et al., 2010; Wagner et al., 2008). TRPM3 channel activation has been correlated with tumorigenesis, vascular smooth muscle secretion and contraction, insulin secretion, glutamatergic transmission of cerebellar Purkinje neurons, and temperature and pain sensation (Hall et al., 2014; Naylor et al., 2010; Vriens et al., 2011; Wagner et al., 2008; Zamudio-Bulcock et al., 2011; reviewed by Thiel et al., 2013, 2017). In the somatosensory system, TRPM3 channels function as chemo- and thermosensors. Accordingly, the avoidance response to noxious heat and the development of inflammatory heat hyperalgesia is significantly impaired in TRPM3-deficient mice (Vriens et al., 2011).

Stimulation of TRPM3 channels triggers a rapid influx of Ca^{2+} ions into the cells and a subsequent rise in the intracellular Ca^{2+} concentration (Islam, 2011; Lesch et al., 2015; Naylor et al., 2010; Wagner et al., 2008). TRPM3 expression is required to induce a Ca^{2+}-dependent intracellular signaling cascade in pregnenolone sulfate stimulated cells, as shown by knockdown experiments. As a result of TRPM3 stimulation, MAP kinases are activated, in particular ERK1/2 (Heiser et al., 2013; Lesch et al., 2015, 2017; Mayer et al., 2011). The activation of MAP kinases is necessary for TRPM3-mediated transcriptional induction (Lesch et al., 2015, 2017; Mayer et al., 2011). In contrast, calcineurin impairs pregnenolone sulfate-regulated transcription (Lesch et al., 2017; Mayer et al., 2011), suggesting that calcineurin may be part of a negative feedback loop.

Insulinoma cells and pancreatic β-cells express TRPM3 channels. Stimulation of insulinoma cells with the TRPM3 ligand pregnenolone sulfate activates the biosynthesis of the transcription factor Egr-1 in insulinoma cells and pancreatic islets (Fig. 3B and C). Accordingly, the transcriptional activity of Egr-1 is upregulated as a result of TRPM3 channel stimulation (Fig. 3D). The modular structure of Egr-1 is depicted in Fig. 3E, showing the DNA-binding domain with three zinc finger motifs, an N-terminal transcriptional activation domain, and an inhibitory domain located between the activation domain and the DNA-binding domain that functions as a

Fig. 3—Cont'd Müller, I., Mannebach, S., Endo, T., Thiel, G., 2011. Signal transduction of pregnenolone sulfate in insulinoma cells. Activation of Egr-1 expression involving TRPM3, voltage-gated calcium channels, ERK, and ternary complex factors. J. Biol. Chem. 286, 10084–10096 with kind permission from The American Society for Biochemistry and Molecular Biology.

binding site for the transcriptional cofactors NAB1 and NAB2. Egr-1 expression is activated by many extracellular signaling molecules such as growth factors, hormones, neurotransmitters, nutrients, and metabolites and Egr-1 function has been correlated with the control of neuronal cell death, neuronal plasticity, proliferation, reproduction, and inflammation (Lee et al., 1996; Mayer et al., 2009; Topilko et al., 1998; Wei et al., 2000; Xie et al., 2011; Yan et al., 2000). In pancreatic β-cells, Egr-1 and related Egr proteins are required to control insulin biosynthesis, glucose homeostasis, and proliferation/cell death of pancreatic β-cells in vivo (Müller et al., 2012). In a search for Egr-1-regulated target genes in pregnenolone sulfate-stimulated insulinoma cells, the gene encoding PDX-1 was identified (Mayer et al., 2011). Egr-1 binds to the PDX-1 gene under physiological conditions. PDX-1 is a major regulator of insulin expression in pancreatic β-cells (Andrali et al., 2008; Brissova et al., 2002). Accordingly, pregnenolone sulfate stimulation enhanced transcription of an insulin promoter-controlled reporter gene and elevated insulin mRNA levels were detected in pregnenolone sulfate-stimulated insulinoma cells (Mayer et al., 2011).

The 5′-flanking region of the Egr-1 gene contains five serum-response elements (SREs). The SRE consists of the CArG box with the consensus sequence $CC[A/T]_6GG$ that binds a dimer of the serum-response factor (SRF). Adjacent to the CArG box, there is often a binding site for ternary complex factors (TCFs) with the core sequence GGAA/T (Fig. 4A). These TCFs bind to their cognate DNA sequence and to the SRF-dimer. Fig. 4B shows that the TCF protein Elk-1 has an N-terminal DNA-binding domain, a C-terminal activation domain, and a binding site for SRF (domain B). Elk-1 activity is regulated by MAP kinases that phosphorylate serine residues S383 and S389 within the transcriptional activation domain. Experiments performed with HEK293 cells expressing TRPM3 channels showed that the transcriptional activation potential of Elk-1 was significantly elevated following stimulation of TRPM3 channels (Fig. 4C) (Lesch et al., 2014), suggesting that stimulation of TRPM3 channels triggered the phosphorylation and activation of Elk-1. Expression of a dominant-negative mutant of Elk-1 completely prevented the induction of Egr-1 biosynthesis in INS-1 cells that had been stimulated with pregnenolone sulfate, indicating that TCF activation is essential for inducing Egr-1 transcription in pregnenolone sulfate-stimulated insulinoma cells (Mayer et al., 2011) (Fig. 4D).

Stimulation of TRPM3 channels with pregnenolone sulfate or the synthetic TRPM3 ligand CIM0216 leads to a significant elevation of AP-1

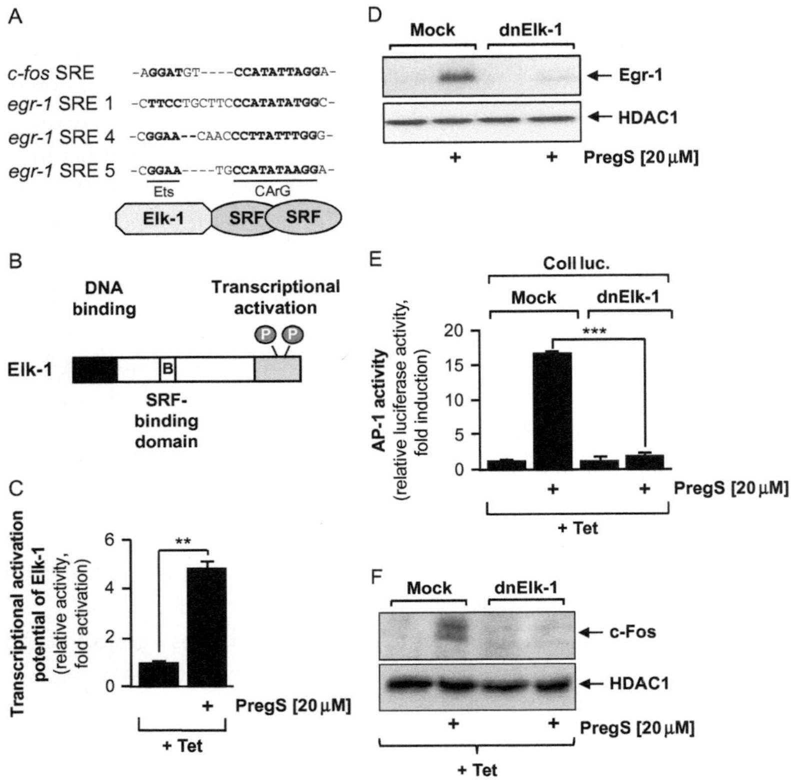

Fig. 4 Stimulation of TRPM3 channels activates the ternary complex factor Elk-1. (A) The serum-response element (SRE) consists of the binding site for a dimer of the serum-response factor (SRF) and an adjacent binding site for Elk-1 or related ternary complex factors. (B) Modular structure of Elk-1. (C) Upregulation of the transcriptional activation potential of Elk-1 in pregnenolone sulfate-stimulated HEK293 cells expressing TRPM3 channels. Cells were treated with pregnenolone sulfate (PregS, 20 mM) (**, $P < 0.01$). (D) Essential role of ternary complex factor activation for the pregnenolone sulfate-induced upregulation of Egr-1 expression in insulinoma cells. Expression of a dominant-negative mutant of Elk-1 blocks the pregnenolone sulfate-induced upregulation of Egr-1 in insulinoma cells. (E) Essential role of ternary complex factors for controlling AP-1 activity in HEK293 cells expressing stimulated TRPM3 channels. Expression of a dominant-negative mutant of Elk-1 blocks pregnenolone sulfate-induced upregulation of AP-1-mediated transcription in HEK293 cells expressing TRPM3 (***, $P < 0.001$). (F) Expression of a dominant-negative mutant of Elk-1 attenuates the upregulation of c-Fos expression in HEK293 cells expressing activated TRPM3 channels. Cells were stimulated with pregnenolone sulfate (PregS, 20 μM). Nuclear extracts were prepared and subjected to Western blot analysis using an antibody directed against the c-Fos protein. The antibody directed against HDAC1 was used as a loading control. *Reproduced with modifications from Mayer, S.I., Müller, I., Mannebach, S., Endo, T.,*

activity in insulinoma cells and HEK293 cells expressing TRPM3 channels (Lesch et al., 2014, 2015, 2017; Müller et al., 2011; Rubil and Thiel, 2017). Recently, the IL-8 gene has been identified as a delayed response gene for the TRPM3-induced signaling cascade (S. Rubil, A. Lesch, N. Mukaida, G. Thiel, manuscript submitted). Stimulation of insulinoma cells with pregnenolone sulfate induced the phosphorylation of c-Jun, which turned c-Jun into an active transcription factor (Müller et al., 2011). Moreover, c-Jun has been shown to be essential for the upregulation of AP-1-mediated gene transcription following stimulation of TRPM3 channels (Lesch et al., 2015). In addition, the TCF Elk-1 functions as a regulator of AP-1 (Kaufmann et al., 2013; Müller et al., 2010; Thiel and Rössler, 2014, 2017; Thiel et al., 2012). Expression of a dominant-negative mutant of Elk-1 almost completely prevented activation of AP-1-regulated gene transcription in TRPM3 expressing HEK293 cells that had been stimulated with pregnenolone sulfate (Fig. 4E) (Lesch et al., 2015). Expression of the dominant-negative Elk-1-mutant reduced c-Fos expression in HEK293 cells expressing activated TRPM3 channels (Fig. 4F) (Rubil et al., 2016), indicating that TCF proteins regulate AP-1 activity via stimulation of c-Fos expression. Thus, TCF activation is essential for connecting TRPM3 activation with transcription of AP-1-controlled genes. The central position of Elk-1 within the TRPM3-induced signaling cascade is depicted in Fig. 5.

Stimulation of TRPM3 channels also induced the phosphorylation of the transcription factor CREB and increased transcription of a CRE-controlled reporter gene (Müller et al., 2011). A transcriptional target of activated CREB is the c-Fos gene, encoding a constituent of the AP-1 transcription factor (Rubil et al., 2016).

In summary, stimulation of TRPM3 channels activates the stimulus-responsive transcription factors AP-1, CREB, Egr-1, and Elk-1, suggesting

Thiel, G., 2011. Signal transduction of pregnenolone sulfate in insulinoma cells. Activation of Egr-1 expression involving TRPM3, voltage-gated calcium channels, ERK, and ternary complex factors. J. Biol. Chem. 286, 10084–10096; Lesch, A., Rubil, S., Thiel, G., 2014. Activation and inhibition of transient receptor potential TRPM3-induced gene transcription. Br. J. Pharmacol. 171, 2645–2658; Rubil, S., Rössler, O.G., Thiel, G., 2016. CREB, AP-1, ternary complex factors and MAP kinases connect transient receptor potential melastatin-3 (TRPM3) channel stimulation with increased c-Fos expression. Br. J. Pharmacol. 173, 305–318 with kind permission from The American Society for Biochemistry and Molecular Biology and John Wiley and Sons.

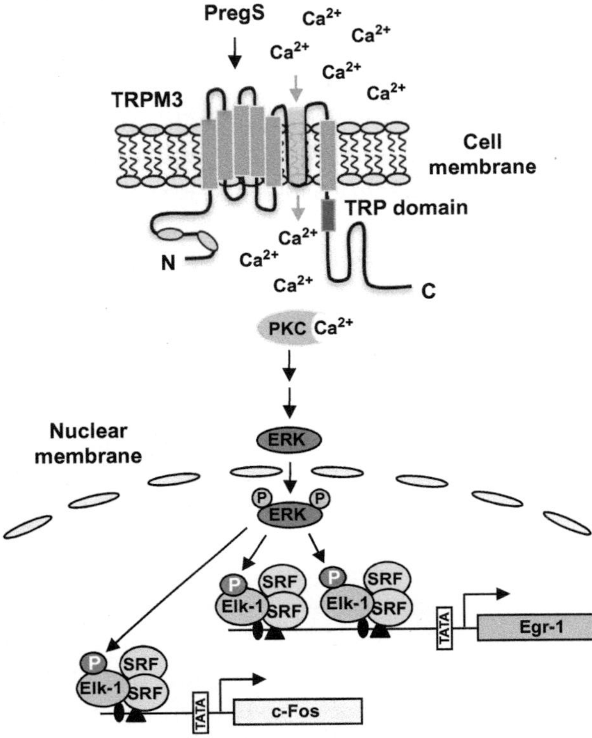

Fig. 5 Central role of ternary complex factor in regulating TRPM3-induced gene transcription. Stimulation of insulinoma cells with pregnenolone sulfate leads to the activation of TRPM3. As a result, Ca^{2+} ions flow through the TRPM3 pore from outside into the cells, increasing the cytosolic Ca^{2+} concentration. The rise in intracellular Ca^{2+} leads to the activation of the extracellular signal-regulated protein kinase (ERK) signaling cascade, most likely via activation of protein kinase C (PKC) that directly or indirectly regulates Raf activity. Phosphorylated and activated ERK translocates into the nucleus and catalyzes the phosphorylation of substrate proteins, including the ternary complex factor Elk-1, an essential component of the serum-response element (SRE) ternary complex. Elk-1 regulates the expression of Egr-1 and c-Fos via SREs in their promoters.

that transcriptional changes may be an important aspect for the biological functions of TRPM3.

2.3 TRPV1

TRPV1 channels belong to the subfamily of transient receptor potential vanilloid (TRPV) channels and show the typical modular structure of TRP channels (Fig. 6) (Bevan et al., 2014). The central ion channel is formed by the transmembrane regions five and six. TRPV1 channels

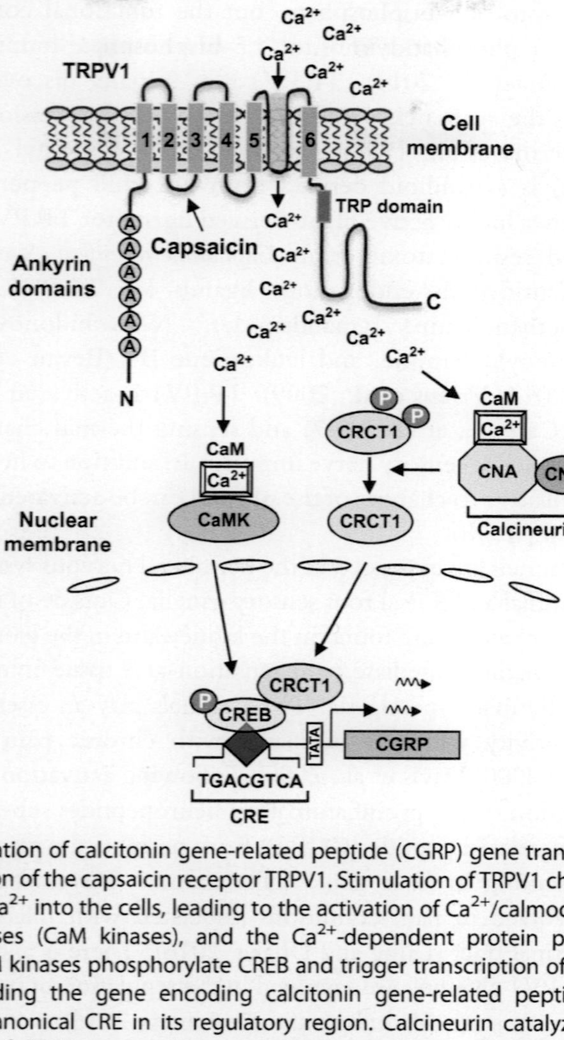

Fig. 6 Regulation of calcitonin gene-related peptide (CGRP) gene transcription following stimulation of the capsaicin receptor TRPV1. Stimulation of TRPV1 channels results in an influx of Ca^{2+} into the cells, leading to the activation of Ca^{2+}/calmodulin-dependent protein kinases (CaM kinases), and the Ca^{2+}-dependent protein phosphatase calcineurin. CaM kinases phosphorylate CREB and trigger transcription of CREB-regulated genes, including the gene encoding calcitonin gene-related peptide (CGRP) that contains a canonical CRE in its regulatory region. Calcineurin catalyzes the dephosphorylation of CREB-regulated transcriptional coactivator-1 (CRTC1), resulting in its nuclear translocation. CRTC1 binds CREB and supports transcription of CREB-regulated target genes.

have six ankyrin repeats at the N-terminal portion of the molecule that represents a multiligand-binding domain for ATP, phosphatidylinositol 4,5-bisphosphate, and calmodulin. A TRP-box is located close to the sixth transmembrane domain that functions as an association domain and regulates channel gating. TRPV1 channels bind calmodulin and phosphatidylinositol 4,5-bisphosphate, but the functional consequences of calmodulin or phosphatidylinositol 4,5-bisphosphate binding is controversial (Bevan et al., 2014). TRPV1 was initially discovered while searching for the capsaicin receptor using an expression cloning approach (Caterina et al., 1997). Capsaicin (*trans*-8-methyl-*N*-vanillyl-6-nonenamide) is a vanilloid derived from hot chili pepper (*Capsicum annuum*). Another highly active plant-derived ligand for TRPV1 channels is the vanilloid resiniferatoxin from *Euphorbia resinifera*. Various lipids have been identified as endogenous ligands for TRPV1, including *N*-arachidonylethanolamine (anandamide), *N*-arachidonoyldopamine (NADA), *N*-oleoyldopamine, and leukotriene B4 (Bevan et al., 2014; Meotti et al., 2014; Vriens et al., 2009). TRPV1 is activated by noxious temperatures (Caterina et al., 1997) and sensing thermal changes of the environment results in sensory nerve impulses. In addition to heat, TRPV1 channels are sensitive to changes of the pH and can be activated by protons (Tominaga et al., 1998).

TRPV1 channels are expressed in the peripheral nervous system, in particular in trigeminal and dorsal root sensory ganglia. Outside of the nervous system, TRPV1 channels are found in the kidney and in the gastrointestinal tract. TRPV1 channels mediate pain sensation and tissue injury-induced inflammatory thermal hyperalgesia. The channels play an essential role in pain hypersensitivity which is associated with chronic pain conditions (Caterina et al., 2000; Davis et al., 2000). Following activation of TRPV1 channels, secretion of the proinflammatory neuropeptides substance P and calcitonin gene-related peptide (CGRP) from peripheral neuronal termini is initiated, resulting in neurogenic inflammation (Devesa et al., 2014). TRPV1 is involved in pain conditions associated with inflammation in the gastrointestinal tract (Laing and Dhaka, 2016). There is additional evidence that TRPV1 channels are involved in the regulation of insulin release and glucose homeostasis (Zsombok and Derbenev, 2016).

TRPV1 channels have a high Ca^{2+} permeability and function as Ca^{2+} permeable cation channels (Caterina et al., 1997). The influx of Ca^{2+} ions into the cells and the subsequent rise in the intracellular Ca^{2+} concentration is often used as an indicator for the activation of TRPV1 channels (Caterina et al., 1997, 2000; Gawa et al., 2005). The rise in Ca^{2+} activates intracellular protein kinases in the cells, including Ca^{2+}/calmodulin-dependent protein kinases and MAP kinases, in particular extracellular signal-regulated protein kinase ERK1/2 and p38 protein kinase (Amantini et al., 2007; T.M. Backes, O. G. Rössler, G. Thiel, unpublished observations; Chen et al., 2009).

Protons, released from infiltrating inflammatory cells, stimulate calcitonin gene-related peptide expression (CGRP) in sensory neurons. CGRP is a neuropeptide of 37 amino acids that is encoded by the calcitonin gene and is generated by alternative splicing. CGRP promotes neurogenic inflammation. The upregulation of CGRP expression is attenuated in the presence of TRPV1 inhibitors and abolished in TRPV1-deficient dorsal root ganglion neurons, indicating that protons directly activate TRPV1 channels, leading to an upregulation of CGRP gene transcription (Nakanishi et al., 2010). The stimulation of TRPV1 channels is accompanied by the phosphorylation of CREB and an upregulation of CRE-mediated gene transcription in dorsal root ganglia. The CGRP gene contains a classical CRE in its regulatory region and experiments confirmed that this motif is essential to couple proton-induced activation of TRPV1 channels with enhanced CGRP gene transcription (Nakanishi et al., 2010). Pharmacological experiments proposed that Ca^{2+}/calmodulin-dependent protein kinases function as signal transducer connecting TRPV1 channels with CREB activation and CGPR transcription (Nakanishi et al., 2010) (Fig. 6). The connection between TRPV1 and CREB is corroborated by the observation that loss of TRPV1 promotes the nuclear exclusion of CREB-regulated transcriptional coactivator CRTC1, also known as transducer of regulated CREB activity 1 (TORC1), in pain sensory neurons. Accordingly, the transcriptional activity of CREB is reduced in TRPV1-deficient dorsal root ganglion neurons, leading to a reduction of gene transcription of CREB-regulated genes (Riera et al., 2014).

In addition to CREB, TRPV1 channel stimulation increased the biosynthesis of c-Jun and c-Fos and activates transcription mediated by the transcription factor AP-1 (T.M. Backes, O.G. Rössler, X. Hui, C. Grötzinger, P. Lipp, G. Thiel, manuscript submitted). The increase in intracellular Ca^{2+} also activates NFAT nuclear translocation via dephosphorylation of NFAT by calcineurin (Jia et al., 2014).

3. CONCLUSION AND FUTURE PROSPECTS

There is ample evidence that stimulation of TRP channels activates stimulus-responsive transcription factors, including AP-1, CREB, Egr-1, Elk-1, and NFAT. In particular, the activation of AP-1 and CREB seems to be a common theme of TRP channel activation. We propose that activation of these proteins explains many of the biological functions connected with TRP channel activation. To corroborate this hypothesis, the delayed response genes of these stimulus-responsive transcription factors need to be identified in different tissues and cell types expressing TRP channels using microarray or next-generation RNA sequencing technologies. Despite the ubiquitous expression of stimulus-responsive transcription factors, cell type-specific target genes have already been identified, for example the PDX-1 gene in insulinoma cells expressing stimulated TRPM3 channels (Mayer et al., 2011), or the CGRP gene in sensory neurons that express activated TRPV1 channels (Nakanishi et al., 2010). Thus, despite their ubiquitous expression, stimulus-responsive transcription factors activate cell type-specific target genes. Epigenetic regulators that induce an open or compact chromatin architecture via histone acetylation or methylation, decide whether a stimulus-responsive transcription factor will have access to a particular target gene. CREB, for instance, has been shown to bind in vivo only to a portion of its cognate sites (Cha-Molstad et al., 2004), indicating that chromatin restricts access of transcription factors to their target genes in a cell-specific manner. Various epigenetic regulators have been identified that determine the biological activities of stimulus-responsive transcription factors. The transcriptional coactivator p300 and its structural homologue CREB-binding protein (CBP) are histone acetyltransferases that regulate the activities of CREB, c-Jun, c-Fos, Egr-1, Elk-1, and other transcription factors (Vo and Goodman, 2001). The histone deacetylase SIRT1 has been described to influence AP-1, c-Jun, and CREB-regulated gene transcription (Gao and Ye, 2008; Monteserin-Garcia et al., 2013; Zhang et al., 2010). The identification and characterization of cell type-specific delayed response genes and their biological functions links two branches of TRP research, the molecular analyses of TRP channel structure, including the elucidation of the conductance properties, performed with easy cellular models, and the research employing whole-animal models with deficiencies in particular TRP channels.

ACKNOWLEDGMENT

We thank Libby Guethlein for critical reading of the manuscript.

REFERENCES

Amantini, C., Mosca, M., Nabissi, M., Lucciarini, R., Caprodossi, S., Arcella, A., Giangaspero, F., Santoni, G., 2007. Capsaicin-induced apoptosis of glioma cells is mediated by TRPV1 vanilloid receptor and requires p38 MAPK activation. J. Neurochem. 102, 977–990.

Andrali, S.S., Sampley, M.L., Vanderford, N.L., Özcan, S., 2008. Glucose regulation of insulin gene expression in pancreatic β-cells. Biochem. J. 415, 1–10.

Badheka, D., Borbiro, I., Rohacs, T., 2015. Transient receptor potential melastatin 3 is a phosphoinositide-dependent ion channel. J. Gen. Physiol. 146, 65–77.

Barco, A., Kandel, E.R., 2006. The role of CREB and CBP in brain function. In: Thiel, G. (Ed.), Transcription Factors in The Nervous System—Development, Brain Function, and Diseases. Wiley-VCH Verlag, Weinheim, pp. 207–241.

Bevan, S., Quallo, T., Andersson, D.A., 2014. TRPV1. In: Nilius, B., Flockerzi, V. (Eds.), Mammalian Transient Receptor Potential (TRP) Cation Channels. In: Handbook of Experimental Pharmacology, vol. 222. Springer-Verlag, Berlin Heidelberg, pp. 207–240.

Bouron, A., Chauvet, S., Dryer, S., Rosado, J.A., 2016. Second messenger-operated calcium entry through TRPC6. In: Rosado, J.A. (Ed.), Calcium Entry Pathways in Non-Excitable Cells. In: Advances in Experimental Medicine and Biology, vol. 898, pp. 201–250.

Brissova, M., Shiota, M., Nicholson, W.E., Gannon, M., Knobel, S.M., Piston, D.W., Wright, C.V.E., Powers, A.C., 2002. Reduction of pancreatic transcription factor PDX-1 impairs glucose-stimulated insulin secretion. J. Biol. Chem. 277, 11225–11232.

Caterina, M.J., Schumacher, M.A., Tominago, M., Rosen, T.A., Levine, J.D., Julius, D., 1997. The capsaicin receptor: a heat-activated ion channel in the pain pathway. Nature 389, 816–824.

Caterina, M.J., Leffler, A., Malmberg, A.B., Martin, W.J., Trafton, J., Petersen-Zeitz, K.R., Koltzenburg, M., Basbaum, A.I., Julius, D., 2000. Impaired nociception and pain sensation in mice lacking the capsaicin receptor. Science 288, 306–313.

Cha-Molstad, H., Keller, D.M., Yochum, G.S., Impey, S., Goodman, R.H., 2004. Cell-type-specific binding of the transcription factor CREB to the cAMP-response element. Proc. Natl. Acad. Sci. U.S.A. 101, 13572–13577.

Chen, Y., Willcockson, H.H., Valtschanoff, J.G., 2009. Vanilloid receptor TRPV1-mediated phosphorylation of ERK in murine adjuvant arthritis. Osteoarthr. Cartil. 17, 244–251.

Chiluiza, D., Krishna, S., Schumacher, V.A., Schlöndorff, J., 2013. Gain-of-function mutations in transient receptor potential C6 (TRPC6) activate extracellular signal-regulated kinases 1/2 (ERK1/2). J. Biol. Chem. 288, 18407–18420.

Chiu, R., Boyle, W.J., Meek, J., Smeal, T., Hunter, T., Karin, M., 1988. The c-fos protein interacts with c-jun/AP-1 to stimulate transcription of AP-1 responsive genes. Cell 54, 541–552.

Davis, J.B., Gray, J., Gunthorpe, M.J., Hatcher, J.P., Davey, P.T., Overend, P., Harries, M.H., Latchman, J., Clapham, C., Atkinson, K., Hugher, S.A., Rance, K., Grau, E., Harper, A.J., Pugh, P.L., Rogers, D.C., Bingham, S., Randall, A., Sheardown, S.A., 2000. Vanilloid receptor-1 is essential for inflammatory thermal hyperalgesia. Nature 405, 183–187.

Devesa, I., Ferrándiz-Huertas, C., Mathivanan, S., Wolf, C., Luján, R., Changeux, J.P., Ferrer-Montiel, A., 2014. CGRP is essential for algesic exocytotic mobilization of TRPV1 channels in peptidergic nociceptors. Proc. Natl. Acad. Sci. U.S.A. 111 (18), 345–18350.

Dietrich, A., Gudermann, T., 2014. TRPC6: physiological function and pathophysiological relevance. In: Nilius, B., Flockerzi, V. (Eds.), Mammalian Transient Receptor Potential (TRP) Cation Channels. In: Handbook of Experimental Pharmacology, vol. 222. Springer-Verlag, Berlin Heidelberg, pp. 157–188.

Dietrich, A., Mederos, Y.S., Gollasch, M., Gross, V., Storch, U., Dubrovska, G., Obst, M., Yildirim, E., Salanova, B., Kalwa, H., Essin, K., Pinkenburg, O., Luft, F.C., Gudermann, T., Birnbaumer, L., 2005. Increased vascular smooth muscle contractility in TRPC6 −/− mice. Mol. Cell. Biol. 25, 6980–6989.

Ding, X., He, Z., Zhou, K., Cheng, J., Yao, H., Lu, D., Cai, R., Jin, Y., Dong, B., Xu, Y., Wang, Y., 2010. Essential role of TRPC6 channels in G2/M phase transition and development of human glioma. J. Natl. Cancer Inst. 102, 1–17.

Gao, Z., Ye, J., 2008. Inhibition of transcriptional activity of c-Jun by SIRT1. Biochem. Biophys. Res. Comm. 376, 793–796.

Gawa, N.R., Tamir, R., Qu, Y., Klionsky, L., Zhang, T.J., Immke, D., Wang, J., Zhu, D., Vanderah, T.W., Porreca, F., Doherty, E.M., Norman, M.H., Wild, K.D., Bannon, A.W., Louis, J.C., Treanor, J.J., 2005. AMG 9810 [(*E*)-3-(4-*t*-butylphenyl)-*N*-(2,3-dihydrobenzo[*b*] [1,4] dioxin-6-yl) acrylamide], a novel vanilloid receptor 1 (TRPV1) antagonist with antihyperalgesic properties. J. Pharmacol. Exp. Ther. 313, 474–484.

Gibon, J., Tu, P., Bohic, S., Richaud, P., Arnaud, J., Zhu, M., Boulay, G., Bouron, A., 2011. The over-expression of TRPC6 channels in HEK-293 cells favours the intracellular accumulation of zinc. Biochim. Biophys. Acta 1808, 2807–2818.

Grimm, C., Kraft, R., Sauerbruch, S., Schulz, G., Harteneck, C., 2003. Molecular and functional characterization of the melastatin-related cation channel TRPM3. J. Biol. Chem. 278, 21493–21501.

Hall, D.P., Cost, N.G., Hegde, S., Kellner, E., Mikhaylova, O., Stratton, Y., Ehmer, B., Abplanalp, W.A., Paney, R., Biesiada, J., Harteneck, C., Plas, D.R., Meller, J., Czyzyk-Krzeska, M.F., 2014. TRPM3 and miR-204 establish a regulatory circuit that controls oncogenic autophagy in clear cell renal cell carcinoma. Cancer Cell 26, 738–753.

Heiser, J.H., Schuwald, A.M., Sillani, G., Ye, L., Müller, W.E., Leuner, K., 2013. TRPC6 channel-mediated neurite outgrowth in PC12 cells and hippocampal neurons involves activation of RAS/MEK/ERK, PI3K, and CAMKIV signaling. J. Neurochem. 127, 303–313.

Held, K., Kichko, T., De Clercq, K., Klaassen, H., Van Bree, R., Vanherck, J.-C., Marchand, A., Reeh, P.W., Chaltin, P., Voets, T., Vriens, J., 2015. Activation of TRPM3 by a potent synthetic ligand reveals a role in peptide release. Proc. Natl. Acad. Sci. U.S.A. 112, E1363–E1372.

Hofmann, T., Obukhov, A.G., Schäfer, M., Harteneck, C., Gudermann, T., Schulz, G., 1999. Direct activation of human TRPC6 and TRPC3 channels by diacylglycerol. Nature 397, 259–263.

Holakovska, B., Grycova, L., Jirku, M., Sulc, M., Bumba, L., Teisinger, J., 2012. Calmodulin and S100A1 protein interact with N terminus of TRPM3 channel. J. Biol. Chem. 287, 16645–16655.

Holendova, B., Grycova, L., Jirku, M., Teisinger, J., 2012. PtdIns(4,5)P2 interacts with CaM domains on TRPM3 N-terminus. Channels (Austin) 6, 479–482.

Islam, M.S., 2011. TRP channels of islets. In: Transient Receptore Potential Channels. Advances in Experimental Medicine and Biology, vol. 704. , pp. 811–830.

Jia, Y., Zhou, J., Wang, Y., 2007. TRPC channels promote cerebellar granule neuron survival. Nat. Neurosci. 10, 559–567.

Jia, X., Zhang, H., Cao, X., Yin, Y., Zhang, B., 2014. Activation of TRPV1 mediates thymic stromal lymphopoietin release via the Ca^{2+}/NFAT pathway in airway epithelial cells. FEBS Lett. 588, 3047–3054.

Johannessen, M., Delghandi, M.P., Moens, U., 2004. What turns CREB on? Cell. Signal. 16, 1211–1227.

Kaufmann, A., Keim, A., Thiel, G., 2013. Regulation of immediate-early gene transcription following activation of Gαq-coupled designer receptors. J. Cell. Biochem. 114, 681–696.

Kuwahara, K., Wang, Y., McAnally, J., Richardson, J.A., Bassel-Duby, R., Hill, J.A., Olson, E.N., 2006. TRPC6 fullfills a calcineurin signaling circuit during pathologic cardia remodeling. J. Clin. Invest. 116, 3114–3126.

Laing, R.J., Dhaka, A., 2016. ThermoTRP and pain. Neuroscientist 22, 171–187.

Lee, S., Sadovsky, Y., Swirnoff, A.H., Polish, J.A., Goda, P., Gavrilina, G., Milbrandt, J., 1996. Luteinizing hormone deficiency and female infertility in mice lacking the transcription factor NGFI-A (egr-1). Science 273, 1219–1221.

Lesch, A., Rubil, S., Thiel, G., 2014. Activation and inhibition of transient receptor potential TRPM3-induced gene transcription. Br. J. Pharmacol. 171, 2645–2658.

Lesch, A., Hui, X., Lipp, P., Thiel, G., 2015. Transient receptor potential melastatin-3 (TRPM3)-induced activation of AP-1 requires Ca^{2+} ions and the transcription factors c-Jun, ATF2, and ternary complex factor. Mol. Pharmacol. 87, 617–628.

Lesch, A., Rössler, O.G., Thiel, G., 2017. Extracellular signal-regulated protein kinase, c-Jun N-terminal protein kinase, and calcineurin regulate transient receptor potential M3 (TRPM3) induced activation of AP-1. J. Cell. Biochem. 118, 2409–2419.

Leuner, K., Kazanski, V., Müller, M., Essin, K., Henke, B., Gollasch, M., Harteneck, C., Müller, W.E., 2007. Hyperforin—a key constituent of St. John's wort specifically activates TRPC6 channels. FASEB J. 21, 4101–4111.

Mayer, S.I., Rössler, O.G., Endo, T., Charnay, P., Thiel, G., 2009. Epidermal growth factor-induced proliferation of astrocytes requires Egr transcription factors. J. Cell Sci. 122, 3340–3350.

Mayer, S.I., Müller, I., Mannebach, S., Endo, T., Thiel, G., 2011. Signal transduction of pregnenolone sulfate in insulinoma cells. Activation of Egr-1 expression involving TRPM3, voltage-gated calcium channels, ERK, and ternary complex factors. J. Biol. Chem. 286, 10084–10096.

Mayr, B., Montminy, M., 2001. Transcriptional regulation by the phosphorylation-dependent factor CREB. Nat. Rev. Mol. Cell Biol. 2, 599–609.

Meotti, F.C., Lemos de Andrade, E., Calixto, J.B., 2014. TRP modulation by natural compounds. In: Nilius, B., Flockerzi, V. (Eds.), Mammalian Transient Receptor Potential (TRP) Cation Channels. In: Handbook of Experimental Pharmacology, vol. 223. Springer-Verlag, Berlin Heidelberg, pp. 1177–1238.

Miehe, S., Crause, P., Schmidt, T., Löhn, M., Kleemann, H.-W., Licher, T., Dittrich, W., Rütten, H., Strübing, C., 2012. Inhibition of diacylglycerol-sensitive TRPC channels by synthetic and natural steroids. PLoS One 7, e35393.

Monteserin-Garcia, J., Al-Massadi, O., Seoane, L.M., Alvarez, C.V., Shan, B., Stalla, J., Paez-Pereda, M., Casanueva, F.F., Stalla, G.K., Theodoropoulou, M., 2013. Sirt1 inhibits the transcription factor CREB to regulate pituitary growth hormone synthesis. FASEB J. 27, 1561–1571.

Morton, S., Davis, R.J., McLaren, A., Cohen, P., 2003. A reinvestigation of the multisite phosphorylation of the transcription factor c-Jun. EMBO J. 22, 3876–3886.

Müller, I., Endo, T., Thiel, G., 2010. Regulation of AP-1 activity in glucose-stimulated insulinoma cells. J. Cell. Biochem. 110, 1481–1494.

Müller, I., Rössler, O.G., Thiel, G., 2011. Pregnenolone sulfate activates basic region leucine zipper transcription factors in insulionoma cells: role of voltage-gated Ca^{2+} channels and transient receptor potential melastatin 3 channels. Mol. Pharmacol. 80, 1179–1189.

Müller, I., Rössler, O.G., Wittig, C., Menger, M.D., Thiel, G., 2012. Critical role of Egr transcription factors in regulating insulin biosynthesis, blood glucose homeostasis and islet size. Endocrinology 153, 3040–3053.

Nakanishi, M., Hata, K., Nagayama, T., Sakurai, T., Nishisho, T., Wakabayashi, H., Hiraga, T., Ebisu, S., Yoneda, T., 2010. Acid activation of Trpv1 leads to an up-regulation of calcitonin gene-related peptide expression in dorsal root ganglion neurons via the CamK-CREB cascade: a potential mechanism of inflammatory pain. Mol. Biol. Cell 21, 2568–2577.

Naylor, J., Li, J., Milligan, C.J., Zeng, F., Sukumar, P., Hou, B., Sedo, A., Yuldasheva, N., Majeed, Y., Beri, D., Jiang, S., Seymour, V.A.L., McKeown, L., Kumar, B., Harteneck, C., O'Regan, D., Wheatcroft, S.B., Kearney, M.T., Jones, C., Porter, K.E., Beech, D.J., 2010. Pregnenolone sulphate- and cholesterol-regulated channels coupled to vascular smooth muscle secretion and contraction. Circ. Res. 106, 1507–1515.

Nilius, B., 2007. TRP channels in disease. Biochem. Biophys. Acta 1772, 805–812.

Nilius, B., Flockerzi, V., 2014. What do we really know and what do we need to know: some controversies, perspectives, and surprises. In: Nilius, B., Flockerzi, V. (Eds.), Mammalian Transient Receptor Potential (TRP) Cation Channels. In: Handbook of Experimental Pharmacology, vol. 223. Springer-Verlag, Berlin Heidelberg, pp. 1239–1280.

Nilius, B., Szallasi, A., 2014. Transient receptor potential channels as drug targets: from the science of basic research to the art of medicine. Pharmacol. Rev. 66, 676–814.

Nilius, B., Owsianik, G., Voets, T., Peters, J.A., 2007. Transient receptor potential cation channels in disease. Physiol. Rev. 87, 165–217.

Reiser, J., Polu, K.R., Möller, C.C., Kenlan, P., Altintas, M.M., Wei, C., Faul, C., Herbert, S., Villegas, I., Avila-Casado, C., McGee, M., Sugimoto, H., Brown, D., Kalluri, R., Mundel, P., Smith, P.L., Clapham, D.E., Pollak, M.R., 2005. TRPC6 is a glomerular slit diaphragm-associated channel required for normal renal function. Nat. Genet. 37, 739–744.

Riera, C.E., Huising, M.O., Follett, P., Leblanc, M., Halloran, J., Van Andel, R., de Magalhaes Filho, C.D., Merkwirth, C., Dillin, A., 2014. TRPV1 pain receptor regulate longevity and metabolism by neuropeptide signaling. Cell 157, 1023–1036.

Rubil, S., Thiel, G., 2017. Activation of gene transcription via CIM0216, a synthetic ligand of transient receptor potential melastatin-3 (TRPM3) channels. Channels (Austin) 11, 79–83.

Rubil, S., Rössler, O.G., Thiel, G., 2016. CREB, AP-1, ternary complex factors and MAP kinases connect transient receptor potential melastatin-3 (TRPM3) channel stimulation with increased c-Fos expression. Br. J. Pharmacol. 173, 305–318.

Schlöndorff, J., del Camino, D., Carrasquillo, R., Lacey, V., Pollak, M.R., 2009. TRPC6 mutations associated with focal segmental glomerulosclerosis cause constitutive activation of NFAT-dependent transcription. Am. J. Physiol. Cell. Physiol. 296, C558–C569.

Shaulian, E., Karin, M., 2002. AP-1 as a regulator of cell life and death. Nat. Cell Biol. 4, E131–E136.

Soni, H., Abebiyi, A., 2016. TRPC6 channel activation promotes neonatal glomerular mesangial cell apoptosis via calcineurin/NFAT and FasL/Fas signaling pathways. Sci. Rep. 6, 29041.

Tai, T., Feng, S., Ge, R., Du, W., Zhang, X., He, Z., Wang, Y., 2008. TRPC6 channels promote dendritic growth via the CaMKIV-CREB pathway. J. Cell. Sci. 121, 2301–2307.

Thiel, G., Rössler, O.G., 2014. Resveratrol stimulates AP-1-regulated gene transcription. Mol. Nutr. Food Res. 58, 1402–1413.

Thiel, G., Rössler, O.G., 2017. Hyperforin activates gene transcription involving transient receptor potential C6 channels. Biochem. Pharmacol. 129, 96–107.

Thiel, G., Lesch, A., Keim, A., 2012. Transcriptional response to calcium-sensing receptor stimulation. Endocrinology 153, 4716–4728.

Thiel, G., Müller, I., Rössler, O.G., 2013. Signal transduction via TRPM3 in pancreatic β-cells. J. Mol. Endocrinol. 50, R75–R83.

Thiel, G., Rubil, S., Lesch, A., Guethlein, L.A., Rössler, O.G., 2017. Transient receptor potential TRPM3 channels: pharmacology, signaling, and biological functions. Pharmacol. Res. 124, 92–99.

Tominaga, M., Caterina, M.J., Malmberg, A.B., Rosen, T.A., Gilbert, H., Skinner, K., Raumann, B.E., Basbaum, A.I., Julius, D., 1998. The cloned capsaicin receptor integrates multiple pain-producing stimuli. Neuron 21, 531–543.

Topilko, P., Schneider-Maunoury, S., Levi, G., Trembleau, A., Gourji, D., Driancourt, M.-A., Rao, C.V., Charnay, P., 1998. Multiple pituitary and ovarian defects in Krox-24 (NGFI-A, Egr-1)-targeted mice. Mol. Endocrinol. 12, 107–122.

Tu, P., Kunert-Keil, C., Lucke, S., Brinkmeier, H., Bouron, A., 2008. Diacylglycerol analogues activate second-messenger-operated calcium channels exhibiting TRPC-like properties in cortical neurons. J. Neurochem. 108, 126–138.

Vo, N., Goodman, R.H., 2001. CREB-binding protein and p300 in transcriptional regulation. J. Biol. Chem. 276, 13505–13508.
Vriens, J., Appendino, G., Nilius, B., 2009. Pharmacology of vanilloid transient receptor potential cation channels. Mol. Pharmacol. 75, 1262–1279.
Vriens, J., Owsianik, G., Hofmann, T., Philipp, S.E., Stab, J., Chen, X., Benoit, M., Xue, F., Janssens, A., Kerselaers, S., Oberwinkler, J., Vennekens, R., Gudermann, T., Nilius, B., Voets, T., 2011. TRPM3 is a nociceptor channel involved in the detection of noxious heat. Neuron 70, 482–494.
Wagner, T.F.J., Loch, S., Lambert, S., Straub, I., Mannebach, S., Mathar, I., Düfer, M., Lis, A., Flockerzi, V., Philipp, S.E., Oberwinkler, J., 2008. Transient receptor potential M3 channels are ionotropic steroid receptors in pancreatic beta cells. Nat. Cell Biol. 10, 1421–1430.
Wang, Y., Jarad, G., Tripathi, P., Pan, M., Cunningham, J., Martin, D.R., Liapis, H., Miner, J.H., Chen, F., 2010. Activation of NFAT signaling in podocytes causes glomerulosclerosis. J. Am. Soc. Nephrol. 21, 1657–1666.
Wei, F., Xu, Z.C., Qu, Z., Milbrandt, J., Zhuo, M., 2000. Role of EGR1 in hippocampal synaptic enhancement induced by tetanic stimulation and amputation. J. Cell. Biol. 149, 1325–1333.
Weissmann, N., Dietrich, A., Fuchs, B., Kalwa, H., Ay, M., Dumitrascu, R., Olschewski, A., Storch, U., Menderos y Schnitzler, M., Ghofrani, H.A., Schermuly, R.T., Pinkenburg, O., Seeger, W., Grimminger, F., Gudermann, T., 2006. Classical transient receptor potential channel 6 (TRPC6) is essential for hypoxic pulmonary vasoconstriction and aveolar gas exchange. Proc. Natl. Acad. Sci. U.S.A. 103, 19093–19098.
Winn, M.P., Conlon, P.J., Lynn, K.L., Farrington, M.K., Creazzo, T., Hawkins, A.F., Daskalakis, N., Kwan, S.Y., Ebersviller, S., Burchette, J.L., Pericak-Vance, M.A., Howell, D.N., Vance, J.M., Rosenberg, P.B., 2005. A mutation in the *TRPC6* cation channel causes familial focal segmental glomerulosclerosis. Science 308, 1801–1804.
Xie, B., Wang, C., Zheng, Z., Song, B., Ma, C., Thiel, G., Li, M., 2011. Egr-1 transactivates Bim gene expression to promote neuronal apoptosis. J. Neurosci. 31, 5032–5044.
Yan, S.F., Fujita, T., Lu, J., Okada, K., Shan Zou, Y., Mackman, N., Pinsky, D.J., Stern, D.M., 2000. Egr-1, a master switch coordinating upregulation of divergent gene families underlying ischemic stress. Nat. Med. 6, 1355–1361.
Zamudio-Bulcock, P.A., Everett, J., Harteneck, C., Valenzuela, C.F., 2011. Activation of steroid-sensitive TRPM3 channels potentiates glutamaterigic transmission of cerebellar Purkinje neurons from developing rats. J. Neurochem. 119, 474–485.
Zhang, R., Chen, H.-Z., Liu, J.-J., Jia, Y.-Y., Zhang, Z.-Q., Yang, R.-F., Zhang, Y., Xu, J., Wei, Y.-S., Liu, D.-P., Liang, C.C., 2010. SIRT1 suppresses activator protein-1 transcriptional activity and cyclooxygenase-2 expression in macrophages. J. Biol. Chem. 285, 7097–7110.
Zhou, C., Tabb, M.M., Sadatrafiei, A., Grün, F., Sun, A., Blumberg, B., 2004. Hyperforin, the active component of St. John's wort, induces IL-8 expression in human intestinal epithelial cells via a MAPK-dependent, NF-κB-independent pathway. J. Clin. Immunol. 24, 623–636.
Zhou, J., Du, W., Zhou, K., Tai, Y., Yao, H., Jia, Y., Ding, Y., Wang, Y., 2008. Critical role of TRPC6 channels in the formation of excitatory synapses. Nat. Neurosci. 11, 741–743.
Zsombok, A., Derbenev, A.V., 2016. TRP channels as therapeutic targets in diabetes and obesity. Pharmaceuticals 9, 50.

CHAPTER SIX

Transcription: Insights From the HIV-1 Promoter

Enrico Ne, Robert-Jan Palstra, Tokameh Mahmoudi[1]
Erasmus University Medical Center, Rotterdam, The Netherlands
[1]Corresponding author: e-mail address: t.mahmoudi@erasmusmc.nl

Contents

International Review of Cell and Molecular Biology, Volume 335
ISSN 1937-6448
https://doi.org/10.1016/bs.ircmb.2017.07.011

Abstract

In this review, we cover transcription regulation of human immunodeficiency virus type 1 (HIV-1) gene expression, focusing on the invaluable contributions, made by HIV research over the years, toward the field of transcription. In this context, the HIV promoter can be considered to be a well-studied model promoter, which although a viral promoter, is subject to the same cellular regulatory mechanisms that modulate the transcriptional control of endogenous host cellular genes. The molecular control of HIV-1 transcription has been well studied and considerable knowledge toward development of alternative strategies for therapies aimed at eradicating both active but also latent HIV-1 has been obtained. Additionally, HIV-1 studies have provided insight into fundamental aspects of transcriptional regulation including transcriptional stochasticity, RNA polymerase II pausing, chromatin regulation of transcription, the role of the +1 nucleosome, the use of an RNA enhancer element, i.e., TAR, the discovery, and essential function of P-TEFb, and the super elongation complex in transcription elongation. These findings have been important not only in deciphering the mechanisms used by HIV-1 to regulate its gene expression and to establish and maintain HIV latency for therapeutic advancement, but were at the same time seminal in pushing the transcription field forward.

1. INTRODUCTION

Human immunodeficiency virus type 1 (HIV-1), the causative agent of the acquired immunodeficiency syndrome or AIDS, destroys the immune system by infecting, replicating in, and depleting predominantly CD4+ T lymphocytes (Burger and Poles, 2003; Epstein et al., 1993). HIV-1 is a lentivirus which enters host cells by binding to the CD4 receptor together with the CCR5 or CXCR4 coreceptors and fuses with the target cell, releasing viral RNA into the cytoplasm. The HIV-1 single-stranded RNA genome is then reverse transcribed into dsDNA by the viral enzyme reverse transcriptase (Hu and Hughes, 2012). The dsDNA preintegration complex then enters the nucleus where it integrates into the host genome aided by the activity of the viral integrase (Craigie and Bushman, 2012). The integrated proviral dsDNA then behaves essentially as a cellular gene and directs transcription of the viral genome (De Crignis and Mahmoudi, 2017). Viral transcripts are differentially spliced and translated into viral proteins, which are cleaved by the viral protease enzyme into maturity. Finally, the assembled virus buds from the host cell and is released into the blood stream (Engelman and Cherepanov, 2012; Freed, 2001, 2015).

In the past decade, the advent of effective combination antiretroviral therapy (c-ART), which targets different steps of the viral life cycle, has proven to be extremely successful in suppressing viral replication. However, due to mechanisms that are not clearly understood, a fraction of the integrated transcriptionally competent proviruses become transcriptionally inactive or latent (Siliciano and Greene, 2011). This reservoir of cells infected with latent HIV-1 is indistinguishable from uninfected cells and masked from c-ART. Because it is transcriptionally competent, latent HIV-1 retains the potential to become transcriptionally activated, which as a consequence leads to a rebound of the infection (De Crignis and Mahmoudi, 2017; Finzi et al., 1999; Siliciano et al., 2003) after termination of c-ART. Therefore, although c-ART has changed HIV-1 infection from a lethal to a chronic disease for people with access to the medication, it is not curative and has to be taken life-long (Maartens et al., 2014).

The shock and kill approach is a curative research strategy that aims to target and eliminate the persistent latent HIV-1-infected cell reservoir that presents the main obstacle to development of a cure. This approach aims to transcriptionally activate the latent viruses such that HIV is expressed and replicates, allowing its elimination by viral cytopathic effects or clearance by an effective immune system (Badalà et al., 2008; Dahabieh et al., 2015; Deeks, 2012; Margolis et al., 2016; Rasmussen et al., 2016). The presence of c-ART during shock and kill would safe-guard against new rounds of infection. The first part of the shock and kill approach relies on derepressing and/or activating the transcriptionally silent HIV-1 promoter (De Crignis and Mahmoudi, 2014; Rasmussen et al., 2016). Therefore, molecular mechanisms that govern the establishment and maintenance of transcriptionally silent HIV-1, and molecular mechanisms responsible for reversal of this process, i.e., transcriptional activation of latent virus, continue to be subjects of intense investigation. These studies were not only instrumental for deciphering transcriptional regulation of HIV-1 gene expression but also provided crucial insights into the regulation of transcription of endogenous genes.

Transcription of HIV-1 is controlled by a promoter located within the 5′ long terminal repeat (5′-LTR) of the integrated provirus. The HIV-1 5′-LTR is a long, promiscuous promoter that hijacks the host cell transcriptional apparatus to tightly control transcription of its viral genome. In the immediate-early phase of transcription, HIV promoter activity critically depends on cellular transcription factors (TFs). The presence of a large number of TF consensus binding sites within the 5′-LTR allows the interaction of a multitude of TFs; transcriptional activators as well as repressors that can modulate promoter activity in a context-dependent manner (Pereira et al., 2000; Rafati et al., 2011; Verdin, 1991; Verdin and Van Lint, 1995). In this way, the HIV-1 5′-LTR has proven to be a useful model promoter system for the identification and characterization of TFs. The context-dependent combinatorial presence of activating and repressive TFs and their associated cofactors creates a balance that determines whether LTR transcription will be silenced, leading to latency, or become active, resulting in a productive infection (Cary et al., 2016; De Crignis and Mahmoudi, 2014; Van Lint et al., 2013).

HIV-1 transcriptional activity is also determined by its chromatin landscape, which provides an additional layer of regulation (Agosto et al., 2015; De Crignis and Mahmoudi, 2017; Lusic and Siliciano, 2017). Within the 5′-LTR, nucleosomes are deposited at specific positions, regardless of the genomic position of virus integration (Mahmoudi, 2012; Verdin et al., 1993). In particular, the nucleosome positioned immediately downstream of the transcription start site (TSS), Nuc-1, which is equivalent to the +1 nucleosome, is repressive to transcription and associated with HIV-1 latency (Rafati et al., 2011; Verdin and Van Lint, 1995). Upon activation, HIV-1 Nuc-1 becomes rapidly and specifically remodeled (Verdin et al., 1993). The well characterized and strict chromatin organization of the HIV 5′-LTR makes it an excellent model system to investigate the role of chromatin structure in transcription regulation. In this context, studies have probed the influence of factors that modulate nucleosome positioning, such as histone-modifying complexes and chromatin remodelers in HIV-1 transcriptional control.

The activity of the HIV-1 promoter is strongly enhanced by the viral transactivator protein Tat. In the absence of Tat, recruitment of cellular TFs to the HIV-1 promoter is sufficient to trigger efficient initiation of transcription. However, very low levels of full-length transcripts are produced at this stage because elongation is terminated prematurely and RNA polymerase II pauses at a region proximal to the TSS (Natarajan et al., 2013; Zhang et al., 2007). Expression of Tat overcomes the block to transcriptional elongation by recruiting the super elongation complex (SEC) to the transactivation response (TAR) region, a short nascent structured mRNA element. This creates a Tat-dependent positive regulatory loop that drastically amplifies HIV-1 transcription, tipping the balance toward productive replication (Flores et al., 1999; Kao et al., 1987; Mousseau et al., 2015b; Ott et al., 2011). In the next sections, we will discuss the factors contributing to transcriptional activation in more detail.

2. STRUCTURE OF THE HIV-1 PROMOTER (5′-LTR)

The integrated HIV-1 provirus, contains two LTRs located, respectively, at the 5′-end and 3′-end of the viral genome. Both LTRs can initiate transcription but the 5′-LTR exerts dominant control and serves as the HIV-1 promoter. Transcriptional interference from the 5′-LTR is likely to repress the promoter activity of the 3′-LTR (Cullen et al., 1984); however, in the presence of a defective 5′-LTR, transcription from the 3′-LTR can be activated (Klaver and Berkhout, 1994).

Like other retroviral LTRs, the HIV-1 LTR is structurally divided into the U3, R, and U5 regions (Fig. 1). Transcription initiates at the beginning of R in the 5′-LTR; the transcripts are then capped and elongated through the U5 region; and the rest of the proviral genome. Transcriptional termination usually occurs within the 3′-LTR, in the R region. At the end of transcription, the polyadenylation signal (poly(A)), encoded by the 3′-LTR U5 region, is added.

The U3 region of the 5′-LTR contains the key regulatory elements for control of transcriptional activity (Gaynor, 1992). The regulatory function of the HIV-1 5′-LTR was probed in early studies that examined the ability of the double-stranded DNA template to act as a promoter for reporter genes. Full-length, mutagenized, and deleted templates of the HIV-1 5′-LTR were fused to chloramphenicol acetyl transferase (CAT) or luciferase reporters and transfected into eukaryotic cell lines and reporter activity was measured (Muesing et al., 1987; Sodroski et al., 1984). These studies identified the main functional regions and *cis*-acting elements in the HIV-1 5′-LTR involved in regulating gene expression and suggested the presence of transacting factors in HIV-infected cells able to enhance the activity of the reporters (Sodroski et al., 1984).

The HIV-1 5′-LTR comprises a region between nucleotides −454 to +188, relative to the TSS, within the approximately 9 kilobases (kb) long viral genome (Fig. 1A) (Ratner et al., 1985; Sanchez-Pescador et al., 1985; Starcich et al., 1985; Wain-Hobson et al., 1985). This region was further divided into four main functional elements based on the effects of each segment on HIV reporter activity or HIV-1 gene expression: the modulatory element (−455, −104), the enhancer region (−109, −79), the core promoter element (−78, −1), and the TAR region (+1, +60) (Fig. 1B). A plethora of (>50) cellular TFs has been reported to bind to the HIV-1 5′-LTR in vitro or predicted to bind the 5′-LTR sequence (Mahmoudi,

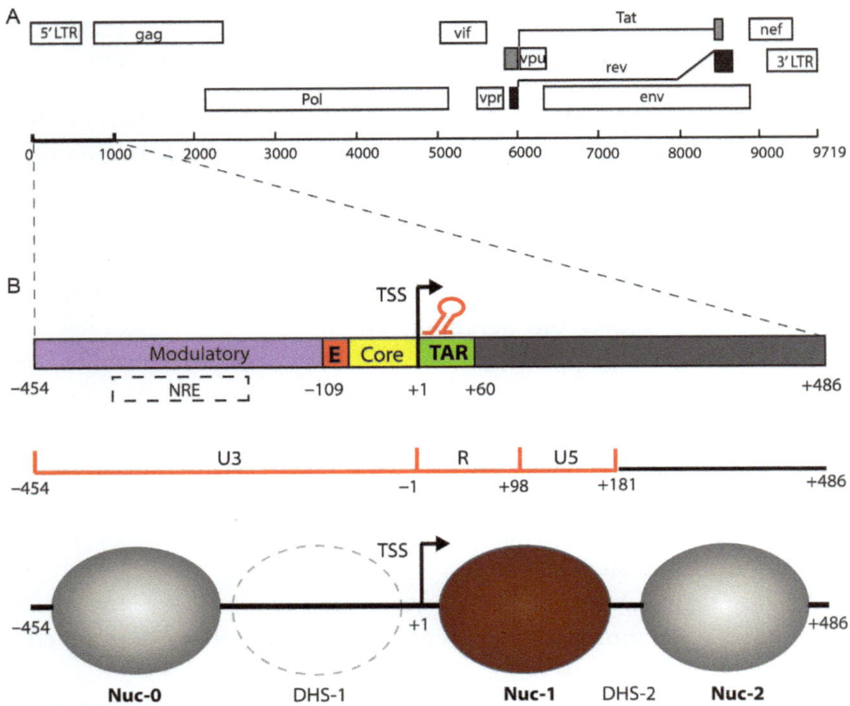

Fig. 1 Schematic representation of the HIV-5′-LTR. (A) Structure of the HIV-1 genome and the 5′-LTR region (~9 kb) that acts as the HIV-1 promoter. The 5′-LTR is divided into the U3, R, and U5 regions. The U3 region can be further divided into the modulatory region (*purple*), enhancer (E) (*red*), basal/core promoter (*yellow*), and TAR (*green*). (B) In vivo observed chromatin landscape of the transcriptionally inactive HIV-1 5′-LTR is defined by the presence of two strictly positioned nucleosomes Nuc-0 and Nuc-1 as indicated connected by a region of accessible DNA, hypersensitive to nucleases, DHS-1 which contains the majority of cellular transcription factor consensus binding sites and which has high propensity for nucleosome formation.

2012; Pereira et al., 2000). However, only a few factors have been shown to be critical for regulation of proviral transcription in vivo: these include nuclear factor (NF)-κB, nuclear factor for activated T cells (NFAT), Sp1, TATA-binding protein (TBP). Other 5′-LTR bound factors are thought to perform a cooperative regulatory function or contribute to the modulation of LTR activity in response to different cellular stimuli. Interestingly, random mutation of the HIV 5′-LTR results in a balance of positive and negative effects on viral fitness, suggesting that the 5′-LTR has not evolved to the most active promoter (van Opijnen et al., 2006).

2.1 The Modulatory Region

Early studies found that deletion of a region between nucleotides −454 to −167 in the HIV 5′-LTR produces a twofold to threefold increase in HIV promoter reporter activity and viral replication (Gaynor, 1992; Lu et al., 1989; Rosen et al., 1985), suggesting that this region behaves as a negative regulatory element. Later this segment was shown to contain multiple *cis*-acting binding sites for activating and repressive factors, supporting a modulatory, rather than repressive function on HIV promoter activity (Pereira et al., 2000) depending on the availability of repressive/activating factors. This region includes three CCAAT/enhancer binding protein (C/EBP) sites (Henderson et al., 1995; Tesmer et al., 1993), an upstream stimulatory factor binding site, the activating transcription factor/cyclic AMP response element binding region (Krebs et al., 1997, 1998), lymphocyte enhancer factor (LEF-1) site, nuclear factor of activated T cells (NF-AT), c-Myb, Ets-1, AP-1, COUP, NF-κB binding sites, and a number of other consensus TF binding sites both shown and predicted to bind (Mahmoudi, 2012; Pereira et al., 2000). Nuclear factor-IB (NF-IB) has also been recently reported to bind to the modulatory region upon HIV-1 infection and to negatively impact viral transcription (Vemula et al., 2015). The 5′-LTR modulatory region is thus thought to regulate the basal activity exerted by the core element and enhancer regions in a context-dependent manner, in response to different cellular signaling pathways. Sequences upstream of the core promoter, between −130 and −201, containing consensus binding sites for LEF-1, NF-AT, and Ets-1 TFs which are highly expressed in T cells, were in particular shown to serve as important elements of regulation of viral expression in peripheral blood lymphocytes and in some T cell lines (Kim et al., 1993).

A recent report using a dCas9 fused to HAT protein p300 has shown that targeting of dCas9-p300 to a segment within the modulatory region that overlaps with Nuc-0 can also drive transcriptional activation supporting a potential regulatory role for this region of the LTR (Limsirichai et al., 2016).

2.2 The Enhancer

Deletion and mutagenesis experiments have identified a region between nucleotides −137 and −17, located directly upstream of the core promoter and the SP binding sites, to be an enhancer element that, irrespective of orientation and distance, is able to activate transcription from a heterologous promoter (Rosen et al., 1985). Consistent with these early studies,

CRISPR/Cas9 studies using a catalytically deficient Cas9 fused to trans-activation domains like VP64 (dCas9-VP64), VP64 arrays (dCas9-SunTag) (Bialek et al., 2016; Ji et al., 2016; Saayman et al., 2016), or SAM (dCas9-SAM) (Bialek et al., 2016; Zhang et al., 2015) determined that the HIV-1 enhancer element behaves as the hotspot for dCas9-mediated activation of proviral transcription, in latently infected T cells. These findings also confirm a general rule of transcriptional activation that requires binding of activating complexes at ~100 bases upstream of the TSS.

It is probably not a coincidence that most HIV-1 subtypes have evolved a tandem repeat of the 10 bp consensus binding sites for NF-κB and related factors (Burnett et al., 2010; Nabel and Baltimore, 1987) right in the hotspot, which constitutes the strongest *cis* activator for the promoter. Different NF-κB subunits have been shown to bind to the HIV-1 enhancer in vitro and in vivo and strongly activate HIV-1 transcription (Duckett et al., 1993; Fujita et al., 1992; Kretzschmar et al., 1992; Lin et al., 1995). The NF-κB sites (Kim et al., 1993; Ross et al., 1991) and NF-κB proteins (Qian et al., 1994) play a key role in regulating viral promoter activity in vivo, in PBMCs and activated T-cells. NF-κB is in fact a crucial activator of immune responses and the cellular acute-phase and it is strongly upregulated in activated T-cells. NF-κB binding is conferred by heterodimer formation of p50 and p65 subunits. Most cell types, including T cells, tightly control NF-κB activity by interaction of p65 with the inhibitory protein I-κB that masks the nuclear localization sequence and keeps the complex sequestered in an inactive state in the cytoplasm. Upon T cell activation by mitogens, cytokines, oxygen radicals, and other agents, NF-κB binding activity is restored: I-κB is phosphorylated by mitogen-activated I-κB kinase, ubiquitinated and sent for degradation by the proteasome. Once released from I-κB, the NF-κB subunits are free to translocate into the nucleus and activate transcription.

NF-κB binds to the kB binding sites, and can cooperatively with Sp1 (Li et al., 1994; Perkins et al., 1993, 1994; Sif and Gilmore, 1994) and other TFs (LEF-1, Ets1, and TFE-3) (Sieweke et al., 1998) synergistically activate HIV transcription, even in the presence of a repressive chromatin environment. The ERK1/ERK2 mitogen-activated protein kinase (MAPK) pathway plays a role in mediating the response to external stimuli in HIV-1 latently infected cells via engagement of the NF-κB sites. Upon MAPK activation AP-1 components c-Fos and c-Jun are reportedly induced and physically interact with NF-κB at the NF-κB sites to synergistically activate the HIV-1 promoter (Yang et al., 1999). The two NF-κB binding sites within the HIV-1 enhancer region can be bound by NFAT factors as well; NFATs were shown to bind the terminal 3′ half of NF-κB motifs located in the U3 region, with similar sequence preference as p50, and modulate HIV promoter activity, increasing infectivity in T-cells (Cron et al., 2000; Kinoshita et al., 1997, 1998; Macian and Rao, 1999), although NFATc2 was shown to also repress transcription in certain contexts via competition with NF-κB for its binding site. Owing to overlap of the consensus sequences, NF-κB–NFAT binding is thought to be mutually exclusive (Chen-Park et al., 2002; Giffin et al., 2003) while binding of NF-κB is enhanced by interactions with Sp1 (Perkins et al., 1993). Although mutation of the NF-κB sites within the HIV-1 enhancer results in only a modest inhibition of viral expression in most cell lines (Liang et al., 1997), the response of the viral enhancer to NF-κB or NF-AT has proven to be essential for reactivation of the latent HIV-1 provirus and to induce viral replication in primary T-cells (Alcamí et al., 1995; Bosque and Planelles, 2009). More recently, HspBP1 was shown to be recruited to the NF-κB sites of the HIV-1 enhancer where it competes with binding of p50/p65 leading to suppression of NF-κB-induced transcription (Chaudhary et al., 2016).

It is important to underline that not all HIV-1 subtypes have two NF-κB sites in their enhancer region. Subtype A/E viruses have a one nucleotide deletion in the upstream NF-κB site that causes a loss of NF-κB binding and increased binding of GABP. The NF-κB to GABP switch in these variants still results in a functional promoter, characterized by an overall reduced basal activity, and enhanced Tat-mediated activation (Verhoef and Berkhout, 1999). Subtype C viruses instead can contain three (Bjorndal et al., 1999; Hunt and Tiemessen, 2000; Naghavi et al., 1999) to four NF-κB sites in their promoter (Bachu et al., 2012b; Hunt and Tiemessen, 2000; Liebert et al., 2002; Papathanasopoulos et al., 2002) and are characterized by moderate enhanced activity (Jeeninga et al., 2000; Montano et al., 1997; Naghavi et al., 1999) and replication fitness (Bachu et al., 2012a).

Upstream of the NF-κB sites, a four-nucleotide AP-1 element was also shown to constitute a latent establishment element as its deletion compromised the ability to establish latent infections, while extension of this site promoted latency (Duverger et al., 2013). c-Jun was shown to suppress HIV-1 promoter activity in the absence of Tat (Duverger et al., 2013). However, c-Jun overexpression also enhances Tat-dependent transcription suggesting that, depending on Tat levels, c-Jun binding to the AP-1 site can mediate repression or enhance HIV-1 promoter activity (van der Sluis et al., 2014).

Interestingly, this AP1 site overlaps with a RBEII binding site, bound by the RBF-2 complex, reported to be necessary for 5′-LTR transcription (Sadowski and Mitchell, 2005), although c-Jun-mediated repression may not solely result from interference with RBF-2 binding (van der Sluis et al., 2014).

2.3 The Core Promoter

The core promoter spans a region of 250 nucleotides and contains the essential elements for HIV-1 promoter basal activity: a TATA box element (Garcia et al., 1989), three tandem GC-rich binding sites for the Sp family of TFs (Jones et al., 1986), and a highly active nonconventional TSS (Zenzie-Gregory et al., 1993). Genetic studies in different HIV-1 subtypes have later revealed that the virus is able to use the overlapping CATA box instead of the canonical TATA box (van Opijnen et al., 2004) and that certain subtypes contain four to five Sp1 sites (Koken et al., 1992; Rosseau et al., 1997).

The TATA box and Sp sites are essential for transcription initiation and basal promoter activity (Jones and Peterlin, 1994) but also necessary for Tat-mediated transactivation as insertions that separate these two regulatory elements compromise transactivation (Huang and Jeang, 1993). HIV-1 basal transcription strictly requires Sp1 and TBP acting together with cellular TAFs (TBP-associated factors) in the general TF complex TFIID (Dynlacht et al., 1991).

Within the complex, Sp1 has been shown to interact directly with TAF-110 (Gill et al., 1994; Hoey et al., 1993). Additionally, Sp1 interacts with additional DNA bound proteins NF-κB (Li et al., 1994; Perkins et al., 1993, 1994; Sif and Gilmore, 1994), YY1 (Seto et al., 1991, 1993), and distal Sp1 monomers (Pascal and Tjian, 1991) and directs the basal machinery to form the preinitiation complex (PIC) via protein–protein interactions (Näär et al., 1998).

TRIM22 was recently reported to interfere with Sp1 binding thereby inhibiting PIC formation and basal proviral transcription (Turrini et al., 2015). Sp1 binding at the core promoter has also been linked to chromatin remodeling as it interacts with histone deacetylases (HDACs) and chromatin modifying factors p300 (Zhao et al., 2003).

The CTGC motifs flanking the TATA box have also been shown to be required for transcription initiation in vitro and Tat-dependent activation in living cells (Wilhelm et al., 2012). Additionally, immediately upstream of the TSS binding sites for the TF ZASC1, have been shown to have a regulatory role as ZASC1 interacts with Tat and positive transcription elongation factor b (P-TEFb) and cooperatively promotes HIV-1 transcriptional elongation (Bruce et al., 2013).

2.4 The TAR Element

Pioneering studies first led to the observation that expression of viral or reporter genes, under the control of the HIV-1 5′-LTR is greatly enhanced in HIV-infected cells (Sodroski et al., 1984, 1985a) or in cells cotransfected with coding sequences of the viral genome (Sodroski et al., 1985b). It quickly became clear that HIV-infected cells must express HIV-specific transacting TFs. Deletion experiments determined that a sequence between −17 and +44, named the TAR element, as well as the presence of an intact enhancer region (Hauber et al., 1989) were essential for the response to transacting factors (Rosen et al., 1985). Given their crucial nature for the regulation of the HIV promoter, the transacting viral sequences were studied extensively to determine their mechanism of action. Cotransfection (Sodroski et al., 1985b) and functional mapping (Arya et al., 1985) studies resulted in identification of a region between nucleotides 5365 and 5607 of the HIV genome responsible for transactivation activity. This region contains the N-terminal segment of the Tat reading frame, which, depending on the subtype, encodes a protein of 86–104 amino acids. Tat was shown to bind to TAR (Dingwall et al., 1989; Roy et al., 1990), a short RNA transcript containing multiple regulatory elements and two copies of the CTCTCGG repeat sequence thereby mediating recruitment of Tat to the HIV-1 promoter (Hauber et al., 1989; Tong-Starksen et al., 1987). In vitro studies later demonstrated that this RNA element has a hairpin-like structure, which acts as a binding site for Tat and is essential for transactivation (Berkhout et al., 1989; Kao et al., 1987; Muesing et al., 1987; Selby et al., 1989). Different subtypes have acquired mutations in the lower stem region of TAR but most changes in the sequence do not seem to significantly affect its secondary structure or sensitivity to Tat activation (Jeeninga et al., 2000; Montano et al., 1997).

2.5 Region Downstream of TSS

Sequences downstream of TSS have an underappreciated role in regulating transcription (Burke and Kadonaga, 1996, 1997; Emanuel and Gilmour, 1993; He et al., 2002; Pelletier et al., 1997). The HIV-1 5′-LTR contains several TF binding sites downstream of the TSS, some of which are well conserved within different HIV-1 strains and subtypes (el Kharroubi and Martin, 1996; el Kharroubi and Verdin, 1994). Important downstream HIV-1 regulatory binding sites are: three AP-1 binding sites located in the R/U5 region (+87, +94) (+118, +125) and (+155, +163) (el Kharroubi and Verdin, 1994; Van Lint et al., 1997), NFATc2 and C/EBP sites (+158, +171) (Dahiya et al., 2014), AP-3 like (+162, +177) (el Kharroubi and Verdin, 1994; Van Lint et al., 1997), NFAT (+162, +177) (Van Lint et al., 1997), DBF-1 (+200, +219) (Liang et al., 1997; Sgarbanti et al., 2002, 2004), and Sp sites (+270, +278) (Dahiya et al., 2014) many of which were identified by in vitro foot printing and DNA binding studies (Dahiya et al., 2014; Rabbi et al., 1997; Roebuck et al., 1996).

Some of these factors including AP1, C/EBP, NFAT contribute to transcriptional activation via recruitment of chromatin remodeling and histone-modifying cofactor complexes to improve accessibility of RNA pol II complex to chromatin (Dahiya et al., 2014). Other factors are likely to contribute to transcriptional activation by influencing transcriptional initiation and elongation (Dahiya et al., 2014). Overall, this region may contribute to modulation of the HIV-1 transcriptional response to activating signals in a context-dependent manner (Dahiya et al., 2014).

3. CHROMATIN LANDSCAPE OF THE HIV-1 PROMOTER

In eukaryotic cells, DNA is tightly packaged into nucleosomes, consisting of a histone octamer core wrapped by 147 bp of DNA (Luger and Richmond, 1998). The nucleosomal core comprises two H2A–H2B and two H3–H4 histone heterodimers (Richmond et al., 1997). Histone H1 locks the DNA at the entry and exit sites of the nucleosomal core, acting as a linker (Richmond et al., 1997). Nucleosome positioning throughout the genome is a very dynamic process, tightly linked with transcription that plays a fundamental role in regulating gene expression. The position of nucleosomes can alter the availability of binding sites for TFs and the recruitment of general transcription machinery in vivo (Radman-Livaja and Rando, 2010), thereby influencing expression. Conversely, binding of specific TFs and chromatin remodelers upon transcriptional activation can also result in destabilization or repositioning of nucleosomes at enhancer or promoter regions (Boyle et al., 2011; Henikoff, 2008).

At least three main cellular mechanisms have been described that can regulate nucleosome occupancy and stability. First, chromatin remodeling complexes, large multisubunit complexes containing a core ATPase domain, can actively reposition or remodel nucleosome using ATP hydrolysis as a source of energy (Clapier and Cairns, 2009). Second, histone-modifying enzymes can add several posttranslational modifications (PTMs) to histones to alter their structural configuration or electrostatic charge, thereby influencing their stability or their ability to recruit regulatory cofactors. The N-terminal tails are unstructured and very accessible, as they protrude from the core nucleosome, and can be subjected to PTMs like acetylation, methylation, phosphorylation, ubiquitination, sumoylation, and poly ADP-ribosylation. Combinatorial patterns of PTMs on histone tails have been proposed to generate a "histone code" that directs chromatin templated processes such as replication, repair, and gene expression (Strahl and Allis, 2000). Finally, DNA methylation at CpG motifs has a well-established regulatory function in transposon silencing, genomic imprinting, X-chromosome inactivation, and also influences the chromatin environment at gene regulatory regions (Li and Zhang, 2014).

3.1 Nucleosome Positioning and HIV-1 Expression

As a naked DNA template, the HIV-1 promoter acts as a strong promoter but it is almost silent in vivo when integrated to the host genome, in the absence of activating stimuli (Huang and Jeang, 1993; Pomerantz et al., 1990). This is because the integrated HIV-1 provirus, like endogenous genes, is packaged into nucleosomes, a structure that adds an extra layer of regulation to the activity of the HIV-1 promoter by limiting access to transcription regulators. Assays probing the in vivo chromatin accessibility of the integrated HIV 5′-LTR has provided invaluable insight into the influence of nucleosome positioning on general transcription. The HIV-1 promoter is embedded into three distinct nucleosomes, Nuc-0 (or −1 nucleosome) (−415/−255), Nuc1 (or +1 nucleosome) (+10/+155), and Nuc2 (+256/+412) which are positioned at defined locations with respect to the enhancer and core regions (Fig. 1B) (Steger and Workman, 1997; Verdin, 1991; Verdin et al., 1993) and delimit two regions hypersensitive to nuclease digestion, DHS-1 and DHS-2 (Verdin et al., 1993). The DHS-1 and DHS-2 regions contain consensus binding sites for multiple TFs that have been shown to bind the LTR in vitro and in vivo (Demarchi et al., 1993, 1996; el Kharroubi and Verdin, 1994; Pereira et al., 2000).

In vitro reconstitution studies of HIV-1 transcription demonstrated that incubation of the template with a nucleosome assembly system, purified from Drosophila embryos, strongly represses transcription (Sheridan et al., 1995), likely by inhibiting the formation of the RNA Pol II PIC (reviewed in Owen-Hughes and Workman, 1994). However, chromatin-mediated repression of the HIV promoter activity could be counteracted by preincubating the DNA, before nucleosome incorporation, with the cellular factors LEF-1, Ets-1, TFE-3, and Sp1 (Sheridan et al., 1995). These data suggested that the presence of cellular factors bound to the HIV 5′-LTR is fundamental to the establishment of nucleosome free regions DHS-1 and DHS-2 (Verdin, 1991; Verdin et al., 1993). In agreement, DNA sequences corresponding to the DHS-1 region, even though accessible, were found to have the highest histone binding score within the LTR sequence, indicating higher propensity for nucleosome formation than LTR sequences encompassing Nuc-0 or Nuc-1 (Rafati et al., 2011). The discrepancy between the observed in vivo chromatin structure and its predicted nucleosomal structure is in agreement with the binding and activity of TFs and cofactors within the DHS-1 region, which keep its nucleosomal state disrupted.

3.2 HIV-1 Nuc-1: The +1 Nucleosome

DNase I foot printing and MNase mapping experiments have demonstrated that Nuc-1, located immediately downstream of the TSS is rapidly removed from the LTR upon transcriptional activation (Verdin, 1991; Verdin et al., 1993). This indicates that the presence of Nuc-1 constitutes a block that needs to be moved or rearranged to allow RNA Pol II elongation (el Kharroubi and Verdin, 1994). Only much later, genome-wide nucleosome mapping studies have demonstrated how widespread and conserved this regulatory mechanism is at endogenous protein-coding genes. Pioneering studies in *Saccharomyces cerevisiae* showed that nucleosomes are generally depleted from active regulatory regions (Bernstein et al., 2004; Lee et al., 2004; Sekinger et al., 2005); higher resolution experiments then clarified that upon gene activation there is an additional depletion of nucleosomes at the promoter regions (Guillemette et al., 2005; Schwabish and Struhl, 2004; Zanton and Pugh, 2006; Zhang et al., 2005).

Genome-wide nucleosome mapping in *S. cerevisiae* and metazoans has shown that nucleosomes at the promoter region of most genes are organized exactly as seen in the HIV promoter: a nucleosome free region, hypersensitive to nuclease digestion which is delimited by −1 and +1 nucleosomes. The gene body instead is packaged by an array of nucleosomes (Barski et al., 2007; Mavrich et al., 2008). Upon transcriptional activation, the +1 nucleosome is evicted but it seems to be rapidly reconstituted after elongation, as highly transcribed genes only show a modest +1 nucleosome depletion (Albert et al., 2007). More recently, genome-wide studies have demonstrated that indeed the +1 nucleosome constitutes a block for RNA pol II transcriptional elongation for essentially all genes, including genes subjected to regulated transcriptional pausing (Weber et al., 2014a).

3.3 Role of ATP-Dependent Chromatin Remodelers

In silico predictions for histone binding point to a lower propensity of the HIV-1 + 1 nucleosome Nuc-1 DNA sequence for nucleosome formation, while the DHS-1 region has the highest propensity for nucleosome formation, suggesting that in vivo, the DHS-1 region is kept accessible, while Nuc-1 is positioned at an energetically less favorable position (Rafati et al., 2011). High resolution MNase mapping demonstrated that the BAF complex, an ATP-dependent chromatin remodeling complex containing BAF250a, actively repositions nucleosomes within the 5′-LTR, positioning Nuc-1 over energetically suboptimal DNA sequences, and hence promotes establishment and maintenance of a repressive state for transcription (Rafati et al., 2011). When BAF complex activity is perturbed, either via siRNA depletion of BAF250 or via inhibition with small molecules, the activity of the HIV promoter is significantly derepressed (Rafati et al., 2011; Stoszko et al., 2016). Other chromatin remodeling factors actively maintain a repressed chromatin state at the HIV promoter thereby repressing transcription. The CHD family, like the BAF family, has a centrally located ATPase domain and two additional tandem bromodomains in the N-terminal region (Marfella and Imbalzano, 2007). CHD1 was shown to promote a repressive nucleosomal configuration in a chimeric yeast–HIV transcription model (Vanti et al., 2009) and in latently infected Jurkat cells (Gallastegui et al., 2011), thereby maintaining latency. More recently, an insertional mutagenesis screen identified CHD1 and CHD2 as positive regulators of HIV-1 expression (Rodgers et al., 2014). The NuRD chromatin remodeling complex combines ATPase activity with HDAC activity. One of its subunits, the methyl binding protein MBD2 was also reported to play a role in maintenance of latency (Blazkova et al., 2009; Kauder et al., 2009). Chromatin remodelers are not only crucial for regulation of basal transcription at the HIV promoter but have also been shown to play a pivotal role in active nucleosome remodeling upon Tat-mediated activation of the promoter (Agbottah et al., 2006; Easley et al., 2010a; Mahmoudi, 2012; Mahmoudi et al., 2006; Tréand et al., 2006; Van Duyne et al., 2011).

3.4 Role of Histone Modifications

PTMs of the N-terminal tail of histones do not only stabilize or destabilize the interaction between DNA and the nucleosomal core but also serve as selective marks, to recruit specific proteins containing specialized domains that can read and bind to the modified histones. These proteins can then act to change chromatin structure, promoting, or repressing transcription. Enzymes that deposit histone marks, referred to as writers, include: histone acetyltransferases (HATs), histone methyltransferases (HMTs/KMTs), protein arginine methyltransferases (PRMTs), histone ubiquitin ligases, and kinases. Proteins containing bromodomains, chromodomains, tudor, PHD fingers, MBT domains, etc., can read the marks (readers) and determine a functional response. PTMs on histone tails are not permanent but can be removed by eraser proteins such as HDACs, lysine demethylases (KDMs), and phosphatases. Acetylation is almost always associated with activation as it masks the positive charge of histones (lowering the affinity for the negatively charged DNA phosphodiester backbone) and helps to loosen the chromatin, thereby facilitating transcription. Methylation can have distinct transcriptional consequences depending on the residue modified. Methylation of H3 lysine residue 4 (H3K4) and 36 (H3K36) is associated with transcriptional activation, while methylation of H3K9 and H3K27 are considered to be repressive.

Reversible acetylation of histones relies on the activity of HATs and HDACs (Tripathy et al., 2011; Van Lint et al., 2013) and represents the best characterized histone modification that regulates HIV-1 promoter activity (Kumar et al., 2015; Van Lint et al., 2013). In the latent state, the HIV-1 promoter is predominantly associated with HDACs (Cary et al., 2016; Donahue et al., 2013; Kouzarides, 2007; Margolis et al., 2016).

Host repressive factors late SV40 factor (LSF), ying-yang 1 (YY1), and NF-κB p50 recruit HDACS to the HIV 5′-LTR thereby causing deacetylation and promoting maintenance of Nuc-1 in a repressive configuration (Williams et al., 2006). The cooperative activity of LSF and YY1 serves to recruit HDAC1 downstream of the TSS (Coull et al., 2000; Romerio et al., 1997), while recruitment to the 5′-LTR is mediated by the interaction with NF-κB p50 homodimers (Williams et al., 2006), with AP-4 (Imai and Okamoto, 2006), CBF-1 (Tyagi and Karn, 2007), and c-Myc-Sp-1 (Jiang et al., 2007). In monocytes and macrophages, Sp1 recruits COUP-TF interacting protein (CTIP2) which in turn recruits HDAC1 and HDAC2 to promote local deacetylation of histone H3 at the HIV-1 5′-LTR (Marban et al., 2005, 2007). Given the well-described role of HDACs in repressing transcription from the 5′-LTR, HDAC inhibitors have been in the spotlight of strategies aiming to reverse latency. Indeed, treatment of latently infected cell lines or CD4+ T cells isolated from HIV-1-infected patients with HDAC inhibitors results in a significant induction of HIV-1 transcription (Barton et al., 2014; Huber et al., 2011; Margolis, 2011; Palmisano et al., 2012; Reuse et al., 2009; Van Lint et al., 1996). HDAC inhibitor treatment was shown to cause reactivation in a broad range of proviruses, including defective ones (Barton et al., 2016). It has been shown that specifically inhibitors of class I HDACs, but not class II HDACs, induce chromatin derepression and efficiently stimulate transcription (Archin et al., 2009; Keedy et al., 2009). Among class I HDACs, specific depletion of HDAC3, but not HDACs 1 or 2 leads to significant activation of latent HIV-1, suggesting a potentially prominent role for HDAC3 in HIV-1 latency (Huber et al., 2011). Interestingly, the HIV-1 accessory protein Vpr was recently found to indirectly counteract latency by inducing degradation of HDAC3 and HDAC1 at the viral promoter (Romani et al., 2016).

Methylation constitutes another important modification of histones that contributes to silencing of the integrated HIV-1 provirus. In the latent state, the HIV-1 promoter is characterized by a predominance of repressive H3K9 dimethyl (Imai et al., 2010) or trimethyl (du Chéné et al., 2007; Marban et al., 2005, 2007) and H3K27 trimethyl (Pearson et al., 2008) histone marks. The association of these marks with transcriptional repression is well-established genome-wide and constitutes a central mechanism for silencing of the HIV-1 promoter as well. The overall effect of these modifications is to promote stabilization of Nuc-1, establishing a repressive environment for HIV-1 gene expression. In cell line models and PBMCs, chromatin repression of the HIV-1 promoter, via H3K9me3, has been shown to be mediated by the histone methyltransferases (HMTs) Suv39H1 and HP1gamma (du Chéné et al., 2007). In microglial cells, CTIP2 recruits Suv39H1 to the HIV-1 promoter resulting in the formation of an H3K9me3 region, which is recognized by HP1-gamma and promotes the formation of a heterochromatic region (Marban et al., 2005, 2007). Another HMT associated with the HIV-1 promoter transcriptional silencing is G9a which promotes the formation of a repressive chromatin environment by deposition of the H3K9me2 mark (Imai et al., 2010). Deposition of the H3K27me3 mark is instead mediated by EZH2, a subunit of the polycomb group repressive complex 2 (PRC2), which is also recruited to the HIV-1 5′-LTR (Friedman et al., 2011). Depletion of EZH2 in latently infected cell lines induces HIV-1 transcription (Friedman et al., 2011). PRC2-mediated H3K27 trimethylation then serves as a recognition mark for recruitment of additional repressive complexes including the polycomb group repressive complex 1 (PRC1), other HMTs, DNA methyltransferases, and HDACs. H3K27me3 is thought to be involved in establishment of latency at an early phase of infection and in maintenance of silencing after integration. Depletion of EZH2 reduced the frequency of latently infected cells after new infections, while inhibitors of PRC2 antagonized repression of the promoter activity in latently infected cells (Matsuda et al., 2015).

3.5 Histone Chaperones and Chromatin Assembly Factors

The composition eukaryotic chromatin is not static and the dynamics of nucleosomes is for a large part determined by the incorporation of histone variants (Venkatesh and Workman, 2015). Histone variants can differ from their canonical counterparts by alteration of a few amino acids or addition of larger domains and their presence influences chromatin-mediated processes such as transcription (Venkatesh and Workman, 2015). Exchange of histones is not only mediated by ATP-dependent chromatin remodelers but also by histone chaperones and assembly factors which save guard the formation of proper nucleosomes (Venkatesh and Workman, 2015).

Several histone chaperones have in recent years been implicated in transcriptional regulation of the HIV-1 5′-LTR. The histone chaperone protein nucleosome assembly protein-1 (NAP-1) was identified as an interactor for flag-tagged Tat (Vardabasso et al., 2008) and as a coactivator of the HIV-1 promoter. The mechanism of NAP-1 5′-LTR activation appeared to primarily involve Tat stabilization and potentially NUP-1-assisted removal of H2A–H2B dimers from nucleosomes which results in an open chromatin environment conductive to transcription (Vardabasso et al., 2008). Early genetic screens, using a minimal HIV-1 transcriptional system in yeast, for factors that modify the transcriptional output of the HIV-1 5′-LTR identified the involvement of chromatin reassembly factors Spt6, Chd1, and FACT as well as the histone chaperones ASF1a and HIRA (Gallastegui et al., 2011; Vanti et al., 2009). Curiously it was found that upon knockdown of these factors the HIV-1 promoter was activated. The authors suggested that chromatin reassembly factors and histone chaperons control the activity of host genes which results in transcriptional interference of the HIV-1 LTR (Gallastegui et al., 2011). A recent functional genomic RNAi screen also identified SUPT16H and SSRP1, both components of the FACT chromatin reassembly complex, as suppressors of HIV-1 transcription (Huang et al., 2015) and demonstrated direct Tat-mediated recruitment of these factors to the viral promoter. This precludes association of the P-TEFb subunit cyclin T1 with the Tat-5′-LTR axis and suggests a direct role for FACT in suppressing the HIV-1 5′-LTR. Two other histone chaperones have been reported to play a role in HIV-1 transcriptional activation: Nucleophosmin (NPM1) and RbAp48 (Gadad et al., 2011; Wang et al., 2016). Acetylated NPM1 is recruited via Tat to DHS-1 of the HIV-1 5′-LTR where it could enhance viral transcription by promoting a dynamic chromatin surrounding. However, NMP1 also regulates the nuclear localization of Tat which potentially significantly contributes to the observed stimulatory effect on HIV-1 transcription (Gadad et al., 2011). RbAp48, a component of several chromatin repressive complexes such as NuRD, NURF, and PRC is also recruited to the 5′-LTR where it represses transcription (Wang et al., 2016). Above studies demonstrate a role for histone chaperones in regulating HIV-1 transcription. Despite the presence of a number of studies implicating a role for histone chaperones in HIV-1 transcription regulation, remains unclear if their mechanism of action involves the deposition of histone variants.

3.6 Role of DNA Methylation

Initial observations for a potential role for DNA methylation in regulation of HIV-1 promoter activity comes from latently infected cell lines, where those harboring heavily methylated promoters were found to be more prone to be highly repressed (Blazkova et al., 2009). The presence of methylation at two CpG islands flanking the TSS was shown to promote HIV promoter binding by the methyl–CpG-binding domain protein (MBD2), where it recruits the repressive NURD complex (Kauder et al., 2009). However, the importance of this regulatory mechanism in vivo is less clear. A role for DNA methylation in maintenance, rather than establishment, of HIV-1 latency was proposed as resting CD4+ T cells isolated from HIV-1-infected patients with undetectable viremia displayed higher levels of DNA methylation at the 5′-LTR compared to cells isolated from viremic patients (Blazkova et al., 2009). Highly methylated DNA has also been observed in PBMCs infected with HIV-1, suggesting that DNA methylation might be one of the mechanisms employed to achieve transcriptional silencing (Maricato et al., 2015). Several studies though have shown that primary cells obtained from c-ART-treated HIV-1-infected patients have low levels of HIV promoter DNA methylation, questioning the importance of this regulatory mechanism in vivo (Blazkova et al., 2012; Ho et al., 2013; Weber et al., 2014b). Recently it was reported that in the latent reservoir of HIV-1-infected patients under prolonged cART, there is an accumulation of DNA methylation at the HIV-1 5′-LTR. This accumulation may arise as a consequence of repeated cycles of activation of the provirus by cellular stimuli and selection of methylated proviruses less prone to activation or alternatively by de novo methylation induced after cell activation (Trejbalová et al., 2016). Thus, methylation of the HIV 5′-LTR may be regulatory in latency control but further investigation, and larger patient cohorts are needed to derive more significant conclusions.

4. BASAL HIV-1 TRANSCRIPTION

4.1 Transcriptional Initiation

RNA polymerase (Pol) II-dependent transcription of protein-coding genes begins with the recruitment of gene-specific regulatory factors to the core promoter. These factors can act directly by recruiting components of the transcription machinery or can mediate the establishment of a transcriptionally permissive chromatin state by recruiting histone modifiers and chromatin remodelers. At the HIV-1 promoter transcriptional initiation occurs at the junction between the U3 and R regions, where three guanosine residues reside, and efficiently starts when cellular factors Sp1, NF-κB, NFAT are recruited (Karn, 2011; Kinoshita et al., 1997; Nabel and Baltimore, 1987; Perkins et al., 1997). Binding of these factors promotes the recruitment of histone-modifying enzymes that cooperatively establish a transcriptionally permissive chromatin environment, thus facilitating the installment of the transcription machinery at the 5′-LTR (Gerritsen et al., 1997; Kilareski et al., 2009; Kobor and Greenblatt, 2002; Peterlin and Price, 2006; Turk and Stoka, 2007). The HIV-1 core promoter can be regarded as a TATA box containing promoter, the most-extensively studied eukaryotic promoter element, present in about 10%–15% of mammalian genes.

Interestingly, it was recently demonstrated that HIV-1 transcription initiation can occur at each of the three G residues at the start site and that depending on which capped guanosine the transcripts begin with, they can be destined for either translation, or packaging as genomic RNA (Kharytonchyk et al., 2016; Masuda et al., 2015). This is in agreement with previous studies which have shown that RNA Pol II can use multiple TSS within TATA-driven promoters (Kadonaga, 2012). The first step for the assembly of the transcription machinery is the binding of the TFIID complex, containing the TBP subunit, to the TATA box. Upon TFIID binding, five more general TFs (TFIIA, TFIIB, TFIIE, TFIIF, and TFIIH), the mediator complex and RNA pol II are recruited to the TATA box, together forming the PIC (for review, Cramer, 2004; Hirose and Ohkuma, 2007; Sikorski and Buratowski, 2009). RNA pol II contains a C-terminal domain (CTD) composed of multiple heptad repeats of the consensus (Tyr1–Ser2–Pro3–Thr4–Ser5–Pro6–Ser7), whose phosphorylation state is crucial to its activity. In the inactive state, the HIV promoter is occupied with hypophosphorylated RNA pol II, TFIID, and mediator which contains the CDK8 repressive module. Upon activation, interaction with CDK8 is lost and the mediator, TFIIH and RNA pol II is set for activation. Concomitantly, the TFIIH-associated kinase CDK7 catalyzes the phosphorylation of Ser 5 of the heptapeptide repeats at the RNA Pol II CTD, required for transcription initiation (Kim et al., 2006). In this context, TFIIH is also necessary to separate the double-stranded DNA and allow RNA pol II access to the single-stranded DNA template.

4.2 RNA Pol II Pausing

Loading of RNA Pol II to promoters has long been considered the main rate-limiting step in gene transcription. Back in the 1987, pioneering work on HIV-1 transcription began to challenge this view (Kao et al., 1987) but only much later genome-wide approaches that tracked RNA Pol II on the DNA and monitored production of nascent transcripts, have revealed that at the majority of genes transcriptional regulation takes place at the level of RNA pol II pausing and its release into productive elongation (Adelman and Lis, 2012) reviewed in Guo and Price (2013) and Jonkers and Lis (2015). Under basal conditions, transcription is initiated but RNA Pol II elongation is restricted (Fig. 2). Shortly after elongation, RNA Pol II pauses and accumulates on the HIV-1 proximal promoter producing only short

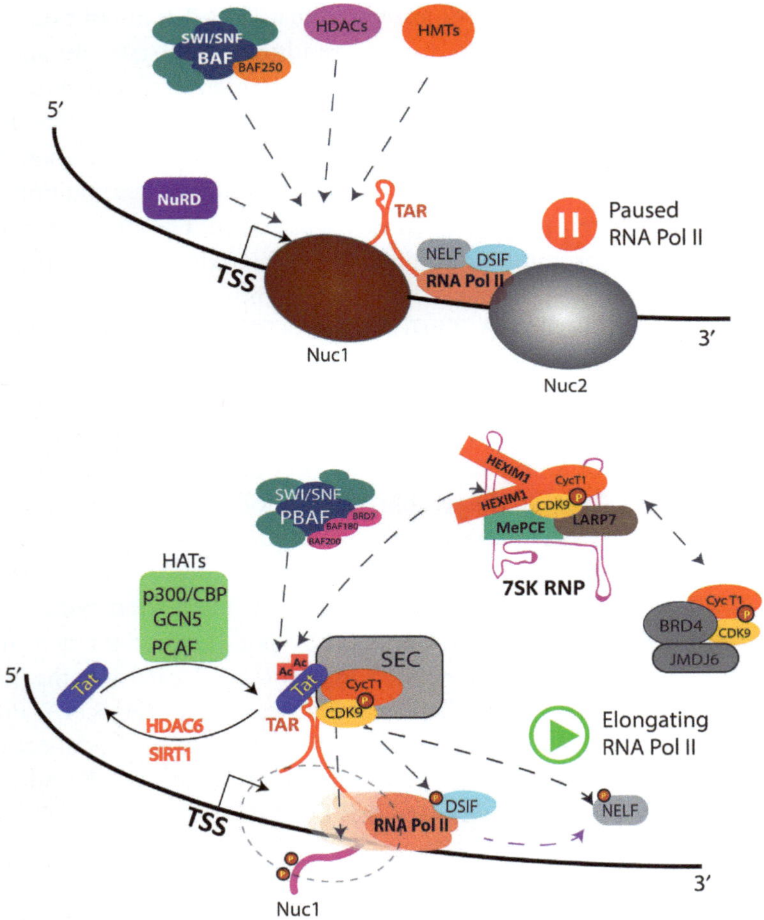

Fig. 2 Regulation of HIV-1 promoter transcriptional activity by Tat and coactivators. HIV-1 Tat recruits pTEFb from its inactive reservoir within the 7SK snRNP–Hexim1 complex to the TAR RNA stem loop structure and together with other components of the SEC phosphorylates the DSIFs and NELFs resulting in release of Pol II pausing and activates transcriptional elongation via phosphorylation of the CTD of RNA Pol II and stimulation of its activity. Tat also associates with various cofactors, which modulate its interaction with TAR and pTEFb via Tat modifications, maintaining its activity and remodeling chromatin to facilitate elongation.

transcripts. Owing to the promoter proximal pausing of RNA pol II, the HIV-1 provirus, is primed for a rapid transcriptional response (Jadlowsky et al., 2014). While RNA pol II pausing initially seemed to be an HIV-1-specific regulatory mechanism it is now known to be common at inducible cellular genes that require fast or synchronous activation (Adelman and Lis, 2012; Hargreaves et al., 2009). In this process two strong inhibitors of elongation (negative elongation factor) NELF and (DRB sensitivity-inducing factor) DSIF cooperatively stabilize the interaction of RNA pol II with DNA, halting elongation of the transcript (Fig. 2). In latent cells, the 3′-end processing protein Pcf11 is then recruited via a direct interaction with NELF which induces premature dissociation of RNA pol II from the DNA template (Cheng and Price, 2008; Jadlowsky et al., 2014; Natarajan et al., 2013; Yamaguchi et al., 1999; Zhang et al., 2007). DSIF, a heterodimer of Spt4 and Spt5, is recruited to the nascent RNA molecules and promotes capping (Cheng and Price, 2008; Missra and Gilmour, 2010; Wen and Shatkin, 1999). Spt5 binds to nascent RNA and in turn recruits NELF (Yamaguchi et al., 2002), generating a configuration in which the transcription complex has limited chance to escape the proximal promoter site, inducing premature termination of elongation within a range of a few 100 nucleotides (Renner et al., 2001). In agreement, depletion of NELF results in enhanced basal HIV-1 transcription and reduced latency (Jadlowsky et al., 2014; Natarajan et al., 2013; Zhang et al., 2007).

Despite many similarities, this regulatory mechanism has some HIV-1-specific features that distinguish it from regulation at endogenous genes. A prominent characteristic is that transcription elongation control at the HIV-1 promoter is much more efficient than typically observed at cellular promoters (Kim et al., 2011). Additionally, the efficiency of elongation in the presence of Tat is also exceptionally high, much higher than typically observed at endogenous genes (Kim et al., 2011). At the HIV-1 promoter, premature termination of elongation produces an abundant 59 nucleotides transcript (Kao et al., 1987). Such high levels of Pol II promoter proximal pausing are not commonly observed at endogenous genes. The nascent RNA forms a unique stem-loop secondary structure referred as TAR element, which is recognized by the transcriptional transactivating protein (Tat) (Fig. 2). The TAR RNA loop region contains an NBE-like element (CUGGGA) similar to the E binding element (CUGAGGA) associated with transcriptionally paused genes. This element may lead to enhanced recruitment of NELF to the HIV-1 5′-LTR (Pagano et al., 2014). Another peculiarity of HIV-1 5′-LTR transcription is that multiple pausing sites exist downstream of the TSS. These features reinforce the pausing signals to ensure silencing of the promoter by decreasing likelihood of RNA Pol II entering into the elongation phase to produce full-length transcripts (Jadlowsky et al., 2014).

5. TRANSCRIPTIONAL ELONGATION

5.1 P-TEFb and the SEC

In order to establish a productive cycle of viral replication, the transcription elongation block has to be removed. Expression of the viral protein Tat at sufficient levels disrupts the NELF- and DSIF-dependent inhibition, induces the recruitment of coactivating factors, the most critical being the SEC (Fig. 2). These events efficiently remove the block in transcription elongation and establish a Tat positive feedback loop in which Tat is rapidly upregulated and strongly pushes HIV-1 transcription toward activation (Karn, 2011; Mbonye and Karn, 2014; Ott et al., 2011). Once more, transcriptional regulation of the integrated HIV-1 provirus proved to be an excellent model system for studying general transcription and was instrumental in elucidating the molecular mechanisms that control RNA pol II elongation. In vitro reconstitution of RNA Pol II transcription elongation demonstrated that Tat strongly activates HIV-1 expression by recruiting the host P-TEFb complex, a heterodimer kinase composed of CDK9 and its regulatory partner cyclin (primarily cyclin T1 and T2a/T2b), to the TAR region of the HIV promoter (Wei et al., 1998). At the TAR region, Tat and P-TEFb form a protein–RNA complex with the TAR RNA stem-loop located at the 5′-end of all nascent viral transcripts. P-TEFb phosphorylates NELF (Fujinaga et al., 2004), releasing it from the complex, and DSIF (Ping and Rana, 2001), converting it to a positive factor, counteracting the inhibitory activity of the complex on RNA pol II elongation (Fig. 2) (Lu et al., 2013; Ott et al., 2011). Additionally P-TEFb phosphorylates Ser-2 on the CTD of RNA pol II, thereby improving its processivity, promoting the coupling between transcription, and RNA processing and the recruitment of chromatin modifying factors (Barboric and Lenasi, 2010; Bentley, 2014; Ping and Rana, 2001; Schüller et al., 2016; Zhou et al., 2012). These key phosphorylation events cause the release of the paused RNA Pol II and induce the elongation phase of transcription.

Later studies have shown that P-TEFb, although essential for Tat-mediated activation of HIV-1 transcription, is not the only player involved in promoting RNA pol II elongation and it is not sufficient to support full Tat-dependent activation (Lu et al., 2014; Suñé et al., 2000). Affinity purification of P-TEFb and Tat, followed by mass spectrometry analysis led to the identification of a multiprotein complex of cellular elongation factors and coactivators, referred to as SEC, that cooperatively stimulate HIV-1 transcription (He et al., 2010; Sobhian et al., 2010). Tat-mediated recruitment of SEC is a potent mechanism to counteract transcriptional pausing and strongly activate transcriptional elongation (Fig. 2). The SEC consists of the scaffold proteins AFF1 to AFF4 (Lu et al., 2015), the elongation factors ELL1 to ELL3 and several additional partners AF9, ENL, and EAF1/2. AFF4 and AFF1 have short hydrophobic domains distributed along a flexible axis that mediates interaction with the other subunits of the complex (Chou et al., 2013; He et al., 2011). AFF4 makes direct contact with Tat and the surface of cyclin T1 (Schulze-Gahmen et al., 2013, 2014) and the cooperative interaction between those subunits, AFF1 and AFF2, which reside in separate complexes, enhance Tat's affinity for P-TEFb with AFF1-SEC predominantly involved in supporting Tat-mediated transactivation of the HIV promoter (Lu et al., 2015).

5.2 Control of P-TEFb Activity

In order to control transcriptional elongation, the activity of P-TEFb has to be tightly regulated in vivo and this is achieved by PTMs of its components CycT1 and CDK9 (Fig. 2) (Budhiraja et al., 2013; Ramakrishnan et al., 2009; Tyagi et al., 2010). The mTOR complex acts as a repressor of HIV-1 transcription in part by interfering with CDK9 phosphorylation (Besnard et al., 2016). Conversely interaction of CycT1/CDK9 with the 7SK small nuclear ribonucleoprotein complex (7SK snRNP) limits P-TEFb activity and availability at gene promoters (Fig. 2) (He et al., 2008; Nguyen et al., 2001; Van Herreweghe et al., 2007; Yang et al., 2001; Yik et al., 2003). The fraction of P-TEFb molecules bound to the 7SK snRNP varies per cell type: in HeLa cells it is about half in normal growing conditions (Nguyen et al., 2001; Yang et al., 2001) while it can reach 90% in Jurkat T cells or PBMCs (Barboric et al., 2007; Kim et al., 2011). Within the complex, the 7SK snRNA is a scaffold for two molecules of P-TEFb complexed with homo or heterodimers of the inhibitory proteins HEXIM1 or HEXIM2 (Li et al., 2005; Peterlin et al., 2012). LARP7 binds to the 3′ uridine-rich terminus of 7SK-snRNA to protect it from degradation by nucleases, while the enzyme MEPCE stabilizes 7SK snRNA by adding a methyl-phosphate cap to its 5′ terminus (He et al., 2008; Jeronimo et al., 2007; Krueger et al., 2008). More recently it was shown that AFF1 and AFF4 are also partners of P-TEFb in the 7SK snRNP complex (Lu et al., 2014). When sequestered within the 7SK snRNP complex, P-TEFb is activated by phosphorylation of Thr186 in the CDK9 T loop, but its association with HEXIM1 blocks ATP binding, hence inhibiting its activity (Li et al., 2005). Within the complex, P-TEFb is also physically restrained, which prevents it from participating in the regulation of gene transcription (Chen et al., 2003). Therefore, in this configuration, the activity of P-TEFb is inhibited but kept in a preactivated form, fully ready to promote elongation upon release from the 7SK snRNP complex (Fig. 2).

5.3 Chromatin Association of the 7SK snRNP

As the 7SK snRNP complex is the major reservoir of transcriptionally inactive P-TEFb, significant effort has been made to determine how it engages with the transcriptional machinery to enable the transition into transcriptional elongation. HIV-1 transcription studies were the first to suggest that the 7SK snRNP complex is anchored to chromatin (Barboric and Lenasi, 2010; D'Orso and Frankel, 2010), with the HEXIM1 and LARP7 subunits cooccupying the HIV-1 core promoter with P-TEFb. This finding was later extended to stimulus-dependent primary response genes (McNamara et al., 2013) and verified genome-wide, through chromatin isolation by RNA purification and sequencing (ChIRP-seq) (Flynn et al., 2016; Liu et al., 2013). In murine embryonic stem cells and human cells, the 7SK snRNP complex has been shown to occupy about 3500 active protein-coding genes along the entire transcribed region with peaks, in accordance with RNA pol II occupancy, at the TSSs and transcriptional elongation sites (TESs). The presence of 7SK snRNA has been detected at active enhancers as well, particularly super enhancers. Considering cooccupancy and interaction studies (Das et al., 2007; D'Orso and Frankel, 2010), it was suggested that RNA pol II acts as a recruitment platform for 7SK snRNP at transcribed genes. More recent findings, however, described the different mechanism by which snRNP is likely to be recruited. The repressive chromatin mark H4R3me2 is recognized and bound by the 5′-terminal half of 7SK at a subclass of enhancers cooccupied by Brd4 and JMJD6 (Liu et al., 2013). Also, Kruppel-associated box-interacting protein KAP1 (TRIM28/TIF1β) could be involved in loading 7SK snRNP to promoter proximal regions containing paused Pol II (McNamara et al., 2016). Additionally, a specific subset of 7SK snRNP contains the transcriptional repressor CTIP2 which together with HEXIM1 and the 7SK enhances inhibition of P-TEFb (Cherrier et al., 2013). Recruitment of this subset of highly repressive 7SK snRNPs to HIV-1 and other promoters is suggested to occur via the nonhistone chromatin protein HMGA1 that interacts with CTIP2 and 7SK within the snRNP complex (Eilebrecht et al., 2013). HIC1 has also recently been shown to be a corepressor of HIV-1 Tat-dependent transcription in complex with HMGA1 and CTIP2 (Le Douce et al., 2016).

5.4 Regulation of P-TEFb Availability

Upon stress-induced response or cellular stimuli that alter cell growth, P-TEFb is rapidly released from 7SK RNP complexes and recruited to active genes. Different signal transduction pathways converge to single components to prevent or disrupt the assembly into 7SK snRNP. In response to T-cell receptor activation, HEXIM1 is phosphorylated at Serine 158 within the 7SK binding region by protein kinase C-ϑ and its ability to bind CycT1 and 7SK is compromised (Fujinaga et al., 2012). Additionally, the unstructured region of HEXIM1 contains Ser/Thr/Tyr target residues, which when phosphorylated, via PI3K/Akt and ERK pathways, abolish HEXIM1–P-TEFb interaction (Contreras et al., 2007; Kim et al., 2011; Mbonye et al., 2015). Expression of HEXIM1 harboring mutations that abrogate phosphorylation of its unstructured C-terminal coiled coil affects the release of P-TEFb from 7SK snRNP and reduces P-TEFb occupancy at the HIV-1 promoter, thereby compromising viral transcription (Contreras et al., 2007; Mbonye et al., 2015). The HAT enzyme p300 also contributes to release of P-TEFb by acetylating lysine residues in the CycT1 coiled-coil domain, inducing dissociation of P-TEFb from HEXIM1 (Cho et al., 2009). Other factors involved in mobilization of P-TEFb, HNRNPA1, HNRNPA2 (Barrandon et al., 2007; Van Herreweghe et al., 2007), SRSF2 (Ji et al., 2013), and DDX21 (Calo et al., 2014) seem to collectively link the release of P-TEFb with elongation and RNA processing.

While the modifications and factors mentioned above contribute to mobilizing P-TEFb without affecting the integrity of the complex, the release of P-TEFb from chromatin-associated 7SK snRNP at enhancers, which are also bound by Brd4-JMJD6, is different. Here, JMJD6 demethylase activity causes the removal of the H4R3me2 repressive chromatin mark and the 5′ 7SK methyl cap, abolishing recruitment of 7SK and alters the stability of 7SK and the 7SK snRNP complex (Liu et al., 2013). Upon release from the 7SK snRNP complex, P-TEFb binds JMJD6 and Brd4 and is loaded on active genes promoters marked by acetylated chromatin, to promote RNA pol II elongation (Fig. 2) (Bisgrove et al., 2007; Chen et al., 2004; Jang et al., 2005; Yang et al., 2005). Formation of the HIV-1 Tat/P-TEFb complex was found to be restrained by P-TEFb association with chromatin-bound Brd4 as knockdown of Brd4 in latently infected cells or treatment with the bromodomain inhibitor JQ1, resulted in increased release of P-TEFb, and a markedly enhanced Tat-dependent activation of HIV-1 (Bartholomeeusen et al., 2012; Boehm et al., 2013; Mbonye et al., 2013; Zhu et al., 2012). Interestingly, Brd2 has also been reported to repress the activity of the HIV-1 promoter, although the mechanism here seems to be independent of Tat as the effect has also been observed in latently infected cell lines that do not express the HIV-1 trans-activator (Boehm et al., 2013).

Tat can also liberate P-TEFb from the 7SK snRNP complex by competing and displacing HEXIM1 binding to CycT1 (Fig. 2) (Barboric et al., 2007; Krueger et al., 2010; Sedore et al., 2007). There is strikingly similarity between the P-TEFb–HEXIM1–7SK binding configuration, within the 7SK RNP complex, and P-TEFb binding to Tat-TAR (Yik et al., 2004). The 7SK binding arginine-rich motif (ARM) of HEXIM1 and the TAR binding ARM of Tat are highly similar (Yik et al., 2004), while 7SK HEXIM1 binding elements are very similar to the minimal consensus structure on TAR (Muniz et al., 2010). Both Tat and HEXIM1 contain a CycT1 binding site and bind CycT1 in a mutually exclusive fashion (Anand et al., 2007; Michels et al., 2004). Tat binding, through its ARM, was shown to displace HEXIM1 from 7SK (Muniz et al., 2010) and via its CycT1 binding domain outcompetes HEXIM1 binding to P-TEFb (Fig. 2) (Barboric et al., 2007). In this process, interaction with AFF1 assists Tat by further increasing its affinity for P-TEFb (Lu et al., 2014).

5.5 Tat Coregulators

In addition to SEC, Tat-dependent transcriptional activation is accompanied by recruitment of transcriptional coactivators and chromatin remodelers which also contribute to strong enhancement of transcription from the HIV-1 promoter (Ott et al., 2011). Tat has been shown to bind and recruit HATs p300/CBP, GCN5, and PCAF to the HIV-1 5′-LTR (Ahmad and Venkatesan, 1988; Benkirane et al., 1998; Col et al., 2001; Deng et al., 2001; Lusic et al., 2003; Marzio et al., 1998; Vendel and Lumb, 2003) which, by acetylation of histone tails, contributes to establishment of a more accessible chromatin environment for transcription (Easley et al., 2010a). Aside from histones, HATs can acetylate and regulate the activity of TFs (Jeng et al., 2015). Upon Tat-dependent recruitment, for example, p300/CBP acetylates the p50 subunit of NF-κB (Furia et al., 2002), thereby increasing its binding affinity and most likely the ability to enhance HIV-1 transcription in the context of p50–p65 heterodimer (Deng et al., 2003).

Tat itself is modified by PTMs including phosphorylation, acetylation, ubiquitination, and methylation, which generates a "Tat-code" that determines its binding partners, most critically its interaction with P-TEFb and TAR. Tat can be acetylated at lysine residues 28, 50, and 51 (D'Orso and Frankel, 2010; Kiernan et al., 1999; Ott et al., 1999), ubiquitinated (Brès et al., 2003; Zhang et al., 2014), or undergo methylation at lysines 50 and 51 and arginine residues 52 and 53 (Ott et al., 1999; Van Duyne et al., 2011; Xie et al., 2007). Additionally, Tat can undergo CDK2-dependent inhibitory phosphorylation on Serines 16 and 46 (Ammosova et al., 2006). Methyl-transferase Set7/9 methylates Tat at Lys51, enhancing HIV-1 transcription, subsequently Tat is demethylated by LSD1 ensuring the recycling of Tat in an activation cycle (Pagans et al., 2010; Sakane et al., 2011). Polyubiquitination of Tat Lys71, which is Hdm2 dependent, promotes Tat-dependent transactivation (Brès et al., 2003). The acetyltransferase P300 acetylates Tat at Lys50 and Lys51 inducing dissociation from TAR. PCAF is subsequently recruited to acetylated lys50 and acetylates Tat at Lys 28 which increases its affinity for P-TEFb during transcriptional elongation while HDAC6-dependent deacetylation allows its release from P-TEFb and TAR leading to the beginning of a new cycle of transcription (Brès et al., 2003; Huo et al., 2011; Kiernan et al., 1999; Pagans et al., 2005). Additionally, acetylated Lys50 allows Tat-mediated recruitment of the PBAF chromatin remodeling complex to the LTR (Agbottah et al., 2006; Mahmoudi et al., 2006; Ott et al., 1999; Tréand et al., 2006). This occurs via the PBAF complex subunits BAF180 and BRG1 contain bromodomains that can recognize and bind acetylated Tat (ac-Tat) (Fig. 2) (Agbottah et al., 2006; Easley et al., 2010b; Mahmoudi et al., 2006; Rafati et al., 2011; Van Duyne et al., 2011). Tat-dependent recruitment of the PBAF complex is likely to enhance productive elongation as in the absence of the catalytic subunit BRM prematurely terminated RNA transcripts have been shown to accumulate at the HIV-1 promoter (Mizutani et al., 2009). In line with this model, it has been reported that depletion of the BAF-specific subunit BAF250 does not affect Tat-dependent activation, whereas depletion of BAF200 and BAF180, PBAF-specific subunits, result in impaired HIV-1 transcription (Easley et al., 2010b; Rafati et al., 2011). ChIP experiments have additionally demonstrated that, upon activation, binding of the BAF250 subunit to the HIV-1 5′-LTR is rapidly abrogated while PBAF subunits BAF200 and BAF180 are recruited at the promoter (Rafati et al., 2011; Van Duyne et al., 2011). Moreover, acetylated Tat Lys 50 and 51 allow interaction with bromodomain containing proteins (Dorr et al., 2002; Mujtaba et al., 2002). Among these proteins, BRD4 was shown to diminish P-TEFb availability thus negatively regulating HIV-1 transcription. BRD4, therefore, competes with Tat for interaction and prevents P-TEFb from elongation of transcription (Bartholomeeusen et al., 2012; Bisgrove et al., 2007; Mbonye et al., 2013).

Given the magnitude of the Tat transcriptional feedback loop and Tat-driven epigenetic changes at the viral promoter, strategies aiming at selective inhibition of Tat activity may promote the establishment of a persistent silencing of the HIV-1 promoter, referred as "deep latency" (Mousseau and Valente, 2012; Mousseau et al., 2015a). This alternative approach has been explored in several studies with encouraging results (reviewed in Mousseau and Valente, 2012; Mousseau et al., 2015a). As a proof of concept, recent studies using didehydro-cortistatin A (dCA), a specific inhibitor of Tat-TAR binding, have shown that dCA treatment in the presence of c-ART leads to the establishment of a stronger transcriptional silencing, with lower propensity to reactivation in in vitro and ex vivo models. Importantly, it has been observed in a primary cell model of latency that treatment with dCA has a long lasting inhibitory effect on transcription even after its removal, suggesting that in the absence of Tat activity the HIV-1 promoter may undergo epigenetic changes that promote stronger repression (Mousseau et al., 2015b).

6. STOCHASTICITY IN TRANSCRIPTION

Throughout this review, we have described gene expression and cell fate as a merely deterministic process. Indeed, in the majority of cases, the complex interplay of regulatory mechanisms we have introduced determines the activity of a promoter and ultimately the cell fate. Similarly, for HIV infections, these layers of regulation cooperatively determine latent vs productive infections. There are cases though where cells seem to choose their fate without an apparent cause, stochastically. The main reason is that gene transcription and ultimately expression do not have a perfect "switch" but it is subjected to a certain degree of noise. This phenomenon is clearly revealed for the HIV-1 promoter by the observation that monoclonal Jurkat cell lines harboring latent HIV-1-derived virus integrated at a single integration site display a nonuniform basal transcriptional state during latency and also in response to activation signals. Thus a population of identical cells shows variability in basal HIV-1 promoter activity as demonstrated by the percentage of cells expressing the GFP reporter gene, and in response to activating stimuli where a percentage but not all of the latent cells undergo transcription elongation to express GFP.

6.1 Noise in Basal Transcription

Noise in transcription originates from transitions between on and off states at the promoter. In the off state, the RNA polymerase II pauses on the HIV-1 proximal promoter but when the block is removed, multiple RNA polymerase units simultaneously read the gene's open reading frame (Tantale et al., 2016), resulting in a burst of RNA transcripts (Dar et al., 2012; Singh et al., 2010). This amplification originates from a mechanism of mediator-dependent reinitiation of transcription. According to this model, after a burst, RNA Pol II molecules rapidly initiate transcription but also can undergo rapid escape from the promoter without fruitful elongation such that the promoter stochastically switches from an on to an off state (Tantale et al., 2016). Oscillations in gene expression also result from random differences in the timing of the process, fluctuations in the availability of RNA polymerases and TFs, differences in cell size, asymmetric distribution, and availability of cellular factors following cell division and other factors (Kaern et al., 2005; Miller-Jensen et al., 2011; Raj and van Oudenaarden, 2008; Raser, 2010; Swain et al., 2002). The chromatin environment at the gene regulatory regions is also considered to be a determinant factor for controlling transcriptional noise (Miller-Jensen et al., 2011). In general, when the gene regulatory regions are accessible and the TSS is free from nucleosomes, transcription can occur relatively freely and more frequently. Conversely, when the chromatin is in a condensed state, transcription is less likely to occur (Raj and van Oudenaarden, 2008). Different studies have suggested that chromatin-related events can alter the grade of stochasticity of gene expression (Dey et al., 2015; Raser and Shea, 2006). It has been observed by electron microscopy in budding yeast that, at the steady state, there is an intrinsically stochastic component that regulates the configurational variation of nucleosomes at gene promoters.

6.2 Modulation of Noise

Transcriptional noise is the driving force of stochastic events but it is not sufficient, per se, to produce an effect. In order for an event to occur stochastically, a mechanism is necessary that stabilizes and possibly amplifies the response to the noise (Losick and Desplan, 2008). HIV-1 has this mechanism. The potent HIV-1 transactivator Tat protein can trigger a positive feedback loop to induce its own expression 50–100-fold (Chavez et al., 2015; Singh and Weinberger, 2009; Weinberger et al., 2005). Even small probabilistic fluctuations in Tat expression can, therefore, by amplification, be determinant and shift the equilibrium from latency to replication. Stochastic fluctuations in the regulatory mechanisms that control HIV-1 transcription can, therefore, influence viral latency. In this light HIV-1 expression is an excellent model system to study the intrinsically stochastic nature of transcription.

A deterministic explanation for the fact that the provirus is more prone to remain silent in resting memory CD4+ T cells, the major reservoir of latently HIV-1-infected cells, is that, due to their resting state, some key cellular TFs such as NF-AT and NF-κB, are sequestered in the cytoplasm and depleted from the nucleus. As a consequence, transcription initiation and elongation do not occur. T cell activation can indeed reverse latency. However, studies show that despite multiple rounds of maximum in vitro activation, latent proviruses remain inactive in a substantial number of cells (Ho et al., 2013) and that the Tat-positive feedback can control latency independently of the cell activation state (Razooky et al., 2015), suggesting that reactivation has a stochastic component. It has been observed in that even in clonal populations of latently infected cell lines there is a probabilistic distribution of cells expressing low to high levels of Tat (Burnett et al., 2009; Miller-Jensen et al., 2013). Transcriptional noise from the HIV-1 promoter is, therefore, likely the cause of stochastic reactivation (Weinberger and Weinberger, 2013). Stochastic variability in HIV expression represents a tremendous obstacle for strategies aiming to reactivate latent HIV to deplete the viral reservoir. It has been estimated that a 10,000-fold depletion of the reservoir is required to prevent viremia rebound in 50% of the patients, after interrupting cART (Hill et al., 2014). An interesting strategy to control HIV-1 expression, taking stochasticity into account, is to look for noise modulating molecules instead of activators only. Transcriptional activators typically activate transcription by improving transcriptional initiation efficiency, hence increasing the frequency of transcriptional events

(Dar et al., 2012; Singh et al., 2010). Noise modulating molecules instead act on elongation, increasing and decreasing the duration of the on state, thereby influencing the magnitude of the burst. Dar et al. have shown that noise enhancing molecules alone, or together with traditional activators, can reactivate latent cells significantly better than existing LRA cocktails while noise suppressing molecules can stabilize latency (Dar et al., 2014).

7. HIV-1 TRANSCRIPTION REGULATION IN THE NUCLEAR 3D ENVIRONMENT

Transcription of genes, and hence the integrated HIV-1 provirus, does not take place on a linear chromatin template but instead involves intricate long range chromatin interactions between regulatory elements (Splinter and de Laat, 2011). At the gene level, the active HIV-1 provirus forms a looped chromatin structure mediated by a direct interaction between the 5′- and 3′-LTRs (Fig. 3) (Perkins et al., 2008). It has long been known that transcription initiation is intimately linked to transcription end processing. The existence of gene loops was first inferred from genetic data obtained from yeast screens. These studies demonstrated a genetic interaction between the promoter bound basal TF TFIIB and Ssu72 (Sun and Hampsey, 1996) that later turned out to be a component of the 3′-end processing complex (He et al., 2003). Similarly, in mammalian cells, a direct interaction between the transcriptional coactivator PC4 and CstF-64, another component of the 3′-end processing complex, was found (Calvo and Manley, 2001). The first promoter–terminator interactions through chromatin looping were described for the yeast FMP27 gene (O'Sullivan et al., 2004) and a direct role for 3′-end processing factors in gene loop formation was demonstrated (Ansari and Hampsey, 2005). The occurrence of gene loops in mammalian cells was first demonstrated for the integrated HIV-1 provirus (Perkins et al., 2008). HIV-1 gene loop formation is linked to transcriptional initiation and elongation since the interaction between the 5′-LTR promoter region and the 3′-LTR polyadenylation signal was induced by activation of the HIV-1 provirus by either TPA treatment or overexpression of the Tat transactivator (Perkins et al., 2008). Conversely, pharmacological inhibition of the P-TEFb with flavopyridol resulted in the collapse of the gene loop suggesting that gene loops are unstable and dependent on ongoing transcription (Perkins et al., 2008). HIV-1 gene loop formation relies on both intact 5′-promoter and 3′-end processing sequences.

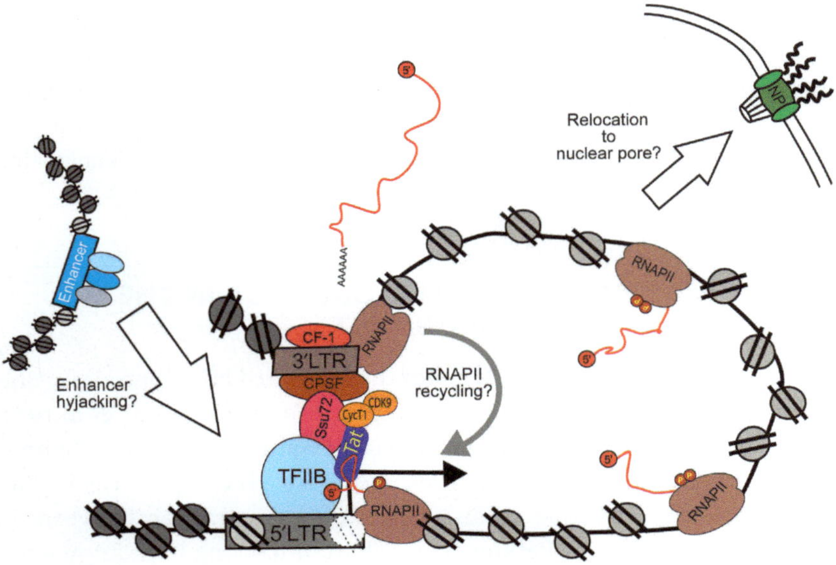

Fig. 3 A model for gene looping of the HIV-1 provirus and its interaction with the 3D nuclear space. The active HIV-1 provirus forms a looped chromatin structure mediated by a direct interaction of factors bound to the 5′- and 3′-LTRs such as the promoter bound basal transcription factor TFIIB and Ssu72 bound to the 3′-LTR. HIV-1 Tat trans-activator bound to the TAR sequence also directly interacts with Ssu72 potentially consolidates gene loop formation. Gene looping facilitated RNAPII recycling could aid transcriptional reinitiation and suppress divergent transcription or stochasticity of expression. It has been suggested that HIV-1 gene loop formation induces relocalization of the HIV-1 provirus to nuclear pores which boosts transcriptional output and stability and facilitates nuclear export of HIV-1 proviral RNA. Endogenous enhancer sequences could potentially modulate HIV1 expression if they are hijacked by the HIV-1 provirus.

However, gene loop formation is not a unique feature of HIV-1 proviruses as replacement of viral promoter and end processing sequences with heterologous elements still supported gene looping (Perkins et al., 2008). This observation suggested that gene loop formation is a universal feature of transcription and indeed it was later demonstrated that the endogenous human BRCA1 (Tan-Wong et al., 2008) and CD68 (O'Reilly and Greaves, 2007) genes also form gene loops. Gene looping in yeast is dependent both on factors bound to the promoter and on 3′-end processing complex subunit components like, e.g., Ssu72 (El Kaderi et al., 2009; Medler et al., 2011). Interestingly, the HIV-1 Tat transactivator directly interacts with Ssu72 and stimulates viral gene expression (Fig. 3) (Chen et al., 2014) but a direct role in HIV-1 gene loop formation remains to be demonstrated.

The functional significance of gene loops is currently not clear especially since gene looping appears to depend on transcription. Conversely, loss of looping does not seem to have a major impact on transcription (Hampsey et al., 2011). It has been suggested that gene looping plays a role in suppression of divergent transcription (Grzechnik et al., 2014; Tan-Wong et al., 2012) and provides "transcriptional memory" by targeting the locus to the nuclear pore (Hampsey et al., 2011; Pascual-Garcia et al., 2017; Tan-Wong et al., 2008). Recent research has also revealed that interactions between the nuclear pore complex, chromatin, and the transcriptional machinery regulates transcriptional activation, elongation, and mRNA processing (Ptak et al., 2014). These mechanisms could all potentially influence latency reversal, stochastic expression, and transcriptional output of the HIV-1 provirus (Fig. 3). Interestingly, the HIV-1 virus tends to integrate into the peripheral nuclear compartment presumably in the vicinity of nuclear pores (Lusic and Siliciano, 2017). Integrated HIV-1 proviruses can efficiently be transcribed when located at the nuclear periphery (Dieudonné et al., 2009) and depletion of the nuclear pore component TPR has a strong impact on HIV-1 provirus expression (Lelek et al., 2015). Therefore it is possible that HIV-1 gene loop formation integrates local chromatin organization with the spatial distribution of the HIV-1 provirus to nuclear pores which results in boosted transcriptional output and stability and additionally facilitates nuclear export of HIV-1 proviral RNA (Fig. 3).

Although HIV-1 preferentially integrates into gene-dense regions and active genes, the site of integration still strongly influences viral expression (Lusic and Siliciano, 2017). Integration in or close to heterochromatic regions or a direct interaction in trans with such regions or other repressive compartments leads to clear position effects on the transcriptional output of the HIV-1 provirus and contributes to latency establishment (Dieudonné et al., 2009; Jordan et al., 2001; Lusic and Siliciano, 2017; Lusic et al., 2013). A recent study using barcoded HIV-1 viruses demonstrated that HIV-1 proviruses integrated proximal to endogenous enhancers are expressed at a higher level than those located in other genomic regions (Chen et al., 2016). The same study also found that latent proviruses are integrated further from enhancers than their active counterparts and that distance to these endogenous regulatory elements also determined the response to different latency reversal agents (Chen et al., 2016). These results suggest that HIV1 expression is not only susceptible to the chromatin environment and availability of TFs but is also subject to regulation by endogenous enhancers (Fig. 3).

The emerging picture is that the 3D chromatin organization and nuclear positioning of the integrated HIV-1 provirus play a crucial part in its transcriptional regulation (Fig. 3). However, direct biophysical evidence supporting a role for gene looping and nuclear positioning in regulation of HIV-1 gene transcription in in vivo-infected patient cells is currently lacking. Also at the mechanistic level questions still remain; e.g., what is the exact role and mechanism of HIV-1 gene loop formation and what is its connection to nuclear positioning? It will also be interesting to know whether the HIV-1 provirus "highjacks" the activity of endogenous regulatory elements by engaging in a direct long range chromatin interaction with enhancers, or do these elements contribute in another way to an increased transcriptional potency of the HIV-1 provirus, e.g., by maintaining a permissive chromatin environment.

8. CONCLUDING REMARKS

The HIV-1 genome does not encode all the factors necessary for its own transcription. Therefore, upon integration into the host cell genome, HIV-1 gene expression, similar to endogenous genes, is regulated by host cellular TFs and the basal transcription machinery. The molecular control of HIV-1 transcription has been intensely studied from a clinical perspective with the aim to cure HIV patients using for example latency reversal therapies. However, the HIV provirus has also been a popular model system to study fundamental aspects of eukaryotic transcriptional regulation. Groundbreaking discoveries related to, e.g., recruitment of TFs, posttranscriptional modification, and remodeling of nucleosomes, recruitment of the basal transcriptional machinery, RNA polymerase II pausing, gene looping, and positioning of gene loci within the nuclear space have been made using this system. It is the unique interplay between clinical driven and fundamental research that has advanced our knowledge of transcriptional regulation and provided valuable insight toward development of alternative strategies for therapies aimed at eradicating both latent and active HIV-1.

ACKNOWLEDGMENTS

This work was supported by two Dutch "Aids Fonds" grants (2014021 and P-12001), the mRACE ErasmusMC grant, and the ERC starting grant (337116—Trxn-PURGE).

REFERENCES

Adelman, K., Lis, J.T., 2012. Promoter-proximal pausing of RNA polymerase II: emerging roles in metazoans. Nat. Rev. Genet. 13 (10), 720–731.

Agbottah, E., et al., 2006. Effect of SWI/SNF chromatin remodeling complex on HIV-1 Tat activated transcription. Retrovirology 3 (1), 48.

Agosto, L., Gagne, M., Henderson, A., 2015. Impact of chromatin on HIV replication. Genes 6 (4), 957–976.

Ahmad, N., Venkatesan, S., 1988. Nef protein of HIV-1 is a transcriptional repressor of HIV-1 LTR. Science (New York, NY) 241 (4872), 1481–1485.

Albert, I., et al., 2007. Translational and rotational settings of H2A.Z nucleosomes across the Saccharomyces cerevisiae genome. Nature 446 (7135), 572–576.

Alcamí, J., et al., 1995. Absolute dependence on kappa B responsive elements for initiation and Tat-mediated amplification of HIV transcription in blood CD4 T lymphocytes. EMBO J. 14 (7), 1552–1560.

Ammosova, T., et al., 2006. Phosphorylation of HIV-1 tat by CDK2 in HIV-1 transcription. Retrovirology 3 (1), 78.

Anand, K., et al., 2007. Cyclin box structure of the P-TEFb subunit cyclin T1 derived from a fusion complex with EIAV Tat. J. Mol. Biol. 370 (5), 826–836.

Ansari, A., Hampsey, M., 2005. A role for the CPF 3′-end processing machinery in RNAP II-dependent gene looping. Genes Dev. 19 (24), 2969–2978.

Archin, N.M., et al., 2009. Expression of latent human immunodeficiency type 1 is induced by novel and selective histone deacetylase inhibitors. AIDS 23 (14), 1799–1806.

Arya, S., et al., 1985. Trans-activator gene of human T-lymphotropic virus type III (HTLV-III). Science 229 (4708), 69–73.

Bachu, M., Mukthey, A.B., et al., 2012a. Sequence insertions in the HIV type 1 subtype C viral promoter predominantly generate an additional NF-κB binding site. AIDS Res. Hum. Retroviruses 28 (10), 1362–1368.

Bachu, M., Yalla, S., et al., 2012b. Multiple NF-κB sites in HIV-1 subtype C long terminal repeat confer superior magnitude of transcription and thereby the enhanced viral predominance. J. Biol. Chem. 287 (53), 44714–44735.

Badalà, F., Nouri-mahdavi, K., Raoof, D.A., 2008. NIH public access. Computer 144 (5), 724–732.

Barboric, M., Lenasi, T., 2010. Kick-sTARting HIV-1 transcription elongation by 7SK snRNP deporTATion. Nat. Struct. Mol. Biol. 17 (8), 928–930.

Barboric, M., et al., 2007. Tat competes with HEXIM1 to increase the active pool of P-TEFb for HIV-1 transcription. Nucleic Acids Res. 35 (6), 2003–2012.

Barrandon, C., et al., 2007. The transcription-dependent dissociation of P-TEFb-HEXIM1-7SK RNA relies upon formation of hnRNP-7SK RNA complexes. Mol. Cell. Biol. 27 (20), 6996–7006.

Barski, A., et al., 2007. High-resolution profiling of histone methylations in the human genome. Cell 129 (4), 823–837.

Bartholomeeusen, K., et al., 2012. Bromodomain and extra-terminal (BET) Bromodomain inhibition activate transcription via transient release of positive transcription elongation factor b (P-TEFb) from 7SK small nuclear ribonucleoprotein. J. Biol. Chem. 287 (43), 36609–36616.

Barton, K.M., et al., 2014. Selective HDAC inhibition for the disruption of latent HIV-1 infection. PLoS One 9 (8), e102684.

Barton, K., et al., 2016. Broad activation of latent HIV-1 in vivo. Nat. Commun. 7, 12731.

Benkirane, M., et al., 1998. Activation of integrated provirus requires histone acetyltransferase: p300 and P/CAF are coactivators for HIV-1 tat. J. Biol. Chem. 273 (38), 24898–24905.

Bentley, D.L., 2014. Coupling mRNA processing with transcription in time and space. Nat. Rev. Genet. 15 (3), 163–175.
Berkhout, B., Silverman, R.H., Jeang, K.T., 1989. Tat trans-activates the human immunodeficiency virus through a nascent RNA target. Cell 59 (2), 273–282.
Bernstein, B.E., et al., 2004. Global nucleosome occupancy in yeast. Genome Biol. 5 (9), R62.
Besnard, E., et al., 2016. The mTOR complex controls HIV latency. Cell Host Microbe 20 (6), 785–797.
Bialek, J.K., et al., 2016. Targeted HIV-1 latency reversal using CRISPR/Cas9-derived transcriptional activator systems. PLoS One 11 (6), e0158294.
Bisgrove, D.A., et al., 2007. Conserved P-TEFb-interacting domain of BRD4 inhibits HIV transcription. Proc. Natl. Acad. Sci. 104 (34), 13690–13695.
Bjorndal, A., et al., 1999. Phenotypic characteristics of human immunodeficiency virus type 1 subtype C isolates of Ethiopian AIDS patients. AIDS Res. Hum. Retroviruses 15 (7), 647–653.
Blazkova, J., et al., 2009. CpG methylation controls reactivation of HIV from latency. PLoS Pathog. 5 (8), e1000554.
Blazkova, J., et al., 2012. Paucity of HIV DNA methylation in latently infected, resting CD4+ T cells from infected individuals receiving antiretroviral therapy. J. Virol. 86 (9), 5390–5392.
Boehm, D., et al., 2013. BET bromodomain-targeting compounds reactivate HIV from latency via a Tat-independent mechanism. Cell Cycle 12 (3), 452–462.
Bosque, A., Planelles, V., 2009. Induction of HIV-1 latency and reactivation in primary memory CD4+ T cells. Blood 113 (1), 58–65.
Boyle, A.P., et al., 2011. High-resolution genome-wide in vivo footprinting of diverse transcription factors in human cells. Genome Res. 21 (3), 456–464.
Brès, V., et al., 2003. A non-proteolytic role for ubiquitin in Tat-mediated transactivation of the HIV-1 promoter. Nat. Cell Biol. 5 (8), 754–761.
Bruce, J.W., et al., 2013. ZASC1 stimulates HIV-1 transcription elongation by recruiting P-TEFb and TAT to the LTR promoter. PLoS Pathog. 9 (10), e1003712.
Budhiraja, S., et al., 2013. Cyclin T1 and CDK9 T-loop phosphorylation are downregulated during establishment of HIV-1 latency in primary resting memory CD4+ T cells. J. Virol. 87 (2), 1211–1220.
Burger, S., Poles, M.A., 2003. Natural history and pathogenesis of human immunodeficiency virus infection. Semin. Liver Dis. 23 (2), 115–124.
Burke, T.W., Kadonaga, J.T., 1996. Drosophila TFIID binds to a conserved downstream basal promoter element that is present in many TATA-box-deficient promoters. Genes Dev. 10 (6), 711–724.
Burke, T.W., Kadonaga, J.T., 1997. The downstream core promoter element, DPE, is conserved from drosophila to humans and is recognized by TAFII60 of drosophila. Genes Dev. 11 (22), 3020–3031.
Burnett, J.C., et al., 2009. Control of stochastic gene expression by host factors at the HIV promoter. PLoS Pathog. 5 (1), e1000260.
Burnett, J.C., et al., 2010. Combinatorial latency reactivation for HIV-1 subtypes and variants. J. Virol. 84 (12), 5958–5974.
Calo, E., et al., 2014. RNA helicase DDX21 coordinates transcription and ribosomal RNA processing. Nature 518 (7538), 249–253.
Calvo, O., Manley, J.L., 2001. Evolutionarily conserved interaction between CstF-64 and PC4 links transcription, polyadenylation, and termination. Mol. Cell 7 (5), 1013–1023.
Cary, D.C., et al., 2016. Molecular mechanisms of HIV latency. J. Clin. Investig. 126 (2), 448–454.
Chaudhary, P., et al., 2016. HSP70 binding protein 1 (HspBP1) suppresses HIV-1 replication by inhibiting NF-κB mediated activation of viral gene expression. Nucleic Acids Res. 44 (4), 1613–1629.

Chavez, L., et al., 2015. HIV latency is established directly and early in both resting and activated primary CD4 T cells. PLoS Pathog. 11 (6), e1004955.
Chen, R., Yang, Z., Zhou, Q., 2003. Phosphorylated positive transcription elongation factor b (P-TEFb) is tagged for inhibition through association with 7SK snRNA. J. Biol. Chem. 279 (6), 4153–4160.
Chen, R., Yang, Z., Zhou, Q., 2004. Phosphorylated positive transcription elongation factor b (P-TEFb) is tagged for inhibition through association with 7SK snRNA. J. Biol. Chem. 279 (6), 4153–4160.
Chen, Y., et al., 2014. A gene-specific role for the Ssu72 RNAPII CTD phosphatase in HIV-1 Tat transactivation. Genes Dev. 28 (20), 2261–2275.
Chen, H.-C., et al., 2016. Position effects influence HIV latency reversal. Nat. Struct. Mol. Biol. 24 (1), 47–54.
Cheng, B., Price, D.H., 2008. Analysis of factor interactions with RNA polymerase II elongation complexes using a new electrophoretic mobility shift assay. Nucleic Acids Res. 36 (20), e135.
Chen-Park, F.E., et al., 2002. The kappa B DNA sequence from the HIV long terminal repeat functions as an allosteric regulator of HIV transcription. J. Biol. Chem. 277 (27), 24701–24708.
Cherrier, T., et al., 2013. CTIP2 is a negative regulator of P-TEFb. Proc. Natl. Acad. Sci. U.S.A. 110 (31), 12655–12660.
Cho, S., et al., 2009. Acetylation of cyclin T1 regulates the equilibrium between active and inactive P-TEFb in cells. EMBO J. 28 (10), 1407–1417.
Chou, S., et al., 2013. HIV-1 Tat recruits transcription elongation factors dispersed along a flexible AFF4 scaffold. Proc. Natl. Acad. Sci. U.S.A. 110 (2), E123–31.
Clapier, C.R., Cairns, B.R., 2009. The biology of chromatin remodeling complexes. Annu. Rev. Biochem. 78 (1), 273–304.
Col, E., et al., 2001. The histone acetyltransferase, hGCN5, interacts with and acetylates the HIV transactivator, Tat. J. Biol. Chem. 276 (30), 28179–28184.
Contreras, X., et al., 2007. HMBA releases P-TEFb from HEXIM1 and 7SK snRNA via PI3K/Akt and activates HIV transcription. PLoS Pathog. 3 (10), e146.
Coull, J.J., et al., 2000. The human factors YY1 and LSF repress the human immunodeficiency virus type 1 long terminal repeat via recruitment of histone deacetylase 1. J. Virol. 74 (15), 6790–6799.
Craigie, R., Bushman, F.D., 2012. HIV DNA integration. Cold Spring Harb. Perspect. Med. 2 (7), a006890.
Cramer, P., 2004. RNA polymerase II structure: from core to functional complexes. Curr. Opin. Genet. Dev. 14 (2), 218–226.
Cron, R.Q., et al., 2000. NFAT1 enhances HIV-1 gene expression in primary human CD4 T cells. Clin. Immunol. 94 (3), 179–191.
Cullen, B.R., Lomedico, P.T., Ju, G., 1984. Transcriptional interference in avian retroviruses—implications for the promoter insertion model of leukaemogenesis. Nature 307 (5948), 241–245.
D'Orso, I., Frankel, A.D., 2010. RNA-mediated displacement of an inhibitory snRNP complex activates transcription elongation. Nat. Struct. Mol. Biol. 17 (7), 815–821.
Dahabieh, M.S., Battivelli, E., Verdin, E., 2015. Understanding HIV latency: the road to an HIV cure. Annu. Rev. Med. 66 (1), 407–421.
Dahiya, S., et al., 2014. CCAAT enhancer binding protein and nuclear factor of activated T cells regulate HIV-1 LTR via a novel conserved downstream site in cells of the monocyte-macrophage lineage. PLoS One 9 (2), e88116.
Dar, R.D., et al., 2012. Transcriptional burst frequency and burst size are equally modulated across the human genome. Proc. Natl. Acad. Sci. 109 (43), 17454–17459.
Dar, R.D., et al., 2014. Screening for noise in gene expression identifies drug synergies. Science 344 (6190), 1392–1396.

Das, R., et al., 2007. SR proteins function in coupling RNAP II transcription to pre-mRNA splicing. Mol. Cell 26 (6), 867–881.
De Crignis, E., Mahmoudi, T., 2014. HIV eradication: combinatorial approaches to activate latent viruses. Virus 6 (11), 4581–4608.
De Crignis, E., Mahmoudi, T., 2017. The multifaceted contributions of chromatin to HIV-1 integration, transcription, and latency. Int. Rev. Cell Mol. Biol. 328, 197–252.
Deeks, S.G., 2012. HIV: Shock and kill. Nature 487 (7408), 439–440.
Demarchi, F., et al., 1993. In vivo footprinting analysis of constitutive and inducible protein-DNA interactions at the long terminal repeat of human immunodeficiency virus type 1. J. Virol. 67 (12), 7450–7460.
Demarchi, F., et al., 1996. Activation of transcription factor NF-kappaB by the Tat protein of human immunodeficiency virus type 1. J. Virol. 70 (7), 4427–4437.
Deng, L., et al., 2001. Enhancement of the p300 HAT activity by HIV-1 Tat on chromatin DNA. Virology 289 (2), 312–326.
Deng, W.-G., Zhu, Y., Wu, K.K., 2003. Up-regulation of p300 binding and p50 acetylation in tumor necrosis factor-alpha-induced cyclooxygenase-2 promoter activation. J. Biol. Chem. 278 (7), 4770–4777.
Dey, S.S., et al., 2015. Orthogonal control of expression mean and variance by epigenetic features at different genomic loci. Mol. Syst. Biol. 5 (5), 806.
Dieudonné, M., et al., 2009. Transcriptional competence of the integrated HIV-1 provirus at the nuclear periphery. EMBO J. 28 (15), 2231–2243.
Dingwall, C., et al., 1989. Human immunodeficiency virus 1 tat protein binds trans-activation-responsive region (TAR) RNA in vitro. Proc. Natl. Acad. Sci. U.S.A. 86 (18), 6925–6929.
Donahue, D.A., et al., 2013. Cellular and molecular mechanisms involved in the establishment of HIV-1 latency. Retrovirology 10 (1), 11.
Dorr, A., et al., 2002. Transcriptional synergy between Tat and PCAF is dependent on the binding of acetylated Tat to the PCAF bromodomain. EMBO J. 21 (11), 2715–2723.
du Chéné, I., et al., 2007. Suv39H1 and HP1gamma are responsible for chromatin-mediated HIV-1 transcriptional silencing and post-integration latency. EMBO J. 26 (2), 424–435.
Duckett, C.S., et al., 1993. Dimerization of NF-KB2 with RelA(p65) regulates DNA binding, transcriptional activation, and inhibition by an I kappa B-alpha (MAD-3). Mol. Cell. Biol. 13 (3), 1315–1322.
Duverger, A., et al., 2013. An AP-1 binding site in the enhancer/core element of the HIV-1 promoter controls the ability of HIV-1 to establish latent infection. J. Virol. 87 (4), 2264–2277.
Dynlacht, B.D., Hoey, T., Tjian, R., 1991. Isolation of coactivators associated with the TATA-binding protein that mediate transcriptional activation. Cell 66 (3), 563–576.
Easley, R., Carpio, L., et al., 2010a. Transcription through the HIV-1 nucleosomes: effects of the PBAF complex in Tat activated transcription. Virology 405 (2), 322–333.
Easley, R., Van Duyne, R., et al., 2010b. Chromatin dynamics associated with HIV-1 Tat-activated transcription. Biochim. Biophys. Acta 1799 (3–4), 275–285.
Eilebrecht, S., et al., 2013. HMGA1 directly interacts with TAR to modulate basal and Tat-dependent HIV transcription. RNA Biol. 10 (3), 436–444.
El Kaderi, B., et al., 2009. Gene looping is conferred by activator-dependent interaction of transcription initiation and termination machineries. J. Biol. Chem. 284 (37), 25015–25025.
el Kharroubi, A., Martin, M.A., 1996. cis-Acting sequences located downstream of the human immunodeficiency virus type 1 promoter affect its chromatin structure and transcriptional activity. Mol. Cell. Biol. 16 (6), 2958–2966.
el Kharroubi, A., Verdin, E., 1994. Protein-DNA interactions within DNase I-hypersensitive sites located downstream of the HIV-1 promoter. J. Biol. Chem. 269 (31), 19916–19924.

Emanuel, P.A., Gilmour, D.S., 1993. Transcription factor TFIID recognizes DNA sequences downstream of the TATA element in the Hsp70 heat shock gene. Proc. Natl. Acad. Sci. U.S.A. 90 (18), 8449–8453.

Engelman, A., Cherepanov, P., 2012. The structural biology of HIV-1: mechanistic and therapeutic insights. Nat. Rev. Microbiol. 10 (4), 279–290.

Epstein, F.H., et al., 1993. The Immunopathogenesis of human immunodeficiency virus infection. N. Engl. J. Med. 328 (5), 327–335.

Finzi, D., et al., 1999. Latent infection of CD4+ T cells provides a mechanism for lifelong persistence of HIV-1, even in patients on effective combination therapy. Nat. Med. 5 (5), 512–517.

Flores, O., et al., 1999. Host-cell positive transcription elongation factor b kinase activity is essential and limiting for HIV type 1 replication. Proc. Natl. Acad. Sci. U.S.A. 96 (13), 7208–7213.

Flynn, R.A., et al., 2016. 7SK-BAF axis controls pervasive transcription at enhancers. Nat. Struct. Mol. Biol. 23 (3), 231–238.

Freed, E.O., 2001. HIV-1 replication. Somat. Cell Mol. Genet. 26 (1–6), 13–33.

Freed, E.O., 2015. HIV-1 assembly, release and maturation. Nat. Rev. Microbiol. 13 (8), 484–496.

Friedman, J., et al., 2011. Epigenetic silencing of HIV-1 by the histone H3 lysine 27 methyltransferase enhancer of Zeste 2. J. Virol. 85 (17), 9078–9089.

Fujinaga, K., et al., 2004. Dynamics of human immunodeficiency virus transcription: P-TEFb phosphorylates RD and dissociates negative effectors from the transactivation response element. Mol. Cell. Biol. 24 (2), 787–795.

Fujinaga, K., et al., 2012. PKC phosphorylates HEXIM1 and regulates P-TEFb activity. Nucleic Acids Res. 40 (18), 9160–9170.

Fujita, T., et al., 1992. Independent modes of transcriptional activation by the p50 and p65 subunits of NF-kappa B. Genes Dev. 6 (5), 775–787.

Furia, B., et al., 2002. Enhancement of nuclear factor-kappa B acetylation by coactivator p300 and HIV-1 Tat proteins. J. Biol. Chem. 277 (7), 4973–4980.

Gadad, S.S., et al., 2011. The multifunctional protein nucleophosmin (NPM1) is a human linker histone H1 chaperone. Biochemistry 50 (14), 2780–2789.

Gallastegui, E., et al., 2011. Chromatin reassembly factors are involved in transcriptional interference promoting HIV latency. J. Virol. 85 (7), 3187–3202.

Garcia, J.A., et al., 1989. Human immunodeficiency virus type 1 LTR TATA and TAR region sequences required for transcriptional regulation. EMBO J. 8 (3), 765–778.

Gaynor, R., 1992. Cellular transcription factors involved in the regulation of HIV-1 gene expression. AIDS (London, England) 6 (4), 347–363.

Gerritsen, M.E., et al., 1997. CREB-binding protein/p300 are transcriptional coactivators of p65. Proc. Natl. Acad. Sci. U.S.A. 94 (7), 2927–2932.

Giffin, M.J., et al., 2003. Structure of NFAT1 bound as a dimer to the HIV-1 LTR κB element. Nat. Struct. Biol. 10 (10), 800–806.

Gill, G., et al., 1994. A glutamine-rich hydrophobic patch in transcription factor Sp1 contacts the dTAFII110 component of the drosophila TFIID complex and mediates transcriptional activation. Proc. Natl. Acad. Sci. U.S.A. 91 (1), 192–196.

Grzechnik, P., Tan-Wong, S.M., Proudfoot, N.J., 2014. Terminate and make a loop: regulation of transcriptional directionality. Trends Biochem. Sci. 39 (7), 319–327.

Guillemette, B., et al., 2005. Variant histone H2A.Z is globally localized to the promoters of inactive yeast genes and regulates nucleosome positioning. PLoS Biol. 3 (12), e384.

Guo, J., Price, D.H., 2013. RNA polymerase II transcription elongation control. Chem. Rev. 113 (11), 8583–8603.

Hampsey, M., et al., 2011. Control of eukaryotic gene expression: gene loops and transcriptional memory. Adv. Enzyme Regul. 51 (1), 118–125.

Hargreaves, D.C., Horng, T., Medzhitov, R., 2009. Control of inducible gene expression by signal-dependent transcriptional elongation. Cell 138 (1), 129–145.
Hauber, J., Malim, M.H., Cullen, B.R., 1989. Mutational analysis of the conserved basic domain of human immunodeficiency virus Tat protein. J. Virol. 63 (3), 1181–1187.
He, X., Fütterer, J., Hohn, T., 2002. Contribution of downstream promoter elements to transcriptional regulation of the rice tungro bacilliform virus promoter. Nucleic Acids Res. 30 (2), 497–506.
He, X., et al., 2003. Functional interactions between the transcription and mRNA 3′ end processing machineries mediated by Ssu72 and Sub1. Genes Dev. 17 (8), 1030–1042.
He, N., et al., 2008. A La-related protein modulates 7SK snRNP integrity to suppress P-TEFb-dependent transcriptional elongation and tumorigenesis. Mol. Cell 29 (5), 588–599.
He, N., et al., 2010. HIV-1 Tat and host AFF4 recruit two transcription elongation factors into a bifunctional complex for coordinated activation of HIV-1 transcription. Mol. Cell 38 (3), 428–438.
He, N., et al., 2011. Human polymerase-associated factor complex (PAFc) connects the super elongation complex (SEC) to RNA polymerase II on chromatin. Proc. Natl. Acad. Sci. 108 (36), E636–E645.
Henderson, A.J., Zou, X., Calame, K.L., 1995. C/EBP proteins activate transcription from the human immunodeficiency virus type 1 long terminal repeat in macrophages/monocytes. J. Virol. 69 (9), 5337–5344.
Henikoff, S., 2008. Nucleosome destabilization in the epigenetic regulation of gene expression. Nat. Rev. Genet. 9 (1), 15–26.
Hill, A.L., et al., 2014. Predicting the outcomes of treatment to eradicate the latent reservoir for HIV-1. Proc. Natl. Acad. Sci. 111 (37), 13475–13480.
Hirose, Y., Ohkuma, Y., 2007. Phosphorylation of the C-terminal domain of RNA polymerase II plays central roles in the integrated events of Eucaryotic gene expression. J. Biochem. 141 (5), 601–608.
Ho, Y.-C., Shan, L., Hosmane, N.N., et al., 2013. Replication-competent noninduced proviruses in the latent reservoir increase barrier to HIV-1 cure. Cell 155, 540–551.
Hoey, T., et al., 1993. Molecular cloning and functional analysis of drosophila TAF110 reveal properties expected of coactivators. Cell 72 (2), 247–260.
Hu, W.-S., Hughes, S.H., 2012. HIV-1 reverse transcription. Cold Spring Harb. Perspect. Med. 2 (10), a006882.
Huang, L.M., Jeang, K.T., 1993. Increased spacing between Sp1 and TATAA renders human immunodeficiency virus type 1 replication defective: implication for Tat function. J. Virol. 67 (12), 6937–6944.
Huang, H., et al., 2015. FACT proteins, SUPT16H and SSRP1, are transcriptional suppressors of HIV-1 and HTLV-1 that facilitate viral latency. J. Biol. Chem. 290 (45), 27297–27310.
Huber, K., et al., 2011. Inhibitors of histone deacetylases: correlation between isoform specificity and reactivation of HIV type 1 (HIV-1) from latently infected cells. J. Biol. Chem. 286 (25), 22211–22218.
Hunt, G., Tiemessen, C.T., 2000. Occurrence of additional NF-kappaB-binding motifs in the long terminal repeat region of south African HIV type 1 subtype C isolates. AIDS Res. Hum. Retroviruses 16 (3), 305–306.
Huo, L., et al., 2011. Regulation of tat acetylation and transactivation activity by the microtubule-associated deacetylase HDAC6. J. Biol. Chem. 286 (11), 9280–9286.
Imai, K., Okamoto, T., 2006. Transcriptional repression of human immunodeficiency virus type 1 by AP-4. J. Biol. Chem. 281 (18), 12495–12505.
Imai, K., Togami, H., Okamoto, T., 2010. Involvement of histone H3 lysine 9 (H3K9) methyltransferase G9a in the maintenance of HIV-1 latency and its reactivation by BIX01294. J. Biol. Chem. 285 (22), 16538–16545.

Jadlowsky, J.K., et al., 2014. Negative elongation factor is required for the maintenance of proviral latency but does not induce promoter-proximal pausing of RNA polymerase II on the HIV long terminal repeat. Mol. Cell. Biol. 34 (11), 1911–1928.

Jang, M.K., et al., 2005. The bromodomain protein Brd4 is a positive regulatory component of P-TEFb and stimulates RNA polymerase II-dependent transcription. Mol. Cell 19 (4), 523–534.

Jeeninga, R.E., et al., 2000. Functional differences between the long terminal repeat transcriptional promoters of human immunodeficiency virus type 1 subtypes A through G. J. Virol. 74 (8), 3740–3751.

Jeng, M.Y., Ali, I., Ott, M., 2015. Manipulation of the host protein acetylation network by human immunodeficiency virus type 1. Crit. Rev. Biochem. Mol. Biol. 50 (4), 314–325.

Jeronimo, C., et al., 2007. Systematic analysis of the protein interaction network for the human transcription machinery reveals the identity of the 7SK capping enzyme. Mol. Cell 27 (2), 262–274.

Ji, X., et al., 2013. SR proteins collaborate with 7SK and promoter-associated nascent RNA to release paused polymerase. Cell 153 (4), 855–868.

Ji, H., et al., 2016. Specific reactivation of latent HIV-1 by dCas9-SunTag-VP64-mediated guide RNA targeting the HIV-1 promoter. Mol. Ther. 24 (3), 508–521.

Jiang, G., et al., 2007. C-Myc and Sp1 contribute to proviral latency by recruiting histone deacetylase 1 to the human immunodeficiency virus type 1 promoter. J. Virol. 81 (20), 10914–10923.

Jones, K.A., Peterlin, M.B., 1994. Control of RNA initiation and elongation at the HIV-1 promoter. Annu. Rev. Biochem. 63 (1), 717–743.

Jones, K.a., et al., 1986. Activation of the AIDS retrovirus promoter by the cellular transcription factor, Sp1. Science (New York, NY) 232 (1966), 755–759.

Jonkers, I., Lis, J.T., 2015. Getting up to speed with transcription elongation by RNA polymerase II. Nat. Rev. Mol. Cell Biol. 16 (3), 167–177.

Jordan, A., Defechereux, P., Verdin, E., 2001. The site of HIV-1 integration in the human genome determines basal transcriptional activity and response to Tat transactivation. EMBO J. 20 (7), 1726–1738.

Kadonaga, J.T., 2012. Perspectives on the RNA polymerase II core promoter. Wiley Interdiscip. Rev. Dev. Biol. 1 (1), 40–51.

Kaern, M., et al., 2005. Stochasticity in gene expression: from theories to phenotypes. Nat. Rev. Genet. 6 (6), 451–464.

Kao, S.-Y.Y., et al., 1987. Anti-termination of transcription within the long terminal repeat of HIV-1 by tat gene product. Nature 330 (6147), 489–493.

Karn, J., 2011. The molecular biology of HIV latency: breaking and restoring the Tat-dependent transcriptional circuit. Curr. Opin. HIV AIDS 6 (1), 4–11.

Kauder, S.E., et al., 2009. Epigenetic regulation of HIV-1 latency by cytosine methylation. PLoS Pathog. 5 (6), e1000495.

Keedy, K.S., et al., 2009. A limited group of class I histone deacetylases acts to repress human immunodeficiency virus type 1 expression. J. Virol. 83 (10), 4749–4756.

Kharytonchyk, S., et al., 2016. Transcriptional start site heterogeneity modulates the structure and function of the HIV-1 genome. Proc. Natl. Acad. Sci. 113 (47), 13378–13383.

Kiernan, R.E., et al., 1999. HIV-1 tat transcriptional activity is regulated by acetylation. EMBO J. 18 (21), 6106–6118.

Kilareski, E.M., et al., 2009. Regulation of HIV-1 transcription in cells of the monocyte-macrophage lineage. Retrovirology 6 (1), 118.

Kim, J.Y., et al., 1993. Replication of type 1 human immunodeficiency viruses containing linker substitution mutations in the −201 to −130 region of the long terminal repeat. J. Virol. 67 (3), 1658–1662.

Kim, M., et al., 2006. Distinct pathways for sno RNA and mRNA termination. Mol. Cell 24 (5), 723–734.
Kim, Y.K., et al., 2011. T-cell receptor signaling enhances transcriptional elongation from latent HIV proviruses by activating P-TEFb through an ERK-dependent pathway. J. Mol. Biol. 410 (5), 896–916.
Kinoshita, S., et al., 1997. The T cell activation factor NF-ATc positively regulates HIV-1 replication and gene expression in T cells. Immunity 6 (3), 235–244.
Kinoshita, S., et al., 1998. Host control of HIV-1 parasitism in T cells by the nuclear factor of activated T cells. Cell 95 (5), 595–604.
Klaver, B., Berkhout, B., 1994. Comparison of 5′ and 3′ long terminal repeat promoter function in human immunodeficiency virus. J. Virol. 68 (6), 3830–3840.
Kobor, M.S., Greenblatt, J., 2002. Regulation of transcription elongation by phosphorylation. Biochim. Biophys. Acta 1577 (2), 261–275.
Koken, S.E., et al., 1992. Natural variants of the HIV-1 long terminal repeat: analysis of promoters with duplicated DNA regulatory motifs. Virology 191 (2), 968–972.
Kouzarides, T., 2007. Chromatin modifications and their function. Cell 128 (4), 693–705.
Krebs, F.C., Goodenow, M.M., Wigdahl, B., 1997. Neuroglial ATF/CREB factors interact with the human immunodeficiency virus type 1 long terminal repeat. J. Neurovirol. 3 (Suppl. 1), S28–32.
Krebs, F.C., et al., 1998. Human immunodeficiency virus type 1 long terminal repeat quasispecies differ in basal transcription and nuclear factor recruitment in human glial cells and lymphocytes. J. Biomed. Sci. 5 (1), 31–44.
Kretzschmar, M., et al., 1992. Transcriptional regulation of the HIV-1 promoter by NF-kappaB in vitro. Genes Dev. 6, 761–774.
Krueger, B.J., et al., 2008. LARP7 is a stable component of the 7SK snRNP while P-TEFb, HEXIM1 and hnRNP A1 are reversibly associated. Nucleic Acids Res. 36 (7), 2219–2229.
Krueger, B.J., et al., 2010. The mechanism of release of P-TEFb and HEXIM1 from the 7SK snRNP by viral and cellular activators includes a conformational change in 7SK. PLoS One 5 (8), e12335.
Kumar, A., et al., 2015. Epigenetic control of HIV-1 post integration latency: implications for therapy. Clin. Epigenetics 7 (1), 103.
Le Douce, V., et al., 2016. HIC1 controls cellular- and HIV-1-gene transcription via interactions with CTIP2 and HMGA1. Sci. Rep. 6, 34920.
Lee, C.-K., et al., 2004. Evidence for nucleosome depletion at active regulatory regions genome-wide. Nat. Genet. 36 (8), 900–905.
Lelek, M., et al., 2015. Chromatin organization at the nuclear pore favours HIV replication. Nat. Commun. 6, 6483.
Li, E., Zhang, Y., 2014. DNA methylation in mammals. Cold Spring Harb. Perspect. Biol. 6 (5), a019133.
Li, Y., Mak, G., Franza, B.R. 1994. In vitro study of functional involvement of Sp1, NF-κB/Rel, and AP1 in phorbol 12-myristate 13-acetate-mediated HIV-1 long terminal repeat activation. J. Biol. Chem., 269 (48), 30616–30619.
Li, Q., et al., 2005. Analysis of the large inactive P-TEFb complex indicates that it contains one 7SK molecule, a dimer of HEXIM1 or HEXIM2, and two P-TEFb molecules containing Cdk9 phosphorylated at threonine 186. J. Biol. Chem. 280 (31), 28819–28826.
Liang, C., et al., 1997. Sequence elements downstream of the human immunodeficiency virus type 1 long terminal repeat are required for efficient viral gene transcription. J. Mol. Biol. 272 (2), 167–177.
Liebert, M.A., et al., 2002. Characterization of the south African HIV type 1 subtype C complete 59 long terminal repeat, nef, and regulatory genes. AIDS Res. Hum. Retroviruses 18 (2), 149–159.

Limsirichai, P., Gaj, T., Schaffer, D.V., 2016. CRISPR-mediated activation of latent HIV-1 expression. Mol. Ther. 24 (3), 499–507.
Lin, R., Gewert, D., Hiscott, J., 1995. Differential transcriptional activation in vitro by NF-kappa B/Rel proteins. J. Biol. Chem. 270 (7), 3123–3131.
Liu, W., et al., 2013. Brd4 and JMJD6-associated anti-pause enhancers in regulation of transcriptional pause release. Cell 155 (7), 1581–1595.
Losick, R., Desplan, C., 2008. Stochasticity and cell fate. Science (New York, NY) 320 (5872), 65–68.
Lu, Y., et al., 1989. Effects of long terminal repeat mutations on human immunodeficiency virus type 1 replication. J. Virol. 63 (9), 4115–4119.
Lu, H., et al., 2013. Viral–host interactions that control HIV-1 transcriptional elongation. Chem. Rev. 113 (11), 8567–8582.
Lu, H., et al., 2014. AFF1 is a ubiquitous P-TEFb partner to enable Tat extraction of P-TEFb from 7SK snRNP and formation of SECs for HIV transactivation. Proc. Natl. Acad. Sci. U.S.A. 111 (1), E15–24.
Lu, H., et al., 2015. Gene target specificity of the super elongation complex (SEC) family: how HIV-1 Tat employs selected SEC members to activate viral transcription. Nucleic Acids Res. 43 (12), 5868–5879.
Luger, K., Richmond, T.J., 1998. The histone tails of the nucleosome. Curr. Opin. Genet. Dev. 8 (2), 140–146.
Lusic, M., Siliciano, R.F., 2017. Nuclear landscape of HIV-1 infection and integration. Nat. Rev. Microbiol. 15 (2), 69–82.
Lusic, M., et al., 2003. Regulation of HIV-1 gene expression by histone acetylation and factor recruitment at the LTR promoter. EMBO J. 22 (24), 6550–6561.
Lusic, M., et al., 2013. Proximity to PML nuclear bodies regulates HIV-1 latency in CD4+ T cells. Cell Host Microbe 13 (6), 665–677.
Maartens, G., Celum, C., Lewin, S.R., 2014. HIV infection: epidemiology, pathogenesis, treatment, and prevention. Lancet 384 (9939), 258–271.
Macian, F., Rao, A., 1999. Reciprocal modulatory interaction between human immunodeficiency virus type 1 Tat and transcription factor NFAT1. Mol. Cell. Biol. 19 (5), 3645–3653.
Mahmoudi, T., 2012. The BAF complex and HIV latency. Transcription 3 (4), 171–176.
Mahmoudi, T., et al., 2006. The SWI/SNF chromatin-remodeling complex is a cofactor for Tat transactivation of the HIV promoter. J. Biol. Chem. 281 (29), 19960–19968.
Marban, C., et al., 2005. COUP-TF interacting protein 2 represses the initial phase of HIV-1 gene transcription in human microglial cells. Nucleic Acids Res. 33 (7), 2318–2331.
Marban, C., et al., 2007. Recruitment of chromatin-modifying enzymes by CTIP2 promotes HIV-1 transcriptional silencing. EMBO J. 26 (2), 412–423.
Marfella, C.G.A., Imbalzano, A.N., 2007. The Chd family of chromatin remodelers. Mutat. Res. 618 (1–2), 30–40.
Margolis, D.M., 2011. Histone deacetylase inhibitors and HIV latency. Curr. Opin. HIV AIDS 6 (1), 25–29.
Margolis, D.M., et al., 2016. Latency reversal and viral clearance to cure HIV-1. Science (New York, NY) 353 (6297), aaf6517.
Maricato, J.T., et al., 2015. Epigenetic modulations in activated cells early after HIV-1 infection and their possible functional consequences. PLoS One 10 (4), e0119234.
Marzio, G., et al., 1998. HIV-1 Tat transactivator recruits p300 and CREB-binding protein histone acetyltransferases to the viral promoter. Proc. Natl. Acad. Sci. U.S.A. 95 (23), 13519–13524.
Masuda, A., et al., 2015. Position-specific binding of FUS to nascent RNA regulates mRNA length. Genes Dev. 29 (10), 1045–1057.

Matsuda, Y., et al., 2015. Epigenetic heterogeneity in HIV-1 latency establishment. Sci. Rep. 5, 7701.
Mavrich, T.N., et al., 2008. A barrier nucleosome model for statistical positioning of nucleosomes throughout the yeast genome. Genome Res. 18 (7), 1073–1083.
Mbonye, U., Karn, J., 2014. Transcriptional control of HIV latency: cellular signaling pathways, epigenetics, happenstance and the hope for a cure. Virology 454–455, 328–339.
Mbonye, U.R., et al., 2013. Phosphorylation of CDK9 at Ser175 enhances HIV transcription and is a marker of activated P-TEFb in CD4+ T lymphocytes. PLoS Pathog. 9 (5), e1003338.
Mbonye, U.R., et al., 2015. Phosphorylation of HEXIM1 at Tyr271 and Tyr274 promotes release of P-TEFb from the 7SK snRNP complex and enhances proviral HIV gene expression. Proteomics 15 (12), 2078–2086.
McNamara, R.P., et al., 2013. Transcription factors mediate the enzymatic disassembly of promoter-bound 7SK snRNP to locally recruit P-TEFb for transcription elongation. Cell Rep. 5 (5), 1256–1268.
McNamara, R.P., et al., 2016. KAP1 recruitment of the 7SK snRNP complex to promoters enables transcription elongation by RNA polymerase II. Mol. Cell 61 (1), 39–53.
Medler, S., et al., 2011. Evidence for a complex of transcription factor IIB (TFIIB) with poly(a) polymerase and cleavage factor 1 subunits required for gene looping. J. Biol. Chem. 286 (39), 33709–33718.
Michels, A.A., et al., 2004. Binding of the 7SK snRNA turns the HEXIM1 protein into a P-TEFb (CDK9/cyclin T) inhibitor. EMBO J. 23 (13), 2608–2619.
Miller-Jensen, K., et al., 2011. Varying virulence: epigenetic control of expression noise and disease processes. Trends Biotechnol. 29 (10), 517–525.
Miller-Jensen, K., et al., 2013. Genetic selection for context-dependent stochastic phenotypes: Sp1 and TATA mutations increase phenotypic noise in HIV-1 gene expression. PLoS Comput. Biol. 9 (7), e1003135.
Missra, A., Gilmour, D.S., 2010. Interactions between DSIF (DRB sensitivity inducing factor), NELF (negative elongation factor), and the Drosophila RNA polymerase II transcription elongation complex. Proc. Natl. Acad. Sci. 107 (25), 11301–11306.
Mizutani, T., et al., 2009. Loss of the Brm-type SWI/SNF chromatin remodeling complex is a strong barrier to the Tat-independent transcriptional elongation of human immunodeficiency virus type 1 transcripts. J. Virol. 83 (22), 11569–11580.
Montano, M.A., et al., 1997. Divergent transcriptional regulation among expanding human immunodeficiency virus type 1 subtypes. J. Virol. 71 (11), 8657–8665.
Mousseau, G., Valente, S., 2012. Strategies to block HIV transcription: focus on small molecule tat inhibitors. Biology 1 (3), 668–697.
Mousseau, G., Kessing, C.F., et al., 2015a. The Tat inhibitor didehydro-cortistatin A prevents HIV-1 reactivation from latency. mBio 6 (4), e00465.
Mousseau, G., Mediouni, S., Valente, S.T., 2015b. Targeting HIV transcription: the quest for a functional cure. Curr. Top. Microbiol. Immunol. 389, 121–145.
Muesing, M.a., Smith, D.H., Capon, D.J., 1987. Regulation of mRNA accumulation by a human immunodeficiency virus trans-activator protein. Cell 48 (4), 691–701.
Mujtaba, S., et al., 2002. Structural basis of lysine-acetylated HIV-1 Tat recognition by PCAF bromodomain. Mol. Cell 9 (3), 575–586.
Muniz, L., et al., 2010. Controlling cellular P-TEFb activity by the HIV-1 transcriptional transactivator Tat. PLoS Pathog. 6 (10), e1001152.
Näär, A.M., Ryu, S., Tjian, R., 1998. Cofactor requirements for transcriptional activation by Sp1. Cold Spring Harb. Symp. Quant. Biol. 63, 189–199.
Nabel, G., Baltimore, D., 1987. An inducible transcription factor activates expression of human immunodeficiency virus in T cells. Nature 326 (6114), 711–713.

Naghavi, M.H., et al., 1999. DNA sequence of the long terminal repeat of human immunodeficiency virus type 1 subtype a through G. AIDS Res. Hum. Retroviruses 15 (5), 485–488.

Natarajan, M., et al., 2013. Negative elongation factor (NELF) coordinates RNA polymerase II pausing, premature termination, and chromatin remodeling to regulate HIV transcription. J. Biol. Chem. 288 (36), 25995–26003.

Nguyen, V.T., et al., 2001. 7SK small nuclear RNA binds to and inhibits the activity of CDK9/cyclin T complexes. Nature 414 (6861), 322–325.

O'Reilly, D., Greaves, D.R., 2007. Cell-type-specific expression of the human CD68 gene is associated with changes in pol II phosphorylation and short-range intrachromosomal gene looping. Genomics 90 (3), 407–415.

O'Sullivan, J.M., et al., 2004. Gene loops juxtapose promoters and terminators in yeast. Nat. Genet. 36 (9), 1014–1018.

Ott, M., et al., 1999. Acetylation of the HIV-1 tat protein by p300 is important for its transcriptional activity. Curr. Biol. 9 (24), 1489–1492.

Ott, M., Geyer, M., Zhou, Q., 2011. The control of HIV transcription: keeping RNA polymerase II on track. Cell Host Microbe 10 (5), 426–435.

Owen-Hughes, T., Workman, J.L., 1994. Experimental analysis of chromatin function in transcription control. Crit. Rev. Eukaryot. Gene Expr. 4 (4), 403–441.

Pagano, J.M., et al., 2014. Defining NELF-E RNA binding in HIV-1 and promoter-proximal pause regions. PLoS Genet. 10 (1), e1004090.

Pagans, S., et al., 2005. SIRT1 regulates HIV transcription via Tat deacetylation. PLoS Biol. 3 (2), e41.

Pagans, S., et al., 2010. The cellular lysine methyltransferase Set7/9-KMT7 binds HIV-1 TAR RNA, monomethylates the viral transactivator Tat, and enhances HIV transcription. Cell Host Microbe 7 (3), 234–244.

Palmisano, I., et al., 2012. Amino acid starvation induces reactivation of silenced transgenes and latent HIV-1 provirus via down-regulation of histone deacetylase 4 (HDAC4). Proc. Natl. Acad. Sci. U.S.A. 109 (34), E2284–93.

Papathanasopoulos, M.A., et al., 2002. Full-length genome analysis of HIV-1 subtype C utilizing CXCR4 and intersubtype recombinants isolated in South Africa. AIDS Res. Hum. Retroviruses 18 (12), 879–886.

Pascal, E., Tjian, R., 1991. Different activation domains of Sp1 govern formation of multimers and mediate transcriptional synergism. Genes Dev. 5 (9), 1646–1656.

Pascual-Garcia, P., et al., 2017. Metazoan nuclear pores provide a scaffold for poised genes and mediate induced enhancer–promoter contacts. Mol. Cell 66 (1). 63–76.e6.

Pearson, R., et al., 2008. Epigenetic silencing of human immunodeficiency virus (HIV) transcription by formation of restrictive chromatin structures at the viral long terminal repeat drives the progressive entry of HIV into latency. J. Virol. 82 (24), 12291–12303.

Pelletier, M.R., et al., 1997. Efficient transcription of an immunoglobulin kappa promoter requires specific sequence elements overlapping with and downstream of the transcriptional start site. Nucleic Acids Res. 25 (20), 3995–4003.

Pereira, L.A., et al., 2000. A compilation of cellular transcription factor interactions with the HIV-1 LTR promoter. Nucleic Acids Res. 28 (3), 663–668.

Perkins, N.D., et al., 1993. A cooperative interaction between NF-kappa B and Sp1 is required for HIV-1 enhancer activation. EMBO J. 12 (9), 3551–3558.

Perkins, N.D., et al., 1994. An interaction between the DNA-binding domains of RelA(p65) and Sp1 mediates human immunodeficiency virus gene activation. Mol. Cell. Biol. 14 (10), 6570–6583.

Perkins, N.D., et al., 1997. Regulation of NF-kappaB by cyclin-dependent kinases associated with the p300 coactivator. Science (New York, NY) 275 (5299), 523–527.

Perkins, K.J., et al., 2008. Transcription-dependent gene looping of the HIV-1 provirus is dictated by recognition of pre-mRNA processing signals. Mol. Cell 29 (1), 56–68.
Peterlin, B.M., Price, D.H., 2006. Controlling the elongation phase of transcription with P-TEFb. Mol. Cell 23 (3), 297–305.
Peterlin, B.M., Brogie, J.E., Price, D.H., 2012. 7SK snRNA: a noncoding RNA that plays a major role in regulating eukaryotic transcription. Wiley Interdiscip. Rev. RNA 3 (1), 92–103.
Ping, Y.H., Rana, T.M., 2001. DSIF and NELF interact with RNA polymerase II elongation complex and HIV-1 Tat stimulates P-TEFb-mediated phosphorylation of RNA polymerase II and DSIF during transcription elongation. J. Biol. Chem. 276 (16), 12951–12958.
Pomerantz, R.J., et al., 1990. Cells nonproductively infected with HIV-1 exhibit an aberrant pattern of viral RNA expression: a molecular model for latency. Cell 61 (7), 1271–1276.
Ptak, C., Aitchison, J.D., Wozniak, R.W., 2014. The multifunctional nuclear pore complex: a platform for controlling gene expression. Curr. Opin. Cell Biol. 28, 46–53.
Qian, J., et al., 1994. Chemically selected subclones of the CEM cell line demonstrate resistance to HIV-1 infection resulting from a selective loss of NF-kappa B DNA binding proteins. J. Immunol. 152 (8), 4183–4191.
Rabbi, M.F., et al., 1997. U5 region of the human immunodeficiency virus type 1 long terminal repeat contains TRE-like cAMP-responsive elements that bind both AP-1 and CREB/ATF proteins. Virology 233 (1), 235–245.
Radman-Livaja, M., Rando, O.J., 2010. Nucleosome positioning: how is it established, and why does it matter? Dev. Biol. 339 (2), 258–266.
Rafati, H., et al., 2011. Repressive LTR nucleosome positioning by the BAF complex is required for HIV latency. PLoS Biol. 9 (11), e1001206.
Raj, A., van Oudenaarden, A., 2008. Nature, nurture, or chance: stochastic gene expression and its consequences. Cell 135 (2), 216–226.
Ramakrishnan, R., Dow, E.C., Rice, A.P., 2009. Characterization of Cdk9 T-loop phosphorylation in resting and activated CD4(+) T lymphocytes. J. Leukoc. Biol. 86 (6), 1345–1350.
Raser, J.M., 2010. Noise in gene expression. Science 2010 (2005), 2010–2014.
Raser, J.M., Shea, E.K.O., 2006. Control of stochasticity in eukaryotic gene expression Jonathan. Science 304 (5678), 1811–1814.
Rasmussen, T.A., Tolstrup, M., Søgaard, O.S., 2016. Reversal of latency as part of a cure for HIV-1. Trends Microbiol. 24 (2), 90–97.
Ratner, L., et al., 1985. Complete nucleotide sequence of the AIDS virus, HTLV-III. Nature 313 (6000), 277–284.
Razooky, B.S., et al., 2015. A hardwired HIV latency program. Cell 160 (5), 990–1001.
Renner, D.B., et al., 2001. A highly purified RNA polymerase II elongation control system. J. Biol. Chem. 276 (45), 42601–42609.
Reuse, S., et al., 2009. Synergistic activation of HIV-1 expression by deacetylase inhibitors and prostratin: implications for treatment of latent infection. PLoS One 4 (6), e6093.
Richmond, T.J., et al., 1997. Crystal structure of the nucleosome core particle at 2.8 |[thinsp]||[Aring]| resolution. Nature 389 (6648), 251–260.
Rodgers, M.J., et al., 2014. CHD1 and CHD2 are positive regulators of HIV-1 gene expression. Virol. J. 11 (1), 180.
Roebuck, K.A., Gu, D.S., Kagnoff, M.F., 1996. Activating protein-1 cooperates with phorbol ester activation signals to increase HIV-1 expression. AIDS (London, England) 10 (8), 819–826.
Romani, B., et al., 2016. HIV-1 Vpr reactivates latent HIV-1 provirus by inducing depletion of class I HDACs on chromatin. Sci. Rep. 6, 31924.

Romerio, F., Gabriel, M.N., Margolis, D.M., 1997. Repression of human immunodeficiency virus type 1 through the novel cooperation of human factors YY1 and LSF. J. Virol. 71 (12), 9375–9382.

Rosen, C.A., Sodroski, J.G., Haseltine, W.A., 1985. The location of cis-acting regulatory sequences in the human T cell lymphotropic virus type III (HTLV-III/LAV) long terminal repeat. Cell 41 (3), 813–823.

Ross, E.K., et al., 1991. Contribution of NF-kappa B and Sp1 binding motifs to the replicative capacity of human immunodeficiency virus type 1: distinct patterns of viral growth are determined by T-cell types. J. Virol. 65 (8), 4350–4358.

Rosseau, C., et al., 1997. Long terminal repeat and *nef* gene variants of human immunodeficiency virus type 1 in perinatally infected long-term survivors and rapid progressors. AIDS Res. Hum. Retroviruses 13 (18), 1611–1623.

Roy, S., et al., 1990. A bulge structure in HIV-1 TAR RNA is required for Tat binding and Tat-mediated trans-activation. Genes Dev. 4 (8), 1365–1373.

Saayman, S.M., et al., 2016. Potent and targeted activation of latent HIV-1 using the CRISPR/dCas9 activator complex. Mol. Ther. 24 (3), 488–498.

Sadowski, I., Mitchell, D.A., 2005. TFII-I and USF (RBF-2) regulate Ras/MAPK-responsive HIV-1 transcription in T cells. Eur. J. Cancer 41 (16), 2528–2536.

Sakane, N., et al., 2011. Activation of HIV transcription by the viral Tat protein requires a demethylation step mediated by lysine-specific demethylase 1 (LSD1/KDM1). PLoS Pathog. 7 (8), e1002184.

Sanchez-Pescador, R., et al., 1985. Nucleotide sequence and expression of an AIDS-associated retrovirus (ARV-2). Science (New York, NY) 227 (4686), 484–492.

Schüller, R., et al., 2016. Heptad-specific phosphorylation of RNA polymerase II CTD. Mol. Cell 61 (2), 305–314.

Schulze-Gahmen, U., et al., 2013. The AFF4 scaffold binds human P-TEFb adjacent to HIV Tat. eLife 2013 (2), 10–12.

Schulze-Gahmen, U., et al., 2014. AFF4 binding to TAT-P-TEFb indirectly stimulates TAR recognition of super elongation complexes at the HIV promoter. eLife 2014 (3), 1–13.

Schwabish, M.A., Struhl, K., 2004. Evidence for eviction and rapid deposition of histones upon transcriptional elongation by RNA polymerase II. Mol. Cell. Biol. 24 (23), 10111–10117.

Sedore, S.C., et al., 2007. Manipulation of P-TEFb control machinery by HIV: recruitment of P-TEFb from the large form by Tat and binding of HEXIM1 to TAR. Nucleic Acids Res. 35 (13), 4347–4358.

Sekinger, E.A., Moqtaderi, Z., Struhl, K., 2005. Intrinsic histone–DNA interactions and low nucleosome density are important for preferential accessibility of promoter regions in yeast. Mol. Cell 18 (6), 735–748.

Selby, M.J., et al., 1989. Structure, sequence, and position of the stem-loop in tar determine transcriptional elongation by tat through the HIV-1 long terminal repeat. Genes Dev. 3 (4), 547–558.

Seto, E., Shi, Y., Shenk, T., 1991. YY1 is an initiator sequence-binding protein that directs and activates transcription in vitro. Nature 354 (6350), 241–245.

Seto, E., Lewist, B., Shenk, T., 1993. Interaction between transcription factors Spl and YY1. Nature 365 (6445), 462–464.

Sgarbanti, M., et al., 2002. Modulation of human immunodeficiency virus 1 replication by interferon regulatory factors. J. Exp. Med. 195 (10), 1359–1370.

Sgarbanti, M., et al., 2004. Analysis of the signal transduction pathway leading to human immunodeficiency virus-1-induced interferon regulatory factor-1 upregulation. Ann. N.Y. Acad. Sci. 1030 (1), 187–195.

Sheridan, P.L., et al., 1995. Activation of the HIV-1 enhancer by the LEF-1 HMG protein on nucleosome-assembled DNA in vitro. Genes Dev. 9 (17), 2090–2104.

Sieweke, M.H., et al., 1998. Cooperative interaction of Ets-1 with USF-1 required for HIV-1 enhancer activity in T cells. EMBO J. 17 (6), 1728–1739.

Sif, S., Gilmore, T.D., 1994. Interaction of the v-Rel oncoprotein with cellular transcription factor Sp1. J. Virol. 68 (11), 7131–7138.

Sikorski, T.W., Buratowski, S., 2009. The basal initiation machinery: beyond the general transcription factors. Curr. Opin. Cell Biol. 21 (3), 344–351.

Siliciano, R.F., Greene, W.C., 2011. HIV latency. Cold Spring Harb. Perspect. Med. 1 (1), a007096.

Siliciano, J.D., et al., 2003. Long-term follow-up studies confirm the stability of the latent reservoir for HIV-1 in resting CD4+ T cells. Nat. Med. 9 (6), 727–728.

Singh, A., Weinberger, L.S., 2009. Stochastic gene expression as a molecular switch for viral latency. Curr. Opin. Microbiol. 12 (4), 460–466.

Singh, A., et al., 2010. Transcriptional bursting from the HIV-1 promoter is a significant source of stochastic noise in HIV-1 gene expression. Biophys. J. 98 (8), L32–L34.

Sobhian, B., et al., 2010. HIV-1 Tat assembles a multifunctional transcription elongation complex and stably associates with the 7SK snRNP. Mol. Cell 38 (3), 439–451.

Sodroski, J.G., Rosen, C.A., Haseltine, W.A., 1984. Trans-acting transcriptional activation of the long terminal repeat of human T lymphotropic viruses in infected cells. Science (New York, NY) 225 (4660), 381–385.

Sodroski, J., Patarca, R., et al., 1985a. Location of the trans-activating region on the genome of human T-cell lymphotropic virus type III. Science 229 (4708), 74–77.

Sodroski, J., Rosen, C., et al., 1985b. Trans-acting transcriptional regulation of human T-cell leukemia virus type III long terminal repeat. Science (New York, NY) 227 (4683), 171–173.

Splinter, E., de Laat, W., 2011. The complex transcription regulatory landscape of our genome: control in three dimensions. EMBO J. 30 (21), 4345–4355.

Starcich, B., et al., 1985. Characterization of long terminal repeat sequences of HTLV-III. Science (New York, NY) 227 (4686), 538–540.

Steger, D.J., Workman, J.L., 1997. Stable co-occupancy of transcription factors and histones at the HIV-1 enhancer. EMBO J. 16 (9), 2463–2472.

Stoszko, M., et al., 2016. Small molecule inhibitors of BAF; a promising family of compounds in HIV-1 latency reversal. eBioMedicine 3, 108–121.

Strahl, B.D., Allis, C.D., 2000. The language of covalent histone modifications. Nature 403 (6765), 41–45.

Sun, Z.W., Hampsey, M., 1996. Synthetic enhancement of a TFIIB defect by a mutation in SSU72, an essential yeast gene encoding a novel protein that affects transcription start site selection in vivo. Mol. Cell. Biol. 16 (4), 1557–1566.

Suñé, C., et al., 2000. An in vitro transcription system that recapitulates equine infectious anemia virus tat-mediated inhibition of human immunodeficiency virus type 1 Tat activity demonstrates a role for positive transcription elongation factor b and associated proteins in the mechanism of tat activation. Virology 274 (2), 356–366.

Swain, P.S., Elowitz, M.B., Siggia, E.D., 2002. Intrinsic and extrinsic contributions to stochasticity in gene expression. Proc. Natl. Acad. Sci. U.S.A. 99 (20), 12795–12800.

Tantale, K., et al., 2016. A single-molecule view of transcription reveals convoys of RNA polymerases and multi-scale bursting. Nat. Commun. 7, 12248.

Tan-Wong, S.M., et al., 2008. Dynamic interactions between the promoter and terminator regions of the mammalian BRCA1 gene. Proc. Natl. Acad. Sci. U.S.A. 105 (13), 5160–5165.

Tan-Wong, S.M., et al., 2012. Gene loops enhance transcriptional directionality. Science 338 (6107), 671–675.

Tesmer, V.M., et al., 1993. NF-IL6-mediated transcriptional activation of the long terminal repeat of the human immunodeficiency virus type 1. Proc. Natl. Acad. Sci. U.S.A. 90 (15), 7298–7302.

Tong-Starksen, S.E., Luciw, P.a., Peterlin, B.M., 1987. Human immunodeficiency virus long terminal repeat responds to T-cell activation signals. Proc. Natl. Acad. Sci. U.S.A. 84 (19), 6845–6849.

Tréand, C., et al., 2006. Requirement for SWI/SNF chromatin-remodeling complex in Tat-mediated activation of the HIV-1 promoter. EMBO J. 25 (8), 1690–1699.

Trejbalová, K., et al., 2016. Development of 5′ LTR DNA methylation of latent HIV-1 provirus in cell line models and in long-term-infected individuals. Clin. Epigenetics 8, 19.

Tripathy, M.K., Abbas, W., Herbein, G., 2011. Epigenetic regulation of HIV-1 transcription. Epigenomics 3 (4), 487–502.

Turk, B., Stoka, V., 2007. Protease signalling in cell death: caspases versus cysteine cathepsins. FEBS Lett. 581 (15), 2761–2767.

Turrini, F., et al., 2015. HIV-1 transcriptional silencing caused by TRIM22 inhibition of Sp1 binding to the viral promoter. Retrovirology 12 (1), 104.

Tyagi, M., Karn, J., 2007. CBF-1 promotes transcriptional silencing during the establishment of HIV-1 latency. EMBO J. 26 (24), 4985–4995.

Tyagi, M., Pearson, R.J., Karn, J., 2010. Establishment of HIV latency in primary CD4+ cells is due to epigenetic transcriptional silencing and P-TEFb restriction. J. Virol. 84 (13), 6425–6437.

van der Sluis, R.M., et al., 2014. Interplay between viral Tat protein and c-Jun transcription factor in controlling LTR promoter activity in different human immunodeficiency virus type I subtypes. J. Gen. Virol. 95 (Pt. 4), 968–979.

Van Duyne, R., et al., 2011. Varying modulation of HIV-1 LTR activity by BAF complexes. J. Mol. Biol. 411 (3), 581–596.

Van Herreweghe, E., et al., 2007. Dynamic remodelling of human 7SK snRNP controls the nuclear level of active P-TEFb. EMBO J. 26 (15), 3570–3580.

Van Lint, C., et al., 1996. Transcriptional activation and chromatin remodeling of the HIV-1 promoter in response to histone acetylation. EMBO J. 15 (5), 1112–1120.

Van Lint, C., et al., 1997. Transcription factor binding sites downstream of the human immunodeficiency virus type 1 transcription start site are important for virus infectivity. J. Virol. 71 (8), 6113–6127.

Van Lint, C., Bouchat, S., Marcello, A., 2013. HIV-1 transcription and latency: an update. Retrovirology 10 (1), 67.

van Opijnen, T., et al., 2004. The human immunodeficiency virus type 1 promoter contains a CATA box instead of a TATA box for optimal transcription and replication. J. Virol. 78 (13), 6883–6890.

van Opijnen, T., Boerlijst, M.C., Berkhout, B., 2006. Effects of random mutations in the human immunodeficiency virus type 1 transcriptional promoter on viral fitness in different host cell environments. J. Virol. 80 (13), 6678–6685.

Vanti, M., et al., 2009. Yeast genetic analysis reveals the involvement of chromatin reassembly factors in repressing HIV-1 basal transcription. PLoS Genet. 5 (1), e1000339.

Vardabasso, C., et al., 2008. The histone chaperone protein nucleosome assembly protein-1 (hNAP-1) binds HIV-1 Tat and promotes viral transcription. Retrovirology 5 (1), 8.

Vemula, S., et al., 2015. HIV-1 induced nuclear factor I-B (NF-IB) expression negatively regulates HIV-1 replication through interaction with the long terminal repeat region. Virus 7 (2), 543–558.

Vendel, A.C., Lumb, K.J., 2003. Molecular recognition of the human coactivator CBP by the HIV-1 transcriptional activator tat. Biochemistry 42 (4), 910–916.

Venkatesh, S., Workman, J.L., 2015. Histone exchange, chromatin structure and the regulation of transcription. Nat. Rev. Mol. Cell Biol. 16 (3), 178–189.

Verdin, E., 1991. DNase I-hypersensitive sites are associated with both long terminal repeats and with the intragenic enhancer of integrated human immunodeficiency virus type 1. J. Virol. 65, 6790–6799.

Verdin, E., Van Lint, C., 1995. Internal transcriptional regulatory elements in HIV-1 and other retroviruses. Cell. Mol. Biol. (Noisy-le-Grand) 41 (3), 365–369.

Verdin, E., Paras, P., Van Lint, C., 1993. Chromatin disruption in the promoter of human immunodeficiency virus type 1 during transcriptional activation. EMBO J. 12 (8), 3249–3259.

Verhoef, K., Berkhout, B., 1999. A second-site mutation that restores replication of a Tat-defective human immunodeficiency virus. J. Virol. 73 (4), 2781–2789.

Wain-Hobson, S., et al., 1985. Nucleotide sequence of the AIDS virus, LAV. Cell 40 (1), 9–17.

Wang, J., et al., 2016. RbAp48, a novel inhibitory factor that regulates the transcription of human immunodeficiency virus type 1. Int. J. Mol. Med. 38 (1), 267–274.

Weber, C.M., Ramachandran, S., Henikoff, S., 2014a. Nucleosomes are context-specific, H2A.Z-modulated barriers to RNA polymerase. Mol. Cell 53 (5), 819–830.

Weber, et al., 2014b. Epigenetic analysis of HIV-1 proviral genomes from infected individuals: Predominance of unmethylated CpG's. Virology 449, 181–189.

Wei, P., et al., 1998. A novel CDK9-associated C-type cyclin interacts directly with HIV-1 Tat and mediates its high-affinity, loop-specific binding to TAR RNA. Cell 92 (4), 451–462.

Weinberger, A.D., Weinberger, L.S., 2013. Stochastic fate selection in HIV-infected patients. Cell 155 (3), 497–499.

Weinberger, L.S., et al., 2005. Stochastic gene expression in a lentiviral positive-feedback loop: HIV-1 Tat fluctuations drive phenotypic diversity. Cell 122 (2), 169–182.

Wen, Y., Shatkin, A.J., 1999. Transcription elongation factor hSPT5 stimulates mRNA capping. Genes Dev. 13 (14), 1774–1779.

Wilhelm, E., et al., 2012. CTGC motifs within the HIV core promoter specify Tat-responsive pre-initiation complexes. Retrovirology 9, 62.

Williams, S.A., et al., 2006. NF-kappaB p50 promotes HIV latency through HDAC recruitment and repression of transcriptional initiation. EMBO J. 25 (1), 139–149.

Xie, B., et al., 2007. Arginine methylation of the human immunodeficiency virus type 1 Tat protein by PRMT6 negatively affects Tat interactions with both cyclin T1 and the Tat transactivation region. J. Virol. 81 (8), 4226–4234.

Yamaguchi, Y.Y., et al., 1999. NELF, a multisubunit complex containing RD, cooperates with DSIF to repress RNA polymerase II elongation. Cell 97 (1), 11.

Yamaguchi, Y., et al., 2002. Evidence that negative elongation factor represses transcription elongation through binding to a DRB sensitivity-inducing factor/RNA polymerase II complex and RNA. Mol. Cell. Biol. 22 (9), 2918–2927.

Yang, X., Chen, Y., Gabuzda, D., 1999. ERK MAP kinase links cytokine signals to activation of latent HIV-1 infection by stimulating a cooperative interaction of AP-1 and NF-kappaB. J. Biol. Chem. 274 (39), 27981–27988.

Yang, Z., et al., 2001. The 7SK small nuclear RNA inhibits the CDK9/cyclin T1 kinase to control transcription. Nature 414 (6861), 317–322.

Yang, Z., et al., 2005. Recruitment of P-TEFb for stimulation of transcriptional elongation by the bromodomain protein Brd4. Mol. Cell 19 (4), 535–545.

Yik, J.H.N., et al., 2003. Inhibition of P-TEFb (CDK9/Cyclin T) kinase and RNA polymerase II transcription by the coordinated actions of HEXIM1 and 7SK snRNA. Mol. Cell 12 (4), 971–982.

Yik, J.H.N., et al., 2004. A human immunodeficiency virus type 1 Tat-like arginine-rich RNA-binding domain is essential for HEXIM1 to inhibit RNA polymerase II transcription through 7SK snRNA-mediated inactivation of P-TEFb. Mol. Cell. Biol. 24 (12), 5094–5105.

Zanton, S.J., Pugh, B.F., 2006. Full and partial genome-wide assembly and disassembly of the yeast transcription machinery in response to heat shock. Genes Dev. 20 (16), 2250–2265.

Zenzie-Gregory, B., et al., 1993. HIV-1 core promoter lacks a simple initiator element but contains a bipartite activator at the transcription start site. J. Biol. Chem. 268 (21), 15823–15832.

Zhang, H., Roberts, D.N., Cairns, B.R., 2005. Genome-wide dynamics of Htz1, a histone H2A variant that poises repressed/basal promoters for activation through histone loss. Cell 123 (2), 219–231.

Zhang, Z., et al., 2007. Negative elongation factor NELF represses human immunodeficiency virus transcription by pausing the RNA polymerase II complex. J. Biol. Chem. 282 (23), 16981–16988.

Zhang, L., et al., 2014. Modulation of the stability and activities of HIV-1 Tat by its ubiquitination and carboxyl-terminal region. Cell Biosci. 4 (1), 61.

Zhang, Y., et al., 2015. CRISPR/gRNA-directed synergistic activation mediator (SAM) induces specific, persistent and robust reactivation of the HIV-1 latent reservoirs. Sci. Rep. 5, 16277.

Zhao, S., et al., 2003. Requirement of a specific Sp1 site for histone deacetylase-mediated repression of transforming growth factor beta type II receptor expression in human pancreatic cancer cells. Cancer Res. 63 (10), 2624–2630.

Zhou, Q., Li, T., Price, D.H., 2012. RNA polymerase II elongation control. Annu. Rev. Biochem. 81 (1), 119–143.

Zhu, J., et al., 2012. Reactivation of latent HIV-1 by inhibition of BRD4. Cell Rep. 2 (4), 807–816.

FURTHER READING

Satinder Dahiya, Y.L., et al., 2014. Role of downstream elements in transcriptional regulation of the HIV-1 promoter. J. Hum. Virol. Retrovirol. 1 (2), 1–7.